National Autonomy, European Integration and the Politics of Packaging Waste

Netherlands School for Social and Economic Policy Research

Netherlands School for Social and Economic Policy Research (AWSB)

The Netherlands School for Social and Economic Policy Research (AWSB) comprises twelve research schools from five faculties spread across four Dutch universities - Utrecht, Tilburg, Rotterdam and Amsterdam (UvA). The School carries out research into questions relating to the welfare state, citizenship, labor and social security. Research undertaken by AWSB is both fundamental and strategic, as well as multi-disciplinary and internationally oriented. The more than 100 reseachers at AWSB are, for the most part, trained in law, sociology and economics. The research school currently supervises about 60 PhD students.

Markets and Regulation

The regulation of markets, and changes therein, is a major theme both in public policy and in recent work in a variety of disciplines: economics, law, political science, sociology, public administration, policy studies. Governments are deregulating, reregulating, privatizing and introducing market elements in the provision of public services. The increased regulatory activities of supra-national organizations (EU, WTO) modify the regulatory capacities of the nation-state. Academics are studying these developments and have (re)discovered the importance of institutions for economic performance.

This series publishes work on these topics, among others done at the AWSB Research School for Social and Economic Policy. Books may cover subjects such as problems of regulation and deregulation of (specific) markets, the economic effects of specific institutions, privatization and new public management, styles of policy implementation and regulatory enforcement, the structure and functioning of regulatory agencies, comparisons of policies and regulations between countries, the effects of internationalization including European regulation on the policies and room for manoeuvre of the nation-states.

Series-editors are Frans van Waarden, Gerrit Faber and Jan Simonis, all researchers at the Netherlands School for Social and Economic Policy Research (AWSB).

The series is open to studies in both English and Dutch.

National Autonomy, European Integration and the Politics of Packaging Waste

NATIONALE AUTONOMIE, EUROPESE INTEGRATIE EN VERPAKKINGSAFVALBELEID
(met een samenvatting in het Nederlands)

Proefschrift ter verkrijging van de graad van doctor aan de Universiteit Utrecht

op gezag van de Rector Maginificus, Prof. dr. H.O. Voorma

ingevolge van het besluit van College voor Promoties in het openbaar te verdedigen

op 9 october 1998 des ochtends te 10.30 uur

door

Markus Haverland

geboren op 1 januari 1967 te Biedenkopf (Duitsland)

Promotor: prof. dr. B.F. van Waarden, Sociale Wetenschappen, Universiteit Utrecht

Preface

This book would have never been written if Frans van Waarden had not talked me into leaving Germany and writing a dissertation under his supervision. A step I have never regretted! I am grateful for his persuasive power, but also for granting me time to socialise in the Netherlands. I also appreciate that he offered me all possible freedom to develop a research project according to my own preferences and I admire the patience with which he read the (too) many pages I gave him over the years.

The Netherlands School for Social and Economic Policy Research and the *Interdisciplinair Sociaal-Wetenschappelijk Onderzoeksinstituut,* Utrecht provided me with an interdisciplinary environment and helped me to maintain a general interest in social science and public policy issues. They also offered a number of useful and very well conducted general courses. I would like to thank Heinze Oost who was instrumental in clarifying my research project and asking narrowly defined research questions.

I had the pleasure to share a room with Kees Le Blansch for two years. He was very helpful in becoming familiar with the 'Dutch case' in a broad respect. Being in the final phase of his dissertation he also offered me also a lot of useful, practical advice.

Further, I am indebted to the hospitality of Wyn Grant who made a three-month visit at Warwick University possible. This not only provided me with a suitable location for field research in the UK but also allowed me to participate in the attractions a British university campus has to offer .

I am thankful to Monique Hoezee and Verina Ingram who were working on similar projects and generously shared information. I owe a special debt of gratitude to the twenty interviewees who made time in their busy schedules and provided me with a lot - often sensitive - information. Without their generosity the study would have had less 'flesh' and 'blood'.

At a crucial phase of my project I had the chance to present preliminary findings at a number of international conferences. I am particularly grateful to the participants of the European Consortium of Political Research Workshop on the Implementation of Environmental Policy held in Bern 1997 for their inspiring discussion of my paper. Their comments were fruitful as was the network of contacts emerging from this sprankling event.

In different, but equally important ways, many people have contributed to the shape of this book. Either by discussing related papers and/or draft chapters, by performing editing tasks, or by doing, in fact, all of this. I would, in particular, like to thank Nynke Deinema, Michaela Drahos, Jan Eberg, Wyn Grant, Johan van der Gronden, Ken Hanf, Monique Hoezee, Dorine Lustig, Dan Keleman, Beate Kohler-Koch, Andrea Lenschow, Astrid van Oorschoten, Sebastiaan Princen, Jaap Tuin and Wil Pansters. Rainer Eising and Andrea Lenschow read the whole manuscript and provided me with many incisive comments. A special thanks to them.

Finally, I am indebted to the many people working for the *Centrum voor Beleid en Management* who warmly welcomed me and often spared a place in their agenda for private activities.

The book is dedicated to the friends and colleagues who made me forget that writing a dissertation is primarily a solitary activity.

Contents

Figures

Tables

Glossary

AGVU	Arbeitsgemeinschaft Verpackung und Umwelt (German industry organization)
AIM	European Association of Industries of Branded Products
ALARA	As Low As Reasonably Achievable (Prevention principle)
AMvB	Algemene Maatregel van Bestuur (Dutch Order In Council)
APME	Association of Plastic Manufacturers in Europe
AOO	Afval Overleg Orgaan (Dutch waste management council)
BBU	Bundesverband Bürgerinitiativen Umweltschutz (German environmentalists)
BCME	Beverage Can Makers in Europe
BDE	Bundesverband der Entsorgungswirtschaft (German association of waste firms)
BDI	Bundesverband der Deutschen Industrie (Federation of German industry)
BEUC	Bureau of European Consumer Unions
BGBl.	Bundesgesetzblatt (German Official Law Gazette)
BMK	Das Bessere Müllkonzept (German citizen's group)
BMU	Bundesministerium für Umwelt, Naturschutz und Reaktorsicherheit (German Federal Ministry of the Environment)
BMWi	Bundesministerium für Wirtschaft (German Federal Ministry of Economic Affairs)
BRC	British Retail Consortium (retailer trade association)
BUND	Bund für Umwelt und Naturschutz Deutschlands (German environmentalists)
CBI	Confederation of British Industry
CDU	Christlich Demokratische Union (German conservative political party)
CEC	Commission of the European Communities
CIA	Chemical Industries Association
COPAC	Consortium of the Packaging Chain (British consortium of six trade associations)
COREPER	Permanent Representatives Committee of the (EU) Council
CRMH	Centrale Raad voor de Milieuhygiëne (Dutch Environmental Advisory Board)
CSU	Christlich Soziale Union (Bavarian conservative party, ' sister party' of CDU)
D	Germany
DG	Directorate General of the Commission of the European Communities
DGM	Directoraat Generaal Milieubeheer (Environmental Division of the Dutch Environment Ministry)
DGMH	Directoraat Generaal Milieuhgiëne (Environmental Division of the Dutch Environment Ministry, from 1982 on DGM)
DIHT	Deutscher Industrie und Handelstag (Peak Association of [statutory] regional boards of trade and industry)
DM	Deutsche Mark (German mark)
DoE	(British) Department of the Environment
DSD	Duales System Deutschland (Industry organization that runs the German recycling system for sales packaging)
DTI	(British) Department of Trade and Industry
EEB	European Environmental Bureau (Brussels based environmentalists)
EC	European Community (until 31.10.1993)

ECJ	European Court of Justice
ECR	European Court Review (Official Gazette of the European Court of Justice)
EP	European Parliament
EPA	(British) Environmental Protection Act (1990)
ERRA	European Recovery and Recycling Association (industry group)
EU	European Union (from 1.11.1993 on)
EUROPEN	European Organization for Packaging and the Environment (industry)
EZ	Ministerie voor Economische Zaken (Dutch Ministry of Economic Affairs)
F	France
FDP	Freie Demokratische Partei (German liberal party)
FEAD	European Federation of Waste Management
FoE	Friends of the Earth
GVM	Gesellschaft für Verpackungsmarktforschung (German organisation for packaging market research)
HM	Her Majesty's
I	Italy
INCPEN	The Industry Council for Packaging and the Environment (British packaging trade association)
IW	Institut der deutschen Wirtschaft (Research institute of German industry)
KWD	Kantoor, Winkel en Diensten Afval (commercial waste)
LARAC	Local Authority Recycling Advisory Committee
LCA	Life Cycle Analysis
LNV	Ministerie voor Landbouw, Natuurbeheer en Visserij (Dutch Ministry for Agriculture, Nature Conservation and Fishing)
MKB	MKB-Nederland (Dutch peak association of small and medium sized enterprises)
MSSD	Most Similar Systems Design (research design)
NL	The Netherlands
NMP	Nationaal Milieubeleidsplan (Dutch National Environmental Policy Plan)
OECD	Organization for Economic Cooperation and Development
PAC	Packaging Advisory Committee (Senior managers from leading British companies from all sectors of the packaging chain, since Jan. 1996)
PCF	Packaging Chain Forum (European industry network)
PET	Polyethyleneterephtalate
PRG	Producer Responsibility Industry Group (Senior representatives of 26 leading British companies, mainly packers/fillers/retailers 1993-1994)
PRN	Packaging Recovery Notes (British system of tradeable permits)
PSC	Packaging Standards Council
RCEP	(British) Royal Commission on Environmental Pollution
RIVM	Rijksinstituut voor Volksgezondheid en Milieuhygiëne (Dutch national institute for public health and environmental protection)
SEA	Single European Act (1987)
SEFEL	European Secretariat of Manufacturers of Light Metal Packaging
SNM	Stichting Natuur en Milieu (Dutch environmentalists)
SPAN	Sustainable Packaging Action Network (environmentalist network)
SPD	Sozialdemokratische Partei Deutschlands (German social democratic party)
SRU	Sachverständigenrat für Umweltfragen (German expert council)

SVM	Stichting Verpakking en Milieu, Foundation of Packaging and the Environment (Dutch industry foundation representing packaging interests of the packaging chain)
TEU	Treaty of the European Union
TK	Tweede Kamer (Second chamber 'House of Commons' of the Dutch Parliament)
UBA	Umweltbundesamt (German federal environment agency)
UK	United Kingdom
US	United States
VALPAK	(Largest) industry organization for running the British collective recovery and recycling scheme
VMK	Vereniging Milieubeheer Kunststofverpakkingen (Dutch plastic packaging industry organization)
VNG	Vereniging van Nederlandse Gemeenten (Union of Dutch Municipalities)
VNO-NCW	Verbond van Nederlandse Ondernemingen - Nederlands Christelijk Werkgeversverbond (Dutch peak organization of employer/industry organizations)
VoMil	Ministerie voor Volksgezondheid en Milieuhygiëne (Dutch Ministry for Public Health and Environmental Hygiene, until 1982
VROM	Ministerie voor Volkshuisvesting, Ruimtelijke Ordening and Milieu (Dutch Ministry for Environmental Protection)
V-WRAG	VALPAK-Working Responsibility Advisory Group (Delegates of some 50 companies representing all sectors of the packaging chain, 1994-1995)
WCA	Waste Collection Authorities (i.e. District Councils in England)
WCED	World Commission on Environment and Development
WDA	Waste Disposal Authorities (i.e. County Councils in England)

1. Introduction

1.1 The research problem: National policy diversity and European integration

This book presents analysis and comparison of policy transformation and regulatory change in the United Kingdom, Germany, and the Netherlands in the context of policymaking processes in the European Union (EU)[1] over the last decade. It is informed by the tensions between policy and institutional diversity on the national level and the process of European integration. The central research question is whether and how European integration brings about a convergence of national policies.

<u>Does European integration bring about a convergence of national policies?</u>

Investigations into convergence, persistence and divergence require a combination of cross-national and longitudinal comparisons. One has to choose countries that differed in the past and to analyse these cross-national variations. Then one has to study how national policies developed in the context of European integration, and finally one has to analyse whether the cross-national policy variation has decreased over time.

The central research question will be analysed for the case of packaging waste policies in Germany, the Netherlands and the United Kingdom. Packaging has been among the most contentious environmental issues in Western Europe in the last decade. A great deal of uncertainty exists about the environmental effects of packaging, and there is political and societal disagreement about the need and the proper design of packaging waste policies. For environmentalists and many consumers, packaging is a symbol of our throw-away society, while for industry packaging is vital to our globalising economy, given its important functions for transport, self-service and marketing. The issue of packaging waste not only divides industry and large parts of the public; there are also conflicting interests between different industrial sectors.

---

[1] The Maastricht Treaty, that came into force on 1 November 1993, has created the European Union. The institutions and processes relevant for this study are, however, still based on that pillar of the European Union to which still the term European Community (EC) applies. Since the term European Union is commonly used also for the EC pillar, it will be employed in this study for all instances in which no specific reference is made to the period preceding the EU.

1.1.1 National policy diversity

The study started with the old but still fascinating question of why people in different countries are governed in different ways. I was struck by the fact that by the end of the 1980s packaging waste had become a top priority on the political agenda in Germany and the Netherlands, but not in the UK. Given similar problems due to common characteristics of production, distribution and consumption patterns in these countries, I wondered: why is the fate of used packaging an important issue in one country but not in another country? A closer look revealed that not only problem perceptions differed between the three countries. Also the policy action taken by those who perceived the problem as serious varied. Germany and the Netherlands adopted political programmes concerning packaging waste in the early 1990s. Yet, different choices were made. In Germany consumers had to sort out all their packaging and throw it in separate dustbins. Retailers had to ensure that 72 per cent of their drinks were sold in refillables, business at large had to ensure that most one-way sales packaging material was recycled. A large part of Dutch industry committed itself to prevent and recycle packaging waste, while nothing similarly ambitious was done in the UK. The German government imposed its policy on business, while the Dutch government engaged in a voluntary agreement with parts of trade and industry, and the British government relied on rather vague industries' codes of practice and some lukewarm commitments.

A review of the literature on comparative public policy reveals that the puzzle of common problems and different solutions is not confined to the issue of packaging waste. Governments choose nationally distinctive policies to deal with problems caused by the internationalization of the economy, unemployment, the rising costs of the welfare state, migration or environmental pollution (see for example: Katzenstein 1985, Scharpf 1987, Vogel and Kun 1987). Even when national governments have similar problem perceptions and ideologies, as in the case of the neo-liberal reform of the health sector and the telecommunication sectors in Germany and Britain, their policies can differ remarkably (see Döhler 1990 and Grande 1989 respectively).

In these studies differences in policy approaches have been identified concerning the degree of legal codification of the programmes, the content of policy objectives, the level and detail of targets, the instruments chosen, the distribution of responsibilities between different levels of government and between government and private actors, and the characteristics of the policy process, such as the number and type of actors involved in policy formulation and policy implementation and the importance of negotiation versus imposition. These cross-national variations in policies have been mainly explained by nationally distinctive patterns of values and interests as well as national specific contingencies, but also by differences in political institutions such as the horizontal and vertical organization of the state, the legal system, and the voting system.

The first empirical part of this study (chapters 3 - 7) is dedicated to the description and the analysis of cross-national differences and similarities in national policies concerning the problem of packaging waste in the early 1990s. What did the national policies look like? Which factors and mechanisms explain the differences and similarities found?

Three countries - the Netherlands, the United Kingdom and Germany - have been chosen for this case study. They were selected according to the most similar systems design: they are all liberal democracies and advanced industrialized countries, with similar levels of technological development, but varying significantly on the dependent variable: their approaches towards the problem of packaging waste.

Research question 1: What were the differences and similarities between the German, Dutch and British packaging waste policies in the early 1990s?

Research Question 2: Which factors and mechanisms explain these differences and similarities of the German, Dutch and British packaging waste policies?

The governments' approach to the problem of packaging waste will be compared when the Commission of the European Community (the Commission) came up with its own official proposal for a comprehensive packaging waste policy in 1992. More concrete: I will look at the context, the history, the policy formulation, the policy content, and the first implementation steps of the German Packaging Ordinance ('*Verpackungsverordnung*', 1991), the Dutch Packaging Covenant ('*Convenant Verpakkingen*', 1991) and the relevant provisions of the British White Book on the Environment (1990), the Environmental Protection Act (1990) and the industries' business plan and Codes of Practise (1992).

1.1.2 European integration

The proposal for a European directive on packaging and packaging waste by the European Commission in 1992 and its adoption by the Council of Ministers and the European Parliament in 1994 indicate that the packaging waste issue transcends the territories of individual nation states. Policies aimed at the prevention or the reduction of million of tonnes of used packaging, often intervene into markets, because certain types of packaging might be discriminated or even banned. The free movement of goods, one of the pillars of the single market, might be affected. Hence, trade aspects but also environmental aspects of packaging and packaging waste have become a European issue and the tensions between environmental protection and the free movement of goods go hand in hand with tensions between European integration and national tradition and autonomy.

As stated above, it is not only the interest in cross-national comparison and analysis that drives this study. In the second part of the study, policy developments - changes and continuities will be investigated by looking at the interrelationship of national and EU policymaking (chapters 8 - 11). The Internal Market Programme, which aims at the free movement of goods, labour, capital, and services, puts pressure on national governments to harmonize regulations. Member States might have to recognize the standards of other countries, allowing importers to market their products on the domestic market and thereby undermining their national policies. They might also be forced by the European Commission and the European Court of Justice to modify or abolish national policies which are not in compliance with the Treaty. Member States also have to implement European regulations and directives, which usually reflect not only their own national preferences but also those of other Member States, the European Commission and the European Parliament. Accordingly, European legislation might be in conflict with national policies. European integration might impinge not only directly but also indirectly on national policies by providing a framework for increased debate about policy problems and solutions, thus facilitating policy learning. What is the relevance and what are the consequences of this development? First, we have to study why and how the Directive was adopted and what it looks like (chapter 8). Then we have to look at its impact on Britain (chapter 9), the Netherlands (chapter 10) and Germany (chapter 11). Is Britain forced to do more? Does the Dutch agreement between government and industry need to be replaced? Can the ambitious German system survive? In this empirical part of the study the political process resulting in the European Directive on Packaging and Packaging Waste will be analysed as well as the substance of the Directive and its implementation - legally and practically - in the Member States. This part will also take into account other developments than those associated with European integration in order to specify the contribution of integration *vis-à-vis* other factors.

Research question 3: How have the German, Dutch and British national packaging waste policies developed since the early 1990s? Have policies changed? Which factors and mechanisms either triggered or blocked changes? What was the impact of European integration?

The mutual impact of national approaches and supranational policymaking are at the centre of the study. By comparing variations in national policies before and after Europeanization and analysing the underlying causes and mechanisms of the policy development, the question can be answered whether policies have converged and whether and how policy developments were related to European integration (chapter 12).

Research question 4: Does European integration bring about a convergence of the German, Dutch and British packaging waste policies?

The tension between European integration and national diversity is of theoretical interest. Comparative studies have revealed that cross-national variations can be partly explained by variations in national institutions, which are embedded in national traditions. The answer to the question of whether European integration results in a convergence of policies will shed light on the strength and stickiness of national institutions and traditions and will allow for a deeper understanding of the mechanisms of policy change. EU packaging waste policymaking has been analysed by a number of scholars (see Gehring 1997, Golub 1997, Porter 1995). These studies focused, however, almost entirely on the European policymaking process. This is the first study looking systematically into the interrelationship of national and European packaging waste policymaking, including the period before and after the EU policymaking process. The study will therefore not simply add yet another case to the growing literature on the impact of European integration on Member States (see, for instance, the contributions in the volumes edited by Majone 1996a, Meny, Muller and Quermonne 1996, and Kohler-Koch and Eising forthcoming). The inclusion of a comparative account of each country's history on packaging waste policy preceding the European policymaking process makes this study rather particular in this context.

Yet, the questions posed have not only theoretical but - implicitly - also political and societal relevance. Taking the perspective from the point of European integration this study reveals whether and to what extent national institutions and traditions hinder this integration. Adherents of a national perspective will see to what degree Member States still pursue nationally distinct policies. Environmentalists will see at what level environmental standards were possible in this case. For all actors it will be interesting to compare the different ways of government and industry interactions, the strategies employed and the - intended and unintended - consequences of policy choices. This includes the opportunities and limitations of voluntary agreements as well as the external effects of interventionist environmental policymaking. More in general, comparative policy analysis carries an essentially critical notion, though the focus is empirical and analytical. Cross-national and historical comparisons reveal that many choices are not necessarily determined by functional requirements, or as it is called in Germany, '*Sachzwänge*'. Divergent solutions have been applied to the same problems; all with their distinctive advantages and disadvantages. To some extent the world is a laboratory in which many things have been tried and much can be learned. Yet, the process of learning is itself a political process, and the study will address the possibilities and limitations of policy learning.

The comparative nature of this study implies that most statements such as '*high level*', '*formalized*', ''*strict standard*', are relative. They derive from cross-national or longitudinal comparisons. The respective frame of reference will always be stated. Therefore, these '*judgements*' derive not from a normative frame. Many people argue, for instance, that the German packaging ordinance is not strict enough. From a cross-national perspective, however, the German approach ranks among the most stringent in the world. According to this frame of reference it has to be explained, why it is so strict, rather than, why it is so soft.

The remainder of this introduction consists of a theoretical and methodological part. In the theoretical part, the four research questions will be elaborated on and substantiated in the light of scientific literature. The first goal is to situate this study within the framework of relevant scientific discussion in order to demonstrate its potential contribution. The second goal is to identify elements of policy approaches that are useful for the disaggregation of the dependent variable '*policy*', and to identify potential explanatory variables, that is the factors and mechanisms that shape (a) national variations and similarities and (b) the persistence and convergence of national policies.

The methodological part starts with a discussion of the opportunities and limitations of the most similar systems design which will be employed in this study. Then, the choice of countries will be substantiated. After having delineated the period and structure of comparison, the methods of data collection will be presented.

This introduction will draw heavily on a review of scientific literature. Such an account is very helpful in a case study strategy. In comparison with other research strategies such as historical accounts, surveys or experiments, case studies in particular benefit from the prior development of theoretical propositions to guide data collection and analysis. Moreover, case studies usually do not represent a 'sample'. The investigator's goal is not to enumerate frequencies, that is statistical generalizations, but to engage in analytical generalizations, that is to expand and generalize theories, which in turn have been informed by earlier studies (Dogan and Pelassy 1990: 122, Yin 1994: 10).

1.2 Theoretical framework

1.2.1 Variations in national policies: empirical insights and conceptual issues

For decades there has been a debate among scholars of policy analysis about the extent of national differences in policies. There is no. doubt that on a general level advanced industrial states have shown similar developments in recent history. In a seminal study, Andrew Shonfield (1965) analysed the development of major industrialized countries in the twenty years following the Second World War. Shonfield observed similarities between the countries under investigation, such as increased influence of public authorities on the management of the economic system, a rising amount of public funds and spending for social welfare, a tendency towards increased regulation and control of competition, and an increasing predilection for long range economic planning (1965: 66-67). More recent studies have also stressed cross-national similarities. According to Freeman (1985) policymaking in all advanced capitalist states has shown a tendency toward technical problem-solving and rational planning. A professionalization of policymaking took place in conjunction with the bureaucratization and institutionalization of policy programmes, based on broad and intensive social scientific data analysis, and formal techniques of planning and analyses to be employed in the definition of policy problems, the development and selection of responses, and evaluation (Freeman 1985, see also Bell 1973).

Underneath these very general observations, however, more subtle variations have been identified. A case in point is the study of Shonfield mentioned above. Despite the similar trends he found in advanced industrialized countries, he also described significant variations in organising the economic system. He compared Germany and France and observed a strong predilection of the French government for centralized economic planning while Germany rejected it. Furthermore, some countries like France successfully introduced systems of indicative planning, whereas countries like the United Kingdom failed to introduce comprehensive planning instruments. In regard to the problems of contemporary capitalism Shonfield notes that '*the essential French view,..., is that the effective conduct of a nation's economic life must depend on the concentration of power in the hands of a small number of exceptionally able people, exercising foresight and judgement of a kind not possessed by the average successful man of business'* (71-72). The British style, as Shonfield describes it, is to '*remove the government officials, suppress the independent person, ensure that there is equality between firms with no advantage for the dynamic enterprise...*' (160).

Studies that focused on the 1970s and 1980s emphasized the cross-national differences of political choices toward economic problems even more. A critical event in this context was the oil crisis in the first half of the 1970s that challenged the trend of steady economic growth. Several studies have shown that countries tended to react differently to this commonly perceived threat. In a book edited by Katzenstein (1978), scholars found that despite a greater entanglement of advanced industrial states in the world market these states developed different strategies to manage this interdependence. The United States adopted a hard line towards OPEC producer states; Britain intensified the development of North Sea oil reserves; West Germany and France countered with a commercial offensive; Italy relied on direct foreign assistance, and Japan accelerated its investment programme in countries exporting raw materials. The Katzenstein volume also comprises other examples of variations of strategies adopted by industrialized states. Examples include topics such as international monetary reform and non-agricultural trade (1978:296).

Differences in national policies were not confined to economic policymaking. Variations have also been found in relatively new areas such as environmental policy. In the 1970s and 1980s a number of cross-national studies were carried out. These studies focused mainly on areas where the transboundary character of pollution has become apparent, first of all air pollution (see for a review: Kern and Bratzel 1996, Knoepfel *et al.* 1987, Pehle 1993, Vogel and Kun 1987). These studies revealed significant cross-national variances in policies devised to cope with similar problems. Examples are Lundqvist (1980), who examined the relationship between the structural organization of Sweden and the US and its choice of policy alternatives to control pollution; Brickman, Jasanoff and Ilgen (1985) who compared policies dealing with chemical regulation and cancer in the US, France, Britain and Germany; Kelman (1981) who studied occupational safety and health policy in the US and Sweden; Knoepfel and Weidner (1985) who investigated SO_2 pollution control policies in EU-countries and the US; and Vogel (1986) who compared a broad array of environmental measures in the US and Britain. These studies reveal variations in the government's approach to problem solving concerning the intensity of intervention, the degree of detail of regulation and prescriptions for implementation, the range of opportunities states provide for non-governmental groups to participate in the policy process, the nature of the relationship between regulatory officials and industrialists, and the strategies employed by government officials for securing compliance.

Based on their findings, scholars of comparative public policy developed typologies for national systems of policymaking. Katzenstein, for instance, discerned three dominant forms of contemporary capitalism: liberalism (weak state) in the United States and Britain; statism (strong state) in Japan and France; and corporatism (strong state-organganized interests relations) in the small European states and to a lesser extent, in West Germany (Katzenstein 1985: 20). Another related typology to characterize state intervention has been suggested by

Dyson, who also compared national industrial policies as a reaction to economic crisis in the 1970s. Dyson distinguished between interventionist and a non-interventionist national industrial cultures. The interventionist culture comprises: discriminatory policy, firm level intervention, proactive government, business cooperation, regular and stable contact and informal consultation. The non-interventionist style is characterized by even handed non-discrimination, industry level intervention, reactive government, business suspicion, irregular and *ad hoc* contact, formal consultation. The interventionist culture can be found in countries like France, Italy and Germany, whereas the non-interventionist culture is typical for Britain and the United States (1983: 31). Katzenstein and Dyson have derived their typologies from a '*weak state/strong state*' dichotomy. The concept of policy styles developed by Richardson, Gustafsson and Jordan (1982) contains two dimensions, namely (a) the government approach to problem solving and (b) the relationship between government and other actors in the policy process. The dimensions are constructed along a) active vs. reactive approach to problem solving and b) consensual- vs. conflict/hierarchy-oriented relations between the state and other actors (1982, 12-13).

These typologies are useful for the general cross-area characterization of national policy approaches, but there are two problems involved: an empirical one and a conceptual one. The empirical problem is that notions such as '*national*' culture or '*national*' policy styles are not very sensitive to cross-sectoral differences in policymaking. Sector-oriented research, however, found problem-solving approaches and government-industry relations that challenged the general typologies. In France traditional expectation of strong dirigiste control and state-leadership was challenged. In the case of consumer electronics one firm dominated the state agencies. The relationship could be described as bargained corporatism verging on agency capture. The United Kingdom, labelled as a weak state, regulated the pharmaceutical industry in a proactive radical way (Wilks and Wright 1987: 284-285, see also Cawson 1985). A number of scholars argued therefore that each sector poses its own problems, sets its own constraints, and generates its own type of conflicts (Freeman 1985: 481-490). They predict differentiation within individual countries across sectors and similarities across nations within sectors, whereas adherents to the concept of national policy styles, or national culture expect similarities within individual countries across sectors and differences across countries within the same sector. Because this study does not have a cross-sectoral focus, I cannot rule out that sector characteristics are more important than national features. Therefore I will not start the empirical exercise with the adjective 'national'.

The conceptual problem with the above mentioned typologies is that they are not meant as an analytical frame for a fine-grained analysis. Although a conceptualization of the dependent variable is needed, which is sufficiently abstract to allow for accurate cross-national and cross-sectoral comparisons, e.g. the comparison of packaging waste regulation with food quality or occupational heath policies, the concept has also to be refined enough to get a detailed picture of the case under study. The general problem for a meaningful conceptualization is that there is no standard definition of concepts like policy, policy style, regulatory style or policy networks (Feick 1992: 261, 274, van Waarden 1992:30).

In recent years, however, several efforts were made to review national typologies and case studies and to arrive at refined and systematic categories. My conceptualization draws especially on Feick (1992) because his concept of policy profile consistently refines the dimension 'problem solving approach' introduced by Richardson, Gustafsson and Jordan. In addition, this study benefits from the growing literature on policy networks (in particular van Waarden 1992, 1995). This literature helps to refine the structural aspects of the second dimension of Richardson, Gustafsson and Jordan, namely the relations between the state and other actors.

Feick (1992) introduced five dimensions for what he called policy profiles: formalization, integration, continuity, intensity and programming. Two of them are not useful for this study. Feick's dimension 'integration' refers to the degree of integration of single policies (e.g. packaging waste policies) in a policy sector or problem field (e.g. environmental policy). This study confines the dependent variable, however, to the single issue packaging and the environment. The dimension 'continuity' refers to the degree to which policies follow or deviate from preceding policies. Continuity is an important notion in this study, but only when it comes to the longitudinal comparison. The dimension does not lend itself to cross-national *static* comparison.

My concept of policies comprises five elements. Three elements ('*formalization*', '*strictness of standards*', '*allocation of responsibilities*') refer to the policy content. One element is devoted to the structural aspect of the policy formulation process ('*mode of interest integration*') The fifth refers to the style of policy formulation and implementation ('*orientation towards target group*').

Formalization: Formalization according to Feick is defined as the degree of legal codification of the policy programme. Is the policy based on public law and generally binding for the target group? Or is it based on a contract between government and other actors (mostly private law), and therefore only binding for those members of the target group that signed? Or is the policy based on general statements which do not have any binding character?

Strictness of standards: This category refers to the intensity (Feick) of the regulatory standards and targets adopted. There may be cross-national variations in the objectives for which targets are set. Are targets set for packaging reduction or for the re-use of packaging, or for recycling packaging waste material? Standards can vary in level and time horizon. They can take the form of minimum targets and maximum targets. They can be more or less binding. The strictness of standards includes 'programming' (Feick), that is, the detail and precision with which implementation is prescribed. This includes the detail of definitions, targets, timetables, guidelines for implementation agencies etc. This category points at the rigidity respectively the flexibility of the programme. Devices that allow for change to the programme in the light of experience and new information are also included in this category.

Allocation of responsibilities: Policy programmes establish obligations for their implementation, monitoring and enforcement, for instance, responsibilities for packaging waste collection, transport, and treatment. Each of these functions can be allocated either to public actors or private actors or a combination of both. In regard to public actors a distinction can be made between national ministries, sub-national actors and independent regulatory agencies.

Mode of interest integration: The mode of interest integration points to the structure of policy networks developed in the course of the programme formulation process (see van Waarden 1992, 1995). It includes the number of actors and type of interests involved, the function of the network and its degree of stability. One can basically distinguish between pluralist interest representation in which a great number of interests are represented in rather loose networks and corporatist interest integration in which a small number of highly aggregated interests are represented in more stable networks.

Orientation towards target group: The orientation towards the target group (Richardson, Gustafsson and Jordan 1982) points to the degree of pressure exerted on the target group, the role of persuasion versus coercion, and incentive mechanisms with which the policymaker tries to influence target group behaviour. An impositional style is associated with coercive interaction with the target group, policy incentives that are based on command and control and on economic instruments that discourage a certain course of action such as taxes and charges, legalistic enforcement practices and high sanctions. A mediating style is indicated by an consensus-oriented approach to target group interests, the stimulation of target group behaviour either by subsidies or by the provision of infrastructure, organizational resources or research and development, pragmatic enforcement practices and low or no sanctions.

Table 1 Dimensions of policies

dimension	***range***
formalization	public law - private law - not binding statements
strictness of standards	strict - modest
allocation of responsibilities	public - public-private - private
mode of interest integration	corporatist - pluralist
orientation towards target group	impositional - mediating

Note that the different elements are closely related to each other and that some characteristics are more likely to cluster together than others. It remains to be seen, whether these clusters can be summarized as a certain type of policy.

1.2.2 Explaining variations: ideas, interests, and institutions

In the previous section, I have sketched a number of findings in regard to cross-national similarities and differences of policymaking. I have also defined the dependent variable '*policy*'. In this section, I will elaborate on the question of why countries differ in their national policies.

To explain differences and similarities in policies is the aim of comparative policy analysis. Comparative policy analysis is not a discipline as such. A number of general theoretical approaches prevalent in political science and sociology were used in the last decades to analyse cross-national variations and similarities, including pluralism, marxism and functionalism. It would go beyond the scope of this chapter to discuss the advantages and disadvantages of these approaches in detail (see for an overview Chilcote 1994, Nassmacher 1991), yet, it is important to note that in the course of empirical-oriented comparative research of political processes, these approaches have shown some shortcomings and limited analytical power. Functionalists explain the shape of policies as a reflex to secular developments such as the globalization of economy, to external contingencies such as an oil crisis, or to the characteristics of a certain problem. They therefore have difficulty in explaining differences between countries which are confronted with similar developments and external shocks. Pluralism and Marxism see policies as the outcomes of political struggles with various, often conflicting ideas and interests as inputs. The outcome is determined by the relative strength of the interests. These approaches fail to explain why policies differ where policy inputs - configurations of ideas and interests - are similar. Given the weaknesses of these approaches, explanations have to go beyond both the analysis of interests and the functional requirements of political systems. Such an approach is provided by the '*new institutionalist*' approach. This approach will form the base of my explanatory framework. After having sketched the context of this approach, I will develop the explanatory framework which provides the background against which the question of why German, Dutch and British packaging waste policies were different in the early 1990s will be answered.

New institutionalism

New institutionalism has its roots in a number of social science disciplines. Some scholars adhere to historical sociology and political science (Hall, Immergut, Skocpol), others have a background in organizational sociology (DiMaggio and Powell, March and Olsen) or in economics (North, Shepsle, Weingast)[2]. The common denominator of these authors is the argument that the '*state and institutions matter*'. While pluralism and Marxism see the state as an agent of societal and class interests respectively, new institutionalists argue that the state itself can be an autonomous actor. '*The state is not only affected by society but also affects it*' (March and Olsen 1989:17, see also Evans, Rueschemeyer and Skocpol 1984: 347-350). Scholars informed by new institutionalism were able to explain the puzzling phenomenon that political outcomes in terms of policy choices differ between countries while the inputs in terms of interests or political ideas were quite similar.

The new institutionalist approach is based on the '*old*' institutionalism. Old institutionalism refers to the dominant political science approach prior to the early 1950's. In those days, scholars, mostly legal scientists and historians, were engaged in detailed studies of peculiar administrative, legal and political structures, such as constitutional arrangements and party systems (Hague, Harrop and Breslin 1992: 31, Hoogerwerf 1995: 94, Keman 1993b: 32). Empirical research linking institutions with policies was lacking. There are several differences between the old institutionalism and the new institutionalism. The former concentrated on the description of formal institutions and simply assumed differences in outputs, whereas adherents of the new variant analyse the processes by which institutions

[2] Overviews are provided by Hall and Taylor 1996, Hemerijck and Verhagen 1994 and Immergut 1997.

influence political choices. New institutionalism avoids any determinism between political institutions and distinct political choices. Institutions, however, structure the political conflict, hence shaping political choices to some extent. Moreover, while the old institutionalism emphasized the constraining character of institutions on interaction, its successor stresses that institutions also enable interaction (Benz 1997: 15, Thelen and Steinmo 1992: 3-6).

Elements of the new institutional framework were already apparent in the studies of Shonfield (1965), and Katzenstein (1978), since the authors emphasized the significance of domestic political structures in explaining cross-national variations. More recently Immergut (1992) did a carefully designed investigation. She studied the health policies of three countries, Sweden, France, and Switzerland, which started out with similar healthcare initiatives and similar government preferences towards public healthcare, and the medical profession for private healthcare systems, but nevertheless ended up with quite different programmes. Sweden's healthcare is based on a public system, whereas the Swiss is privately run and the French system lies somewhere in between. This variation could be explained neither by the preferences and strength of political parties or state executives, nor by the preferences and resources of the medical profession or the trade unions. She persuasively explained this puzzling political outcome by highlighting the role of political institutions: '*By changing the probability of veto of executive proposals, institutions changed the balance of power amongst interest groups and between interest groups and the state, and hence accounted for these different policy results*' (Immergut 1994: 10). The Swiss referendum provided the medical profession with an effective point of access to the political process to advance their preference for a private system, while the executive in Sweden was, for a number of institutional reasons, more remote from pressure from the medical profession and could pursue its public health programme more or less unchallenged. The French public-private mix was the result of a parliamentary history with changing coalitions and changing institutional rules.

Other comparative studies came to similar conclusions about the relevance of domestic structure. A seminal study that clearly shows the selection function of institutions is Scharpf (1987). He compared full-employment policies of different governments and showed that institutions either left different options open to these governments, or modified those options. Peter Hall (1986) examined the impact of Keynesianism on state intervention in France and the UK. He found that, although British and French institutions relevant to industrial policy did not fully determine outcomes, they nevertheless structured the flow of ideas and the clash of interests in ways that had a significant impact on these outcomes (see also Döhler, 1990, for health policy in Britain and German, and Grande, 1989, for telecommunication in the same countries).

Studies of political economy and the welfare state have a longer tradition and were theoretically more sophisticated than research in comparative environmental policy, the policy field which is empirically more relevant for my study. Only a few environmental studies of the 1970s and 1980s tried to explain their findings in a theoretically satisfying manner. '*Individual studies almost never attempt to link their theoretical and conceptual ideas into overarching theories that transcend the specific problem area*' (Knoepfel *et al.* 1987: 172).

There were, however, a few theoretically more ambitious studies which employed institutional explanations for cross-national variations in policies. Lundqvist, for example (1980), examined the relationship between the structural organization of Sweden and the US and their choice of policy alternatives to control pollution. He claimed that, though the impacts of air pollution were the same in both countries and the same technological and economic means were available, the two countries adopted different clean air policies. Lundqvist concluded that the U.S. and Swedish systems of government provide clean air policymakers with different political incentives and constraints, resulting in different policies. In times of a large upsurge in public opinion, the constitutionally built-in competitiveness of the American political system coupled with the peculiarities of the electoral and party systems will provide policymakers with tremendous incentives for selecting drastic and escalatory policies. '*Because they are so visible in the eyes of their constituencies, but with no responsibility for actual policy implementation, the United States Congressman will see much merit in selecting radical alternatives*' (1980: 117). Swedish parliamentarians were more remote from public opinion. They have more to gain from going along with the party leadership than from establishing themselves as champions for a particular group or issue. The party leadership in turn must be aware that the party may become part of the government. It would then become responsible for the implementation of selected or recommended policy alternatives. '*This certainly provides an incentive for preferring the practicable to the desirable*' (Lundqvist 1980: 122). In particular the dynamic of policy change varied. The United States is the hare and Sweden is the tortoise.

D. Vogel (1993), building on earlier work (Vogel 1986) sketched the development of environmental policy in of the United States, Britain and Japan over three time periods (1945-67, 1968-73, 1975-88) from a neo-institutionalist perspective. He found that the American separation of power played a crucial role in maintaining the representation of environmental interests. Environmental groups were able to use their access to the courts and their influence in Congress and on the state level and local level to reduce the negative environmental side-effects of the Presidential deregulation programme. In Britain, environmental interests had no effective point of access to counter the relaxation of enforcement of environmental regulations instituted by the Thatcher government (Vogel 262-263). The multiple veto points in the US make it difficult to alter the status quo, mechanisms of access and substantive policies favouring environmental interests are unlikely to be repealed or to fall into disuse. Environmental policymaking is therefore more likely to be a ratchetlike phenomenon (advances that are difficult to reverse) rather than a see-saw like process (cyclical advances followed by equal declines) that often characterizes party government parliamentary systems (Vogel 1993: 267). Extending the scope of his comparison, Vogel also argues that parliamentary systems characterized by proportional representation and coalition governments offer advocates of environmental interests greater opportunity for access and responsiveness than systems with a majority voting system. He alludes in this context to the Netherlands, Sweden and the Federal Republic of Germany who have the proportional voting system in common (Vogel 1993: 269). He emphasized in his article the intermediary character of institutions. Institutional effects are not uniform or unidirectional, but are *'contingent upon the presence or absence of other conditions and variables across countries and time periods'* (Vogel 1993: 266).

Institutions as intermediate variables

The explanatory framework of this study is based on the idea that institutions are intermediate variables linking micro features such as actors' interests with macro features such as socio-economic conditions. In other words, institutions provide room to manoeuvre by mediating the political process framed by micro factors and macro conditions and contingencies (Thelen and Steinmo 1992, 6, 32-33). The micro features involved are essentially the interests, strategies and resources of political actors. Macro conditions are, for instance, the basic distribution of natural resources, i.e. the geophysical conditions in various countries or the socio-economic conditions. Contingencies are, for instance, rather unique historical events such as German re-unification.

As the selection of already presented studies indicates, this study is based on the political science/historical sociological branch of neo-institutionalism. Accordingly, the concept of institutions is rather thin[3]. The study does not use the thick definition commonly used by organizational sociologists where a) institutions are almost everything, including symbols, strategies, technologies, beliefs, paradigms, codes, and knowledge, and b) almost no room is left for the autonomous and strategic action of actors (see March and Olsen 1989: 22, see for a critique Hoogerwerf 1995: 96, Mayntz and Scharpf 1995: 45-46).

Political institutions will be defined as formal and informal rules that structure the relationship between individuals in various units of the polity (see Hall 1986: 19, Windhoff-Héritier 1991: 38). The definition of institutions as the formal and informal rules of the game stresses their intermediary character. The basic idea is that political institutions such as the voting system, the vertical structure of the state, the internal organization of government, or the legal system, function as filters which select interests that enter the political arena and determine their relative power. Institutions provide points of access and veto, and select and thereby privilege certain interests. Hence, institutions constrain political choices available to actors on the one hand. On the other hand, however, they stabilize mutual expectations of actors and thus provide a relatively stable context for strategic action (Windhoff-Héritier 1991: 40-43, see also Immergut 1992, Mayntz and Scharpf 1995).

[3] The distinction between thin and thick institutions is borrowed from Checkel (1998).

Figure 1 Institutions and the policy process

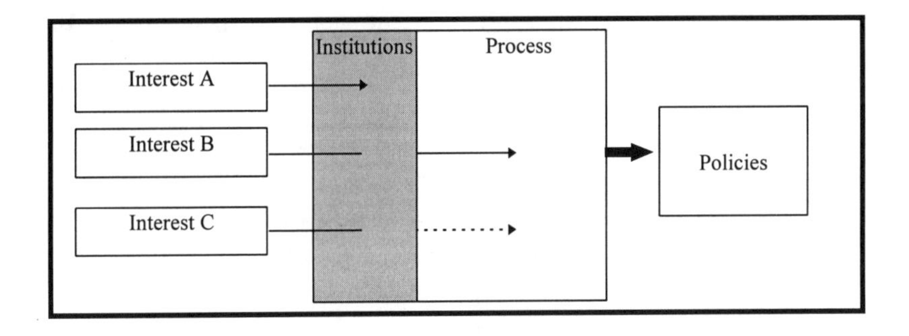

The concept of institutions is, however, not as thin as the definition of hardcore rational choice institutionalists, who deny any effect of institutions on the definition of interests of political actors. This study will assume that political institutions do not only provide the structure for *given* interests. Institutions might also contribute to the construction of interests. In his book about the power of Keynesianism Hall argues that institutions, by providing routines linked to processes of socialization and incentives for certain kinds of behaviour, contribute to the very terms in which the interests of critical political actors are constructed. Furthermore, in many instances the routines that have been institutionalized into the policy process filter new information and affect the force with which new ideas can be expressed (Hall 1992: 91). In other words: institutions are seen not only as a physical frame influencing the distribution of resources and structuring power and authority relations but also as a cognitive and normative frame. '*Institutions affect the substantive outcomes of choices by regulating the access of participants, problems and solutions to choices, and by affecting the participants' allocation of attention, their standards of evaluation, priorities, perceptions, identities and resources*' (Olsen 1991: 73).[4] Hence, the study is in line with the organizational sociologists branch of neo-institutionalism as far as possible *effects* of institutions are concerned, though it uses a thinner *definition* of institutions. Institutions do not *determine* interests. Actors actively seek to accomplish their goals by strategic interaction and the exchange of resources, such as legitimacy, expertise, or money, but their interests, strategies and resources are shaped partly by institutions. This is partly because these institutions also

[4]This quote and the following section provide examples of recent developments in new institutionalism which are to be seen in the broader context of the so called "cognitive" turn of a number of social science disciplines. Whereas the notion of constructed preferences has a rich tradition in sociology it is a rather new focus in political studies (see for an overview Nullmeier 1997, Fischer and Forrester 1993). Examples are for policy science, Hajer (1995), Majone (1989) Sabatier (1993), and for international relations theory, Goldstein and Keohane (1993), Haas (1992), Jachtenfuchs (1993), Risse-Kappen (1994). This list should not suggest, however, that these authors share a common set of epistemological or ontological assumptions, definitions and methods. The opposite is true.

affect the perception of the nature and significance of conditions and contingencies, such as the economic situation or the type and intensity of an environmental problem. To give an example: it is quite likely that the allocation of competencies for the environment within the government matters. 'Waste' as a political problem will probably be perceived in a different way by a civil servant socialized within a ministry of public health, a ministry of trade and industry, or a ministry for the countryside.

Figure 2 Institutions as cognitive and normative frames

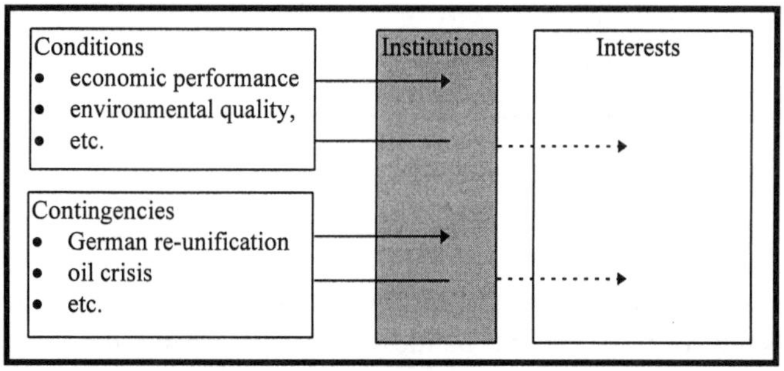

Conditions
- economic performance
- environmental quality,
- etc.

Contingencies
- German re-unification
- oil crisis
- etc.

Institutions

Interests

Ideas as frame and force

It has been said that institutions provide not only a physical frame but also a normative and a cognitive one. Neo-institutionalists have argued that not only the formal and informal rules of the game have this mediating function but ideas have this function as well. As Weir (1992) emphasizes, ideas can play an independent role in causing existing groups to rethink their interests and form alliances that would not be possible under an older system of ideas (Weir 1992: 190). Ideas in this sense can conceptualize as policy paradigms. Policy paradigms are issue-area doctrines. They are an '*overarching set of ideas that specify how the problems facing them are to be perceived, which goals might be attained through policy and what sorts of techniques can be used to reach those goals*' (Hall 1992: 91). Like institutions, ideas or policy paradigms also are '*cognitive maps*' (Axelrod 1976) that structure the very way in which policy-makers see the world and their role within it (Hall 1992: 92). The perception of interest, conditions and contingencies is therefore mediated not only by the formal and informal rules of the game, it is also structured by major ideas. Examples for such ideas are, for instance, the macroeconomic doctrines of monetarism and Keynesianism (Hall 1986, 1992). In the area of environmental protection the most important set of ideas is probably formed by the ideas of human centred environmental policy vs. ecological-oriented environmental policy and - more important in this case study 'economic feasibility' vs. 'ecological modernization'. The adherent of an 'economic feasibility frame' assumes that environmental protection is a burden on economic performance, while the ecological modernist believes that both can be reconciled or even, that environmental protection is a precondition for economic performance.

Figure 3 Ideas as frames and forces

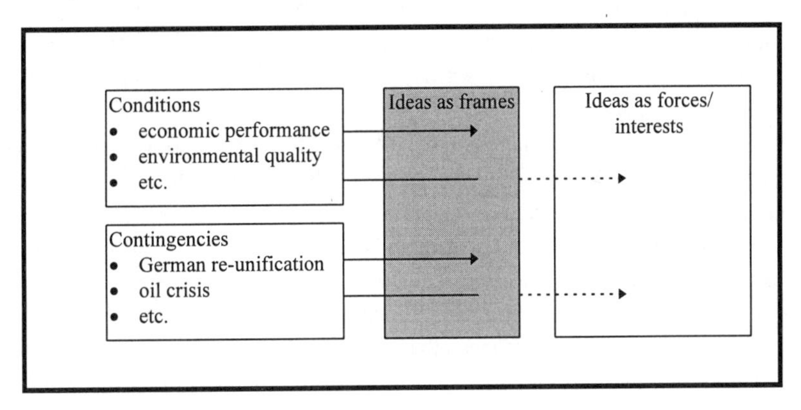

That the two frames - institutions and ideas - are presented in two different figures should not suggest that they function independently of each other. As a matter of fact, ideas, interests and institutions are interacting in many ways. Putting all the possible interactions in one table would contradict the purpose of simplifying the argument.

1.2.3 Convergence or persistence of national policies?

The previous sections provided the conceptual and theoretical framework for answering the research questions whether, how and why German, Dutch and British packaging waste policies differed in the early 1990s. The review of literature from the 1970s and 1980s has revealed that though advanced industrialized societies share a number of secular political developments, they differ in the answers they provide to common contingencies. In the 1990s, the image prevailed that national diversity was increasingly put under pressure due to technological change, increased trade between advanced industrialized countries, and political integration. Technological innovations facilitate and lower the costs of information exchange, travel and transport. As a consequence, international trade in goods and services intensifies and capital and labour become more mobile. These developments are amplified by international treaties and organizations such as the World Trade Association, NAFTA and the EU, which are dedicated to market integration through the reduction of trade barriers (Paqué 1995, Unger and van Waarden 1995). The general picture is that these developments result in fewer cross-national differences and more cross-national similarities in policies. Moreover, this convergence is supposed to result in a low level of regulations. National governments are restricted in their autonomy (Strange 1995) and continuously forced to abolish regulations or at least to lower their standards (Scharpf 1995).

This image is disturbed by recent empirical research which does not confirm a general trend towards deregulation and convergence. S. Vogel (1996), for instance, looked at regulatory developments in telecommunications and financial services. Both sectors are particularly exposed to technological change and economic globalization. Examining Britain and Japan in more detail, he concludes: '*the governments of the advanced industrial countries have not converged in a common trend toward deregulation, but have combined liberalisation and reregulation in markedly different ways. These governments have achieved different degrees of liberalisation, adopted particular types of reregulation, and developed distinctive new styles of regulation*' (S. Vogel 1996: 4). A book edited by Unger and van Waarden comprised a number of case studies in the area of economic, social and environmental policy. In the conclusion the editors state that '*most of our chapters find tendencies towards convergence as well as divergence. The majority expect divergence*' (Unger and van Waarden 1995: 29, see also Berger and Dore 1996 and Hollingsworth and Boyer 1997). Even in cases where there is convergence, this does not mean that the convergence will be on a low level. D. Vogel observed in a number of cases in the area of environmental policy and consumer protection a race to the top rather than a race to the bottom (D. Vogel 1995). This finding is confirmed for a case of international banking analysed by Genschel and Plümper (1997).

The patchy picture of convergence and divergence emerges also in findings of European Union research[5]. Several scholars have suggested convergence like a development towards a regulatory state in Europe (Majone 1994, 1996a, 1996b) or the emergence of a distinct European policy style (Mazey and Richardson 1996). Still, the results are rather patchy. A review of the findings reveal mixed patterns of the influence of European integration on national capacity for autonomous action and in particular, the extent of convergence of national policies in Europe. Eichener (1993), for instance, analysed health and safety at work-place regulations and argued that there has been a strong impact of European regulation on national regulatory policies and practices. Adaptation pressure has been particularly strong because the level of norms and standards adopted on the European level went far beyond the least common denominator across the Member States. More recently, a number of case studies have been presented in an edited volume by Majone (1996a), including not only social regulation but also economic regulation. These cases also show an '*extraordinary impact of European laws, policies and judicial decisions on the action and behaviour of the Member*

[5] Until recently, the EU was less researched than one might expect. The impact of this extreme example of political integration on national policies challenged the discipline of comparative policy analysis rather late. European integration research has been for a long time the domain of formal institutional analysis mostly done by lawyers and by macro integration research informed by international relations theory. When empirical case studies were done then, as Jachtenfuchs criticised '*weitgehend theorielos...(und) ohne systematischen Zusammenhang*' (Jachtenfuchs 1993: 183). Moreover these studies did not look systematically into the interrelationship of European and national policymaking. In recent years, however, the interaction of national diversity and European integration has been more systematically examined. Scholars of international relations and comparative public policy have been increasingly engaged in in-depth analyses of policy developments and regulatory change in the EU context (for an overview: Jachtenfuchs and Kohler-Koch 1996, Wallace and Wallace 1996).

States' (Majone 1996b: 265). Other studies, however, suggest that the impact of European integration on domestic politics is rather modest. Golub, looking at a number of environmental directives, emphasizes that national policy preferences have not been seriously affected by European integration (Golub 1997). And Kurzer, analysing a case in the area of social policy, concludes that '*national governments have lost none of their important decision making capacity*' (quoted in Kohler-Koch 1996: 365).

The review of literature therefore reveals a mixed picture of convergence and persistent diversity of national policies in a decade of globalization and increased European integration. In the following section, potential mechanisms of convergence - both related and unrelated to EU integration will be discussed more in detail. In addition, factors will be identified that may retard or prevent convergence and cause a persistence of national approaches. This section will benefit from the new institutional framework developed earlier, since this theory helps to identify and clarify the forces and obstacles of policy transformation and regulatory change. These new mechanisms are added to the existing explanatory framework in order to explain the extent and the direction of policy transformation in the case of packaging waste and the role of European integration therein (research questions three and four).

1.2.4 Mechanisms of convergence and persistent diversity

Scholars who analysed the similarities in the development of advanced industrial countries in the 1950s and 1960s explained this convergence mainly by the logic of industrialization. They claimed that the characteristics of the similar technological and economic developments pose a set of common political challenges and problems to advanced industrial states. The logic of industrialization would result in the convergence of both structural characteristics of the political system and of public policies (Bell 1973, Galbraith 1967, see Freeman 1985: 467).

Goldthorpe argued in 1984, however, '*that the convergence thesis was very much a product of its time - that is, of the "long boom" of the post-war period...In other words, there was no place in the scenarios of convergent development that were elaborated in the 1950s and 1960s for the severely troubled phase in the economic history of the Western world which actually began in the early 1970s*' (1984:317). The divergence tendencies in the 1970s were related to the decrease in influence of the United States and the increase in importance of domestic structures. Katzenstein, for instance, explained the convergence of policies of advanced industrial states like Germany, Britain, France, Italy and Japan in the early post-war years with the great political and economic predominance of the US. Its presence and preferences deeply affected the political choices of these countries (1978: 296). In the 1970s, however, differences in domestic structures became more important and resulted in a divergence of industrial policy (297).

In the more recent discussion on convergence two basic forces have been identified: competition and trans-national political cooperation (Unger and van Waarden 1995, Vogel 1995). It has been argued that as a consequence of economic internationalization nation states '*have to compete for the same pool of production factors, goods, and markets. They are under pressure to attract or keep capital, labour, and markets for their goods and services*' and countries are also under pressure to compete for the location of such enterprises (Unger and van Waarden 1995: 16). The general image is that this competitive pressure results in convergence on a very low level. Countries will try to surpass each other not only with low wage levels, and lower costs of social security, but also - and this is important for the purpose of this study - with less regulatory restrictions to business[6]. There is, however, an alternative convergence thesis which is based on the idea that strict regulation need not be a competitive disadvantage. High regulatory standards may favour domestic producers. Once a rich country with a large market has adopted a higher standard, foreign producers are forced to adopt it if they do not want to lose access to this market. Other countries may react by raising their own level of regulations, thus starting a regulatory race to the top (Vogel 1995: 5-6, 249pp.)

The second reason for convergence lies in political cooperation. The most substantial forms of political cooperation come about by legal and political integration, in particular by the conclusion of international treaties such as the WTO, NAFTA and the EU. The goal of these forms of political cooperation is to break down the barriers to trade and factor mobility. Unlike the WTO and NAFTA, the EU has the competence to adopt supranational legislation. These rules often compensate for the breakdown of national non-tariff trade barriers (Unger and van Waarden 1995: 18, see also Genschel and Plümper 1997). Both developments, the increasing invalidation of national measures and the adoption of supra-national legislation may be factors for convergence of national policies. Therefore, forces for convergence may be particularly strong in the EU. It is the most institutionalized form of international cooperation and it is dedicated to increased competition through market integration.

European integration

With the treaties on the European Community for Coal and Steel (ECCS 1951), the European Atomic Community (EURATOM 1957), and the European Economic Community (EEC 1957) a process started in which European nation states transferred legislative, executive and judiciary powers to intergovernmental and supra-national institutions. The European Communities were founded because of the experience of the Second World War. They were, however, designed as economic communities rather than political ones. The main objective of the EEC was the creation and maintenance of a common market enabling the free movement of persons, goods, capital and services. The history of the EEC reveals, however, that European integration has gone beyond pure market integration. In the last decade, European legal, political and economic integration has increased in speed, scope and depth. Milestones have been the Internal Market Programme (1985), the Single European Act (1986), the Treaty on the European Union (Maastricht 1993), and the Treaty of Amsterdam (concluded in 1997).

New institutionalism relates political choices to national political institutions. These institutions are deeply rooted in national history, and show remarkable persistence over time.

---

[6] Until the 1970s the convergence and divergence debate centred in particular around industrial and to a lesser extent social policy. This reflected the redistributive and distributive character of most public policies in these decades. In the recent discussions, however, regulatory policymaking gains prominence, such as the regulation of certain economic sectors and social regulation, including standards from the environmental, consumer or labour protection.

European integration, however, resulted in substantial institutional change. The European Treaties established competencies and procedures on a supranational level. They created a new set of rules that define and structure a new political arena above the nation states. The new rules are hierarchically above national legislation. They therefore restrict national sovereignty. Moreover, countries like the UK and the Netherlands who never experienced judicial review of their legislation have to accept that an extra-national body may invalidate parliamentary acts. The institutional change means for domestic actors that '*the game in the principal arena is nested inside a bigger game, ... the set of available options is considerably larger than in the original one' (Tsebelis 1990: 8).* The political conflict is not only structured by new rules of the game, it is also played by more actors, most notably other Member States, the Commission, the European Parliament and the European Court of Justice.

When institutions shape political choices, what is the impact of institutional change? Does the new level of access change the relative power of ideas and interests? And in doing so does it affect national policies? Does this institutional change result in a convergence of national policies? These questions can only be answered empirically.

There are a number of direct and indirect effects of European integration on Member States. The direct effects come about by negative and positive integration. Though both institutional mechanisms point to a convergence of national policies, there are a number of possibilities for Member States to safeguard their nationally distinct approaches. An indirect effect of European integration that may result in convergence is policy-oriented learning. But there are also forces that may prevent learning and convergence. A closer look is therefore necessary. The following part will detail the new mechanisms that either stimulate or prevent the convergence of national policies.

Negative integration

The EEC Treaty and its successors established rules of the game primarily aimed at market integration. Member States are not allowed to erect barriers to trade and have to abolish regulations that do so. In the absence of community law, Member States may adopt national standards for production and marketing. But Article 30 of the Treaty, for instance, stipulates - among other things - that quantitative restrictions on imports and all measures having equally effect are prohibited. Member States are not allowed to prevent the import of goods which are lawfully produced in other countries, even if these do not comply with its national standards. The Court of Justice has explicated this principle of mutual recognition in its famous Cassis de Dijon Case (120/78, ECR 1979: 649-675; see for details chapter 8). Several authors have argued that mutual recognition may safeguard national diversity in the short run. In the long run, however, mutual recognition forces Member States into a race to the bottom, hence to a downward convergence of national policies due to increased competition pressure (Rehbinder and Stewart 1985: 7, Scharpf 1995).

The common market provisions are guarded and enforced by the Commission and ultimately by the European Court of Justice. Member States and the Commission can bring another Member State before the Court if they perceive an infringement of the Treaty's obligations (Article 169, 170 of the Treaty). If the Court decides that a national measure is not in compliance with the Treaty, Member States are required to take the necessary steps to comply with the judgement of the Court (Article 171).

The EU also provides Member States with institutional opportunities to avoid or circumvent the mechanisms of negative integration. Member States which are accused of not complying with Treaty provisions dealing with the single market, have two possibilities to prevent the modification or abolition of their policies. First, they can ask for an exemption. These legitimate exemptions are partly provided by the Treaty itself - see, for example, Article 36 for exemptions to the free movement of goods - and partly by national policy motives such as environmental protection, if they constitute mandatory requirements, also called the Rule of Reason. The European Court of Justice has increasingly shown in this context that it accepts national arguments (see for details chapter 8).

The second possibility to protect national policies against deregulation pressures is to make it a subject of positive integration. Dehousse claimed that '*those Member States that favour high protection levels increasingly prefer to press for a decision at Community level, rather than seeking an escape clause which might harm the interests of their own producers*' (Dehousse 1992: 397). In other words, Member States may try to translate their own policies into European regulations and directives.

Positive integration

Member States are not only required to adapt to the Treaties, the primary legislation. The European Union has adopted thousands of binding regulations and directives to create the common market or to pursue other policy objectives. These manifestations of positive integration have to be incorporated in national law and must be implemented and enforced. The pressure to implement and enforce European directives speedily and properly has been strengthened by ECJ case law (see Prechal 1995). Many directives confer rights to individuals. These individuals can rely on these rights before national courts (Direct Effect, see for example 148/78 [Ratti] ECR 1979, p.1629). Moreover, the ECJ held that in cases where directives confer rights to individuals, Member States can be held liable if individuals suffer damages because of inaccurate implementation (Case C-6/90 [Francovich], ECR 1991, p. 5357). Finally, national legislation has to be interpreted in accordance with relevant directives (Consistent Interpretation, see for example 14/83 [von Colson en Kamann] ECR 1984, p.1189).

As in the case of negative integration, Member States have a number of possibilities to mitigate adaptive pressure of positive integration. They can first do so by trying to prevent the adoption of certain directives or regulations. They can veto initiatives in cases where unanimity is required. Under the conditions of qualified majority voting, they can seek to organize a minority large enough to block the decision. When a blockage is not possible, or from the point of view of the Member State not preferable, Member States can try to shape the draft according to their own preferences and practices.

One of the most comprehensive and systematic pieces of research that focused on the interrelationship between national and EU policymaking has found that Member States indeed try to defend their national practices on the EU level (Héritier *et al.* 1994, 1996). The research group focused on clean air policy in three European countries: the United Kingdom, Germany and France. The authors maintain that policymaking in the European arena is shaped by a competition of existing policies. '*In diesem regulativen Wettbewerb wirkt einmal der eine oder der andere Mitgliedstaat als Schrittmacher der europäischen Politik, während sich die übrigen Länder einem stärkeren oder schwächeren Anpassungsdruck ausgesetzt sehen*' (1994: 2). The logic behind this regulatory competition, in which the Member States try to transform their own policies into European law, is to maintain national problem-solving strategies and institutions, to minimize legal adaptation costs and to create or to secure competitive advantages. The research group also emphasized that political pressure works in both directions. European policies and strategies clash with established national problem-solving approaches, institutional traditions and inter-organizational relations. These national features correspond to a different degree with supranational strategies implying varying degrees of adaptation pressure. The European legislation is a result of regulatory competition including vertical interaction between European institutions and Member States on the one hand, and horizontal interaction between Member States on the other hand. Héritier *et al.* argue that those nations who act as pace setters and who are able to form a coalition with the Commission are more successful in this regulatory competition than other countries. As a consequence, the new policies resemble some national preferences and practices more than others. Policies of the pace setters are more likely to persist than those of the laggards (Héritier *et al.* 1994: 2-3).

Secondly, Member States can weaken the adaptation pressures caused by positive integration in the implementation stage. European legislation, in particular directives, provides discretionary space for Member States to determine the manner in which the directives will be implemented. This national leeway allows room for national preferences. Moreover, and less elegantly, Member States can weaken adaptation pressure by lax and delayed implementation (see Eichener 1996), though often legal risks are involved (see above). Since policy implementation is primarily done by the Member States, national institutions and traditions are likely to shape the implementation to a certain extent. One can therefore expect variations in the implementation of European rules in the Member States, given apparent national implementation styles as they are embedded in national institutions and underpinned by cultural and ideological traditions (Siedentopf and Ziller 1988).

The adaptation pressure produced by positive integration depends also on the degree of harmonization which is intended. In the case of minimum harmonization, Member States are permitted to adopt measures that are more but not less stringent than European directives (Rehbinder and Stewart 1985: 7-8)

It is not always the case, however, that Member States want to defend existing policies and practices. Governments may use the new strategic context to change domestic programmes either to escape from constraints imposed by domestic actors or as a result of a learning process, or as a combination of both. Both rationales point towards a convergence of national policies. Cases where national governments used the European arena to circumvent domestic constraints have been described by Milward 1992 who referred to the Belgian privatization and the Italian monetary discipline case. The role of policy learning as mechanism for policy change is theoretically and empirically so important that it will be elaborated on in more detail below.

Policy learning

Policy change may come about as a reaction to pressure from competition, as an adaptation to supra-national legislation and decisions, or as an outcome of a bargaining process. But policy change may also be the result of learning processes. In the following part of this study the concept of policy learning will be elaborated on and its significance in the context of European integration will be argued.

Approaches focusing on policy learning have been applied for more than two decades. In recent years, however, such approaches have been more frequently used and have been refined, by new institutionalists such as Hall (1993). A pioneering study was done by Heclo (1974), who analysed British and Swedish welfare policy during the initial decade of the century. Heclo viewed policy change as a product of (1) social, economic and political changes and (2) the interaction of people within a political subsystem, involving both competition for power and efforts to develop more knowledgeable means of addressing various aspects of the policy problem. In the words of Heclo: '*Governments not only "power" (or whatever the verb form of that approach might be); they also puzzle. Policymaking is a form of collective puzzlement on society's behalf; it entails both deciding and knowing*' (Heclo 1974: 305-306).

The notion of policy learning has caused a lively debate in political science about its heuristical and explanatory strengths and limits, and its normative connotations. It has to be stressed that in political science policy learning is used in analogy to individual learning. Analogy treats group learning as autonomous, determined by group-level causal processes that correspond to the processes shaping individual learning (Parson and Clark 1997: 15).

Paul Sabatier, one of the main advocates of the policy learning approach, defines policy-oriented learning as '*the relative enduring alteration of thoughts or behavioural intentions that result from experience and are concerned with the attainment (or revision) of policy objectives*' (1993: 19). For the purpose of this study, however, it not relevant whether a political actor has learned as such. It is more important to know whether policy change is the result of policy learning. According to Hall, learning is indicated when policy changes as the result of '*a deliberative attempt to adjust the goals or technique of policy in response to past experience and new information*' (Hall 1993: 278).

The definitions of policy-oriented learning include a number of features that help to distinguish learning from other mechanisms of policy change. First, policy learning refers to '*relatively enduring*' (Sabatier) alterations of beliefs and preferences. Only and insofar as the shift of beliefs and preferences sustain for a longer period is it more than a opportunistic adaptation to a new situation and can one speak of learning. There must be a '*deliberative*' (Hall) attempt to adjust the policies in the light of experience and new information. It has to be shown that there was such an attempt and not a unintentional adaptation to a new situation.

Hall and Sabatier emphasize the role of experience. Policy change is not induced by imposed regulations or by the presence of powerful actors, but by voluntary lesson-drawing from experience. As another author emphasized in this context '*the central characteristic of emulation is the utilization of evidence about a programme or programmes and a drawing of lessons from that experience*' (Bennet 1991: 221). Learning is possible from success as well as from failure. It is important to note that learning is not used in a normative sense. Policymakers can learn the '*wrong*' things, just as a child can learn bad habits (Hall 1993: 293). *Any* changes in policies to which the previously mentioned characteristics apply are a result of policy learning. Moreover, whether an experience is a success or a failure is left to the interpretation of the actor (see Bovens and t'Hart 1996). In fact, political actors can learn different things from the same issue. The selection and interpretation of experience is structured by frames and institutions (see above). And learning is instrumental: members seek to better understand the world in order to further their policy objectives (Sabatier 1993: 19).

Learning from experience can take different forms. Rose (1993) developed a typology of lesson-drawing based on the degree to which policymakers use experience. Policymakers may use programmes elsewhere as an intellectual stimulus to develop a novel programme (inspiration). They may combine familiar elements from programmes in a number of different places to create a new programme (synthesis), or they may combine the elements of two programmes (hybrid). Policymakers can also adjust for contextual differences a programme already in effect in another jurisdiction (adaptation), or enact more or less intact a programme already in effect in another jurisdiction (copying, Rose 1993: 30).

Policy learning and European integration

Policy learning takes place on the basis of own and others' experience. In the past, however, drawing lessons from own experience was more important than learning from abroad. In fact, as Hall emphasizes, policy responds even less to social and economic conditions than it does to the consequences of past policy. Paraphrasing Weir and Skocpol, he says that the interests and ideals that policymakers pursue at any moment in time are shaped by 'policy legacies' or 'meaningful reactions to previous policies' (Hall 1993: 277).

Developments in information and communication technology, however, may help getting lessons diffused more often and more rapidly across space. '*Worldwide information exchange allows for voluntary mutual learning and imitation between nations, people and their organizations and governments*' (Unger and van Waarden 1995: 2). Political globalization is one important stimulus for diffusion and application of knowledge generated somewhere else. In particular European integration may play a role for at least two reasons. First, the EU is an organization whose members share rather similar socio-economic and technological developments, roughly the same degree of industrialization and basic values. These factors may help policymakers to perceive experience of other countries as being useful to their own problems. Secondly, European policymaking is fragmented. The development of legislation takes place in issue-specific subsystems, in which expertise plays an important role and issues are frequently depoliticized. Depoliticization and the emphasis on expertise may facilitate the diffusion of ideas and lessons learned from experience elsewhere.

Policy learning is facilitated in the EU by a number of institutions. To give just two examples: First, under the conditions of majority voting, Member States try to build either passing or blocking coalitions. This generally occurs after the extensive exploration of all kinds of options suggested by Member States or the Commission, and careful studies of existing national programmes. Secondly, Member States have to notify any draft law which might actually or potentially restrict intra-EU free trade in goods (Directive 83/189/EEC). In the Committee concerned with the notified drafts (Committee 83/189) some 550 national laws, decrees etc. are discussed in depth, and well over 300 a year lead to comments or detailed opinions, or both (Pelkmans 1997: 4). '*Nowhere in the world is there such an intense, routine-based process of understanding one another's regulations. Learning is likely to be considerable here*' (Pelkmans 1997: 4).

Information about problems and solutions may not only influence European decision-making but also national political choices. Freeman, for instance, argues that '*traditional policymaking styles may lose their distinctiveness because of the tendency of policy makers in one country to imitate what they perceive to be the successful policies of their neighbours*' (Freeman 1985: 481). In other words, and referring to EU environmental policy: '*Gerade die intensive Interaktion in der EG läßt es als wahrscheinlich erscheinen, daß sich im Zuge der fortschreitenden Integration auch nationale Konzepte der Umweltpolitik auflösen bzw. verändern*' (Jachtenfuchs, Hey and Strübel 1993: 140-141). One should keep in mind, however, that even in cases when dominant ideas are on the international agenda, they get differential access to the political arena in the various countries and result in different substantive and procedural choices, owing to different state institutions.

Path dependency and national institutional traditions

Policy learning mainly occurs in the form of drawing lessons from ones own experience. However, existing policies not only provide the major stimulus for policy learning, they might also be the major obstacle to draw lessons. Once a policy has been established, it is often quite difficult to change, because '*new policies create a new politics*' (Schattschneider 1960). In the words of Rose: '*The precedence of inheritance before choice blocks lesson-drawing insofar as it is not possible to alter programmes already in place*' (Rose 1993: 39). The change of programmes is difficult for a number of reasons. Existing policies provide structures which create resources and incentives that shape the formation and activity of new groups. '*Public policies often create "spoils" that provide a strong motivation for beneficiaries to mobilize in favour of programmatic maintenance or expansion*' (Pierson 1995: 40). Hence, public and private interests may try to block policy change. More generally, policies once established may encourage actors to adapt in ways that lock in a particular path of policy development. The key concept in this respect is path dependency, a notion that swept over to political science from economic historians interested in the development of technology (see Arthur 1989, North 1990 pp. 93-104).

Policies may encourage individuals and organizations to make certain investments, to devote time and money to certain organizations, to stimulate co-ordinated action with other actors, to develop specialized skills or certain habits. Elaborate social networks emerge which, even when they do not lobby to prevent change, '*will greatly increase the cost of adopting once-possible alternatives and inhibit exit from a current policy path*' (Pierson 1995: 42).

Path dependency does not only hamper policy learning. Even in instances where policy changes are likely due to competitive pressures or threatening enforcement of supranational legislation, policymakers have to take into account interests benefiting from the status quo as well as sunk costs. Moreover, not only policies but also institutions create politics. Given the fact that national policies are embedded in national institutions and traditions it may be that pressures for change could meet strong resistance. National state-institutions may be so persistent that EU-involvement cannot change differences in domestic patterns, or it even produces further divergence. How much national policies will be influenced by convergence pressures depends not only on the preferences of governments and policy-induced path dependency, but also on '*how solidly national institutions are ingrained in the various societies*' (Unger and van Waarden 1995: 3).

Figure 4 Mechanisms of convergence and persistent diversity

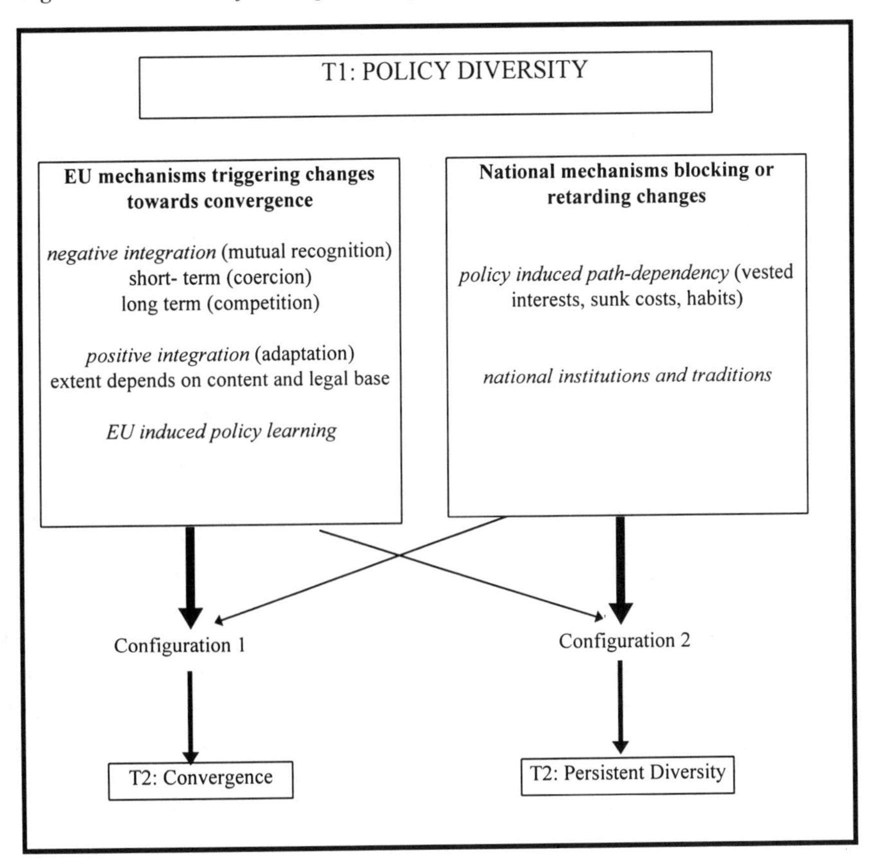

Summary
There are a number of mechanisms that either stimulate or frustrate policy change in the context of European integration. These mechanisms interact which each other in many ways. A simplified model is presented in figure 4.

Given contradictory forces and mechanisms of convergence and persistence it is not surprising that research has shown different degrees of convergence in different policy fields. These cross-sectoral variations suggest that the properties of the policy problem and the sector concerned matters. In the following section, therefore, the issue of this case study will be delineated and a number of its features will be sketched that point either towards convergence or to persistent diversity. More information on the issue of packaging waste will be provided in chapter 2.

1.2.5 Convergence, diversity and packaging waste

This study looks at convergence processes and tries to identify causes and mechanisms of policy transformation and regulatory change. Such a process-oriented research has to confine itself to a small slice of empirical reality and to focus the research on a manageable number of actors, political programmes and other aspects of policymaking. Concentration on a specific issue is facilitated by the fact that policymaking in advanced industrial societies is often organized in political sub-systems. This case will show, however, that policy issues also cut across policy fields, not least because the definition of the issue can change.[7]

This study deals with packaging waste policies. At the centre are policy programmes of authorities who have the formal power to adopt generally binding decisions, though in practice they might prefer more voluntary action. Packaging waste policies deal with the reduction and/or proper treatment of packaging waste, either by prevention, re-use, material recycling, energy recycling, incineration, or landfill. This study will be confined to the quantitative dimension of packaging waste. Qualitative prevention, that is the reduction of hazardous substances in packaging and packaging waste including PVC and heavy metals, will not be covered. Though this qualitative aspect has been addressed by some policy programmes, it remains a minor part of the general story and will be left out in order to simplify the intricate issues. The same goes for policies dealing with marking of packaging, such as labels for recycling or re-use.

The mixed picture of convergence and persistence of policies found by comparative studies suggests that specific properties of the policy issue and the relevant economic sectors might influence the relative importance of the independent variables and therewith the degree of convergence of policies. A cross-national, cross-sectoral analysis would probably be more elegant in accounting for structured variations caused by both national and sector-characteristics. This will not be done in this study. It is, however, possible to describe

[7] For methodological problems and research strategies concerning the identification and definition of policy fields/ policy sub-systems see Freeman 1985, König 1994.

characteristics of this case that might be to some extent issue-specific from a cross-sectoral perspective, and to indicate whether these characteristics serve as a bias towards convergence or persistent diversity and towards certain factors and mechanisms.

In the case of packaging waste, convergence might be hampered by the fact that the issue touches values and substantial economic interests. For many consumers, packaging is a symbol of our throw-away-society, while for large parts of trade and industry, packaging is the key to our modern self-service-oriented consumer society. Moreover, the prevention of packaging waste is not possible by end-of-pipe measures. While pollution into air, soil and water can be prevented by filters and other techniques, the prevention of used packaging or other types of solid waste is only possible by changes of products or production processes. Prevention therefore requires changes in behaviour of companies and/or consumers. Packaging waste policies, therefore, often intervene in economic processes. Packaging waste issues therefore touch deep core values concerning the role of the state in the economy. They may result in '*deep core conflicts*' (Sabatier) between policy actors, which might make policy-oriented learning more difficult, and therefore prevent policy change.

There are, however, at least three related factors that suggest a convergence of policies. Environmental policy in general and packaging waste policy in particular is:

a) a rather new policy field, which is less institutionalized than older policy fields such as agriculture;

b) a field in which policy ideas about problems and solutions have substantially changed and where scientific uncertainty and ambiguity is high and

c) a field in which policies might have effects on the free movement of goods between countries.

A new policy field

Environmental policy in general and packaging waste policy in particular is a rather new policy field. It is less than thirty years old. Pollution control regulations to protect public health have been enacted at least since the Middle Ages. In the late 1960s and early 1970s, however, the '*environment*' was redefined in a more holistic manner. What were once disparate topics began to be recognized as more or less linked. As environmental concern grew, so did environmental organizations. Governments adopted general environmental programmes and a wide range of environmental laws, and created new state and semi-state organizations to deal with them (see Goverde 1993, Hucke 1992, Jordan 1993). From a neo-institutionalist perspective one can argue that environmental policy is less embedded in institutions than old policy fields like health policy or agricultural policy, which are characterized by long established policy networks with well-entrenched interests and stabilized relationships (Naßmacher 1991: 205). Lehmbruch maintained in this context: '*Institutionelle "constraints" sind um so ausgeprägter, je deutlicher die jeweilige Konfiguration durch Merkmale struktureller Verfestigung charakterisiert ist, denen bestimmte Aktionsmuster korrespondieren*' (1988: 255). One can therefore hypothesise that the stronger the institutionalization, the more national policies are likely to resist.

Shifting ideas and scientific uncertainty

The '*age*' of a policy field is important not only in terms of institutional rigidity. In a less institutionalized policy field it is likely that ideas float more freely. Policymakers are more flexible and more open to policy-oriented learning. And indeed, reviews of the history of environmental policy point in this direction, showing a rapid change in the prevailing ideas about problems and solutions. In the 1970s, policymakers thought (a) that the character of environmental problems was well understood; (b) that environmental problems could be managed by a specialist branch of the machinery of government, focusing on specific environmental media (e.g. water, air) and employing end-of-pipe technologies; and (c) that in the setting of pollution control standards a balance had to be struck between environmental protection and economic growth and development. Since the 1980s, however, the prevailing ideas of problems and solutions have been substantially changed. Environmental policymakers now argue that (a) serious environmental problems are frequently not obvious and the link from cause to effect is often long and indirect; (b) that fundamental problems of environmental protection cannot be dealt with by end-of-pipe technologies, but need to be tackled at the source; (c) that the policy strategies adopted characteristically resulted in problem displacement, across time and medium, rather than problem solution; and (d) that environmental protection is a cross-sectional government task. Moreover, a number of policymakers also believe that environmental protection is not a burden upon the economy but a potential source of economic growth. The latter aspect is of particular importance. As Weale pointed out, '*it is in reconceptualizing the relationship between economy and environment that marks the most decisive break with the assumptions that informed the first wave of environmental policy*' (Weale 1992b: 76, see also Goverde 1993, Jänicke 1993, Jordan 1993, WCED 1987).

The dynamic in the field of environmental policy is also fuelled by scientific uncertainty and technical complexity of many environmental issues. This is particulary true of the issue of packaging waste. The relative environmental advantages of alternative types of packaging, packaging materials and waste management options are difficult to assess. There is much to be learned in these relatively new areas of regulatory policymaking (see chapter 2.5 for details).

Shifting ideas and scientific uncertainty point at a large potential for policy-oriented learning and lesson-drawing, and therefore towards policy change and possibly convergence.

Threat to the free movement of goods

Packaging waste policies focus either on processes such as recycling or incineration or on products, such as requirements for refillable packaging. In the latter case the free movement of goods might be threatened. Indeed, many of the conflicts surrounding packaging waste can be interpreted as conflicts between a high level of regulation and the principle of the free movement of goods. Market interventions in the course of preventive waste policy might threaten economic integration or more specifically the common market programme, which is still the essential feature of European integration. Political activities to harmonize national regulations are therefore likely in the European arena, implying policy change at least in some countries and thereby convergence (see for details chapter 8).

1.3 The comparative case study method

'*What know they of England, who only England know*!' (Kipling)

1.3.1 Introduction

The four research questions will be investigated by the comparative case study method. In this introduction the underlying logic and the opportunities and limitations of this method will be discussed.

Most scholars of comparative political analysis derive their mode of explanation - often implicitly - from Mill' s System of Logic (1843/1970)[8]. Two of Mill's methods are relevant for the purpose of this study: the method of difference and the method of concomitant variations.

a) *Method of difference*. If an instance in which the phenomenon under investigation occurs, and an instance in which it does not occur, have every circumstance in common save one, that one occurring only in the former; the circumstances in which alone the two instances differ is the effect, or the cause, or an indispensable part of the cause, of the phenomenon.

b) *Method of concomitant variations*. Whatever phenomenon varies in any manner whenever another phenomenon varies in some particular manner, is either a cause or an effect of the phenomenon, or is connected with it through some fact of causation (Mill cited in Faure 1994: 310, see also Dreier 1997: 554, King, Keohane and Verba 1994: 93, Skocpol l984b: 379).

Mill developed his system for experiments. Experiments, however, are rarely possible in political science, for ethical and practical reasons. Mill himself rejected the suitability of his methods for the social sciences. Still, the underlying logic of experiments helps to design social research. A general problem of comparative public policy analysis, however, is that the number of cases is rather restricted, while the potentially independent variables are many, too many to make meaningful statistical correlations (King, Keohane and Verba 1994: 118). Overdetermination is particular a problem in the study of political processes. The dynamic nature of its social and political reality cannot be captured by means of controlled experiments. There are always more potentially explaining variables than cases. '*(N)o matter*

[8] This does not imply, that comparative politics is a unified discipline within political science. On the contrary, disagreement prevails about what, when why and how to compare. Consult for a first overview of the debate, the chapters of Daalder and Keman in (Keman 1993a).

how perfect the research design, no matter how much data we collect,...,no matter how much experimental control we have, we will never know a causal inference for certain' (King, Keohane and Verba 1994: 79, see also Dogan and Pelassy 1990: 133, Faure 1994: 313, Keman 1993b: 50).

It is therefore important to note that case studies should be seen as analytical generalizations rather than statistical generalizations. Case studies try to expand and generalize theories which in turn have been informed by earlier empirical studies. In the practice of sciences, cases are either implicitly or explicitly viewed in the theoretical mould of a larger number of cases that guide the collection and analysis of data material (Lijphart, 1975: 160 see also Dogan and Pelassy 1990, Yin 1994: 10). In addition to the important role of theory for case studies, there are a number of designs and techniques that make causal inferences more certain. Three of them will be employed in this study: a) the '*most similar systems design (MSSD)*', also known as the '*comparable cases design*'; b) the technique of process tracing, and c) the application of counterfactual reasoning.

Most similar systems design

This study will employ the most similar systems design. The MSSD is based on the '*premise that systems as identical as possible in regard to as many constitutive features as possible represent the optimal samples for comparative research*' (Przeworski and Teune, 1970: 32). Common systemic characteristics are regarded as 'controlled for' while differences of an intersystemic nature are viewed as explanatory variables. Relating this back to Mill's logic of difference, the following theoretical implications are at stake (see Faure 1994: 312): (1) common characteristics are irrelevant in relation to the differences to be explained, because they obtain in cases or systems that share these characteristics, and (2) any set of variables that differentiates these cases may be regarded as independent if the said differentiation corresponds with the observed differences of the dependent variable. Hence, by choosing countries carefully, the number of potentially causal variables can be controlled. Those variables can often be identified by the review of earlier research. For example: the level of economic development may have an impact on national environmental policies. By choosing three countries with a comparable level of economic development, this factor cannot explain variations in the dependent variable, for instance, variations in the strictness of environmental standards. The strategy of holding as many background variables constant as possible is also called '*matching*' (King, Keohane and Verba 1994 205).

Process tracing

Though a most similar systems design controls for a number of potential explanatory variables, it is usually not sufficient to avoid the problem of overdetermination in a three-country study.[9] A research technique that increases the quality of causal inference in this situation is the historical method also known as '*process tracing*', that is, '*to connect the*

[9] One of the few examples where it was possible to identify the distinctive variable is provided by Immergut (1992). She could show that the variations in the public-private mix of the Swedish, French, and Swiss health system can be fully explained by differences in the formal institutions.

phases of the policy process and enable the investigator to identify the reasons for a particular kind of decision through the dynamic of events' (Tarrow 1995: 472 referring to George and McKeow 1985). This method is advocated by historically-oriented new institutionalists for the analysis of policy change. '*The way to do this [investigating policy change, M.H.] is by tracking the development and paths to influence that ideas and material interests take within the institutional context of policymaking*' (Weir 1992: 188). Thus, by a detailed reconstruction of European and domestic policy process important factors and causal mechanisms can be singled out (see also King, Keohane and Verba 1994: 85-87, 224-228, Skocpol 1984a).

Counterfactual analysis
Another technique that is employed to identify causal factors in a '*small N*' situation is counterfactual analysis. That is according to Max Weber '*the mental construction of a course of events which is altered through modification in one or more conditions*' (Weber 1905, quoted in King, Keohane and Verba 1994: 11). Generally speaking, one has to imagine what the world would be like if all conditions up to a specific point were be fixed and then the rest of history were rerun. The researcher shows that Y would have not occurred if X had not been given, other things being equal. In this study I will try to establish what would have happened in the three countries under review, without the European Packaging Directive. I will try to demonstrate, for instance, that without this Directive, the British government would not have introduced a regulation based on public law for packaging. By doing this the importance of the factor 'Directive' can be illuminated. It goes without saying that one can never be certain whether this causal effect exists or not, since one deals with a *counter*factual situation.

In short: It is difficult to single out the explanatory variables or to determine the relative explanatory power of a range of factors with a 'small *N*' setting. But by examining the case of packaging waste in the light of the increasing body of theoretical literature, by using a most similar systems design, by tracing processes in a detailed way, and by applying counterfactual analysis, national specific configurations of factors and mechanisms that can be identified that help to understand the variations, dynamics and inertia of such a complex phenomenon as policy transformation and regulatory change in the context of European integration.

1.3.2 The choice of countries

A careful choice of countries which will be compared can increase the reliability, validity, and certainty of conclusions. In this study Germany, the Netherlands and the United Kingdom have been chosen. What was the rationale for this choice? First, the focus of the study on convergence or persistence demands a certain variation on the dependent variable '*policies*' at the starting point (t_1). Without such a variation in the first period, it is meaningless to look for convergence, since they are already similar. Countries have to be chosen from those who varied in their policies prior to the initiative for comprehensive European packaging waste policy. This is the case for the Netherlands, Germany and the United Kingdom. Secondly, following the logic of the MSSD, countries have to be chosen which show similarities on some potential explanatory variables, to reduce the problem of overdetermination. Germany, the Netherlands and the United Kingdom have been chosen because they have an advanced economy, a democratic system, and a relatively high level of technological development in common. The character of public policy problems, such as the problem of packaging waste and the capacities to solve them, should therefore be quite similar. Thirdly, though the three countries are all parliamentary democracies, they vary significantly in their macro political institutions. This diversity will help to highlight the role of institutions in public policymaking. The following section elaborates on the rationale for the choice of the three countries.

Variations in the dependent variable

It was known from the outset of the study, that Germany, the Netherlands and the United Kingdom differed in their national approaches to packaging waste. It was well publicized that the German government imposed a far-reaching producer responsibility on industry. Coming to the Netherlands, I realized that the Dutch policy was somehow different: industry had agreed voluntarily to meet quite ambitious packaging prevention and recycling targets. This was done in the form of a covenant. Looking abroad, I realized that the issue of packaging waste was a rather minor topic in the British political debate.

The variation in policy approaches fitted with general research findings. The three countries under investigation differ generally in their approaches to environmental policy and also in the point of time at which they were policy innovators in the history of environmental policy (see Héritier *et al.* 1994, 1996, Weale 1992b). Until the 1960s, Britain had a leading position in environmental policy in terms of organization and legal framework. Between 1960 and 1990, however, it fell behind and acquired the image of the 'dirty man of Europe'. Yet, the Environmental Protection Act of 1990 suggests a substantive change towards acceleration and modernization, decentralization in terms of devolution of competence to local authorities (but also partly centralization), industrial self-regulation and public participation. Germany was considered as a pace setter in many respects, in particular concerning the reduction of emissions from industry between 1980 and 1988. In recent years, however, it has developed resistance against European initiatives concerning change of administrative and procedural innovations such as Integrated Prevention and Pollution Control, Eco-Audit, Environmental Impact Assessment, Freedom of Information. The Netherlands has served a specific function in the development and diffusion of new political concepts concerning environmental protection in the late 1980s. The Dutch National Environmental Policy Plan (NEPP, 1988) has been internationally applauded as a progressive strategy for environmental protection, in particular because of its long-term and consensus-oriented approach (Weale 1992b chapter 5). The ENDS-Report a specialized environmental monthly reported, for instance, that '*[British] Industry calls on Government to go Dutch on legislation*' (ENDS Report 1993, 222: 3). Dutch civil servants have also strongly influenced the general concepts of European environmental policy laid down in the Fifth Action Programme.

The longitudinal variation in countries which are policy innovators has the additional advantage that it can encourage the flow of ideas and stimulate substantial cross-national lesson-drawing between policy pioneers and policy laggards. Moreover, given the intensive entanglement of the Dutch economy with that of Germany, the Dutch case may provide useful insights into processes of learning and adoption among neighbours.

Controlled variables: economic and technological level and general type of the political system

Comparative studies have indicated that national approaches to public policy problems are partly shaped by their economic and technological situation (see for example Schmidt 1989a, Wilensky 1975) Since the amount of packaging waste is positively related to material welfare, the 'objective' problem pressure is higher in rich countries than in poor countries. Moreover, the strength of post material values is also positively related to an increase in income. And postmaterialists are more likely to become environmentally active than materialists (Inglehart 1995). Hence 'objective' and subjective problem pressure reinforce each other and sharp differences might be the result which is likely to overrule other explanatory variables. Hence, comparing developed countries with developing countries would produce so many substantial differences that no meaningful and specified conclusions could be reached. Even within the EU, the economic and technological gap between, for instance, the Netherlands and Greece is too big to neglect this factor in making comparisons (see Jänicke 1990).

Therefore, EU countries have been chosen which have no great differences in this respect. Germany, the Netherlands, and the United Kingdom belong to the economically and technologically advanced countries. Their gross domestic product per capita calculated as purchasing power parity was above the average of the EU Member States in the 1970s and 1980s (Brettschneider, Ahlstich, and Zügel 1992: 510). Moreover, as all other EU Member States, these three countries have democratic systems. But in contrast to Portugal, Spain and Greece, the democratic systems have existed for a long time, at least for as long as environmental policy has been on the political agenda. Therefore national variations in policies cannot be explained by changes in the general type of the political system.

Variations in national political institutions

Though the three countries under review have a parliamentary democratic system in common, they vary significantly in the institutional arrangements which structure the political conflict. Cross-national variations in political institutions have been identified as a potential explanation for differences in national policies and nationally distinctive paths of policy transformation and regulatory change. On a general level the countries exhibit different degrees of horizontal and vertical division of formal political power.

Table 2 Variations in national political institutions

		Horizontal division of power		
		High		Low
Vertical division of power	High	D		
			NL	
	Low			UK

The degree of division of power is related to the number of potential barriers to the power of a single party majority in government (Schmidt 1993: 387). The higher the division of power the greater is the number of barriers. These barriers also establish the opportunity structure which provide points of access to the decision-making process for political and societal interests (Immergut 1992, Kitschelt 1983).

The **United Kingdom** is the prototype of a majoritarian government system (Lijphart 1984). Political power is concentrated in the executive and parliamentary branches on the central level. The two are only weakly separated, partly because ministers have to be members of parliament. Moreover, the majority voting system favours a two-party system and makes coalition governments unnecessary. The party in government usually holds a large majority in parliament. The executive machinery is made up of a life-time career and usually bi-partisan civil service of personnel educated as generalists. The power of government is restricted neither by judicial review nor by plebiscites. There is no written constitution against which the courts may review acts of the executive, nor any tradition of the courts playing such a role. The centralized character of the UK is reflected in the fact that central government's share in the overall tax receipts has been some 87% in the 1970s (Lijphart 1984: 178). Though local authorities perform important functions, they are instated by central government and their powers are not constitutionally guaranteed. The role of local authorities was diminished under the Thatcher government by increasing control on their income and spending and by creating alternative local agencies to deliver policies (King 1993; see for a general overview Sturm 1994, Kavanagh 1996).

Germany is, according to the typology of Lijphart, a majoritarian-federal type of government (Lijphart 1984).). The central government is exposed to more checks and balances. It has therefore been labelled as a '*semi-sovereign state*' (Katzenstein 1987). Germany has a proportional voting system, with a 5 per cent threshold. There have always been more than two parties represented in the national parliament. The national government is usually made up of a coalition of parties, which force the dominant party into compromises. The individual ministries enjoy a comparatively high degree of autonomy due to the '*Ressortprinzip*' (Dyson 1992: 16). This autonomy of ministers as the head of departments is strengthened further by the practise of government, where parties consider 'their' departments as domains that are normally 'off limits' to the partners in the coalition (Grande 1987, referred to by Lehmbruch 1992: 37).

The power of the national government is restricted by a number of constitutional devices. Due to the presence of judicial review, legislative acts of parliament can be challenged in the Constitutional Court. In addition, there is an elaborate system of legal devices that allow individuals to challenge the implementation of policies in administrative courts. The power of the central government is also restricted by the division of legislative and executive competencies between the national level and the *Länder* level. In contrast to US federalism, however, the German constitution stipulates in many instances the need for joint decisions of representatives from both levels of the political system. Therefore federal legislators depend to a large extent on the cooperation of the state governments to adopt and implement policies

(Schmidt 1989b: 77). In addition, the constitution guarantees local authorities a certain degree of autonomy. The executive machinery is driven by a civil service which is made up of professionals rather than generalists. The federal government's share of the total tax receipts was 57% in the 1970s and therefore considerably less than in the case of the Netherlands or the UK (Lijphart 1984: 178; see for a general overview Ellwein and Hesse 1987, Katzenstein 1987).

The **Netherlands** belongs to the consensus democracies in the typology of Lijphart. Parliament and the executive branch are more separated in the Netherlands than in the UK partly because ministers are not allowed to be members of parliament. Due to a proportional voting system usually a large number of parties, five to ten, are represented in parliament, and the government is made up of two or more parties. The Dutch civil service is a professional life-time career service. The ministries and their policy divisions are relatively independent of each other. Each ministry is said to have its own legal culture, partly due to different legislative traditions and a lack of uniform recruitment requirements for civil servants (Rosenthal 1989: 239, van Waarden 1995: 358). The Dutch political system does not provide direct political participation through referenda and there is no judicial review of parliamentary acts. The courts cannot invalidate parliamentary acts. Citizens, however, can contest administrative acts under certain conditions and interest groups can force the compliance of organizations with existing laws via the court. Central government is sovereign but has legally delegated certain functions to provinces and local authorities. Though virtually all tax revenues (1970s: 98%) are under the control of central government (Lijphart 1984: 178), decentralized bodies perform important public tasks. There is in particular a large intermediary sector (see for a general overview: van Deth en Vis 1995, van de Heijden 1994, Gladdish 1991).

In addition to the institutions that determine the allocation of formal power within the political system, there are a number of institutions which also potentially shape the political conflict. One of them is the legal system. While the English system is based on case law, the German and the Dutch system have codified law. Hence, while in English case law tradition the legal system developed inductively on a case by case basis, German and Dutch law developed deductively by starting from principles elaborated in general rules, which were then applied to specific cases. Another institutional device which is important is the allocation of tasks within government. For example, this study will show that it makes a difference whether recycling policy is the domain of the Ministry of Trade and Industry as in the case of the UK, or the responsibility of the Ministry of the Environment, as in the case of Germany and the Netherlands.

These national political institutions provide the formal rules of the game. They structure the flow of ideas and the conflict of interests, and they shape the ideas, interests, strategies and the resources of political actors to some extent. There are, however, also informal rules of the game. Moreover, the national institutions are affected by European institutions. British parliamentary sovereignty, for instance, has been restricted by European law. It would go beyond the purpose of this section to detail the informal and European aspects of the institutional arrangements. Aspects will be discussed as they become relevant in the case study.

Table 3 MSSD based on logic of difference applied to the choice of countries

variable	***design***
policies in t_1 (dependent variable)	variation
general level of economic development	controlled-constant
general level of technological development	controlled-constant
basic political system	controlled-constant
national political institutions (one potential explanatory variable)	variation

The choices above do not imply that the control (matching) is complete. Not all factors commonly proposed as independent in comparative public policy theories have been treated so far. Still, this exercise increases the plausibility of causal inferences (King, Keohane and Verba 1994: 206). The arguments given for the choice of the Netherlands, Germany and the United Kingdom also do not imply that the choice was made only on the ground of theoretical and methodological considerations. As a matter of fact, France, as an economically important country with a quite distinctive institutional setting, and Denmark, also regarded as a environmental policy pioneer, would also fit the criteria mentioned above. Yet, in accordance with the theoretical framework a deep understanding - through a detailed reconstruction of policy processes and its context - is required. Given this requirement as well as time constraints, linguistic skills, and lack of financial resources I could 'only' choose those three countries.

1.3.3 Periods and structure of comparison

The analysis of the policy process and regulatory change has been organized along four research questions. The last and essential question of whether European integration brings about a convergence of policies demands for a combination of two modes of comparison: cross-national and longitudinal. An important question is, where to draw the line between the two periods which will be subject to the longitudinal comparison. Since the rationale underlying the study is the influence of EU integration, a distinction is made between the period in which packaging waste policymaking has been primarily a national domain and the period in which packaging waste policies have become Europeanized. Europeanization means that the European Commission issued an official proposal to adopt a packaging waste policy that covers all types of packaging and packaging materials in 1992. This distinction between the two periods is artificial, however. Packaging waste policies were discussed on a European level as early as the mid 1970s. In 1985, a European directive on containers of liquids for human consumption was adopted, and in 1988 the European Court of Justice decided on a case dealing with a national packaging waste regulation. However, the case studies will reveal that neither of these two instances played a decisive role in the development of national packaging waste policies. Two periods will be therefore distinguished: the national processes that finally led to the packaging waste policies in the early 1990s (t_1) and the transformation of these policies and the regulatory change hereafter, until sommer 1998 (t_2).

In order to identify convergence and divergence processes first, cross-national comparisons have to be made for both periods (t_1 and t_2), in addition longitudinal comparisons are required for each country (t_1->t_2). The policy elements of one country can be labelled '*persistent*' when the longitudinal comparison between both periods reveals '*many*' similarities and '*few*' differences. Policy change is indicated by '*many*' differences and '*few*' similarities. A convergence of policies has occurred when a comparison of both cross-national comparisons shows '*more*' similarities and '*less*' differences between the countries in 1997/1998 than in 1991. A divergence of policy styles has occurred when the comparison of cross-national comparisons has shown '*less*' similarities and '*more*' differences between countries in 1997/1998 than in 1991. The structure and object of comparison can be summarized as follows:

Table 4 Structure of comparison

Country **Period**	**Netherlands**	**Germany**	**United Kingdom**	**Mode of Comparison**
Policies until early 1990s (t_1)	Packaging Covenant (1991)	Packaging Ordinance (1991)	Environmental Protection Act (1990), Codes of Practise(1992)	*cross-national/ (synchronic)*
Policies 1991 to 1997/1998 (t_2)	Packaging and Packaging Waste Regulation (1997) Packaging Covenant II (1997)	Draft Amended Packaging Ordinance (1998)	Producer Responsibility Obligations Regulations (1996)	*cross-national/ (synchronic)*
Mode of Comparison	*intranational and (diachronic)/ longitudinal*	*intranational and (diacronic)/ longitudinal*	*intranational/ (diacronic)/ longitudinal*	***convergence/ divergence?***

1.3.4 Methods of data collection

The research questions demand intensive investigations of European and national policy processes over a decade. This cannot be done by experiments, surveys or statistical methods. The study applies the case study strategy for data collection and analysis. According to the seminal definition of Yin, a case study '*investigates a contemporary phenomenon within its real-life context, especially when the boundaries between phenomenon and context are not clearly evident*' (Yin 1994: 13). Since case studies '*cope with the technically distinctive situation in which there will be more variables of interests than data points...[a case study] relies on multiple sources of evidence, with data needing to converge in a triangulating fashion...*' (Yin 1994: 13). The use of multiple sources of evidence allows one to address a broad range of historical, attitudinal, and behavioural issues which are involved in the process of policy transformation and regulatory change. Another advantage of using multiple sources of evidence is the development of a '*converging line of inquiry*' (Yin 1994: 92). Any finding or conclusion is likely to be more convincing and accurate if based on several different sources of information, following a corroborative mode.

The data collection has relied on secondary sources, such as specialized journals and scientific literature and primary sources, in particular official documents and semi-structured interviews. The data have been analysed by qualitative content analysis. Since packaging waste is a controversial issue, a large amount of data has been available, including not only governmental ordinances, programmes, statements, letters, memoranda, administrative documents and the written evidence of speeches and hearings but also many articles from newspapers such as the Financial Times (FT), The Economist, Handelsblatt (HB), NRC, specialized journals such as the Environmental Network Data Service Report (ENDS), European Packaging and Waste Law, Ökologische Briefe (ÖB), and regular publications as well as specific statements of interest groups (a list of anonymous refereed written sources is annexed). In order to collect the written evidence specialized libraries and archives were visited including the one of the Afval Overleg Orgaan (Utrecht) and the Dutch Ministry of Environment (Den Haag), the Government Documentation Centre and Corporate Business Library of the University of Warwick, the House of Lords Library (London) and the Library of the Wissenschaftszentrum Berlin.

In addition, 20 interviews were carried out with key persons involved in the political processes on which I focused. The interviews were semi-structured, mostly taped and took one to three-and-a-half hours. Since a lot of information about the content of policies and the ideas and interests of the relevant actor has been yielded by written evidence, these interviews could be restricted to the key players in the process. In these interviews, the emphasis lay on the strategies of the different actors and the general features of the political processes. These key players have been selected according to the relevance of the organization they represented and the variety of perspectives they offered. The interviewees include representatives of national and subnational administrations, environmental interest groups, and business interests as well as experts (list of interviewees is annexed). Indications for the relevance of the actor have been the frequency with which they appear on hearings and conference lists, in newspaper articles and in interviews. In addition, I have attended a number of specialized conferences and had frequently contact with German and Dutch actors through my function as correspondent for the journal European Packaging and Waste Law, between August 1996 and July 1997.

2. The Politics of Packaging Waste

National policies and regulatory styles may be influenced not only by national institutions and traditions but also by properties of the problem at stake and the structure of the economic sector concerned. In order to facilitate comparison across policy fields, I will elaborate on characteristics of the policy field packaging waste. This chapter also provides a background to the political processes analysed in this study.

Like other problems of advanced industrial societies, the issue of packaging and packaging waste is complex in the sense of being influenced by plural interests, scientific uncertainty and dynamic developments. This chapter will discuss different aspects of the case packaging waste. It will first show that the increase in packaging waste and its complex composition is strongly linked to the development from local traditional agricultural-oriented societies into a global and modern industrialized one. Packaging waste is a very visible consequence of modernity (2.1). In some environmental areas such as air and water pollution and energy consumption one can observe an uncoupling of the amount of pollution produced and economic growth in advanced industrialized societies. This is not the case with regard to municipal waste in general and packaging waste in particular. Growth in production and consumption and an increased amount of packaging waste used to go hand in hand. According to many critical consumers the problem of used packaging belongs to the most serious environmental problems (2.2). Since the prevention of waste cannot take place end-of-pipe as in the case of water pollution or air pollution, policymakers who want to prevent waste have to intervene into production and/or consumption processes. But packaging has important functions for our modern and global society. Changes in policies can have strong redistributive effects, especially in the very competitive food market and for small packaging producers who occupy market niches. Strong and divergent industry and trade interests are at stake in such cases. And in contrast to other environmental issues such as FCKW, many sectors of industry and hundred of thousands of companies in all three countries under investigation are affected (2.3). In addition to strong and plural economic interests, the environmental impact of packaging waste and the different ways to deal with it are still poorly understood. These issues are rather technical and the degree of scientific uncertainty is high (2.4). It is therefore no surprise that the issue of packaging waste has become one of the most controversial environmental topics in the 1980s and 1990s in advanced industrial societies.

2.1 Packaging and modernity

It is a myth that all ancient and traditional societies lived in harmony with their natural environment. The pollution of air, water and soil, the depletion of resources and the reduction of biodiversity is universal across time. Archaeologists have revealed that the generation of waste is no eception to that (Rathje and Murphy 1994: 41-68). But without doubt environmental problems in general and the problem of waste in particular have increased in the course of the industrialization process of modern societies. Packaging waste is one of the most visible and most typical consequences of modernity. If we understand the process of modernization in the sense of Giddens, as a process of the dislocation of time and space, this argument become obvious (Giddens 1990: 17-29). The increasing time span between the production and the consumption of goods as well as the increasing distance between production and consumption made it necessary to create provisions in order to enable and facilitate transport, to protect goods and to keep them fresh. Related developments as urbanization and globalization increased this tendency. The development is probably most visible in the evolving structure of supermarkets as the dominant pattern of the distribution of consumer goods, also known as fast movable goods. The emergence of supermarkets has destroyed more local and more sustainable consumption patterns. Packaging is used to enable the shelving of goods, to enable self service, to hinder the theft of goods and to attract consumer attention. The latter is quite a challenge given the more than 18,000 goods offered at the same time at the same place (van Kempen 1992: 182). The rise of supermarkets in the distribution systems has, for instance, resulted in a decrease of refillables. Supermarkets satisfy the wish for easy access to a broad range of products and the convenient consumption and disposal of them. *'Supermarkets provide shelves of products beyond the wildest dreams of the shopper of the 1950s - from out-of-season vegetables flown in from around the world to time-saving processed or pre-cooked foods*' (Lang and Raven 1994: 124). Packaging also has the function of providing an image of luxury. Examples are lavishly packed chocolate candies or the one-way packaging by fast food chains which provide a feeling of abundance and affluence.

One might argue that the upsurge of post-materialistic values might reduce the problem. But new values do not automatically result in less packaging. For health and safety reasons information is needed, which is almost always provided on packaging. The more conscious consumers want more rather than less information. A case in point is the field of pharmaceutical products. Ironically, labelling for environmental protection reasons is also mostly placed on packaging. Therefore consumer protection goals and environmental goals are once again in conflict (Klages 1990: 47). More important yet for the amount of waste is the process of individualization and the problem of an ageing society. Both developments result in a decrease of the average number of persons per household. Since small portions need relatively more packaging than large portions, the amount of used packaging generated in smaller households is relatively larger than the amount generated in larger households, and packaged micro-wave and frozen foods are consumed more often in smaller households. The OECD summarized the factors contributing towards an increase in the quantity of packaging and packaging waste as follows: growing number of consumers, rising trend in consumer purchasing power, increasing movement of goods over long distances, rising trend in quality requirements (for example hygiene), trends towards small households, convenience needs (fast food, microwave and frozen foods, disposable beverage packaging), increased role of packaging in new marketing and advertising techniques to make goods more attractive (OECD 1995b: 21).

2.2 Packaging and the environment

'*Appropriate places for (refuse) are become scarcer year by year, and the question as to some other method of disposal...must soon confront us*' (US Federal Official 1889 (sic!), quoted in The Economist 29.5.1993).

2.2.1 Figures, not facts

'*Decisions about waste management take place in a statistical vacuum. Few governments have much idea of how much waste their citizens produce and what it consists of, let alone where it can be put*' (The Economist 29.5.93). The environment editor of the Economist exaggerates only slightly. One should be very careful with data available on the amount, composition and disposal routes of waste. There are substantial statistical problems involved. Generally the data situation has always been poor and slowly improving in the 1990s. The OECD stated in 1992 that '*a strong argument can be made that policy-makers do not have adequate data on waste generation and recycling*' (OECD 1992, quoted in Porter 1994: 13). The OECD listed those factors which vary and therefore affect the measurement of packaging waste and its recycling. The value of the figures depends, among other things, on a) whether one includes industrial and post-consumer waste or simply the latter; b) whether one measures material collected for recycling or simply the material used in producing a new product; c) whether one subtracts from the amount recycled contaminants or other materials that are removed in the production process; d) how one accounts for imports and exports in collected materials (OECD 1992, 82-83, referred to in Porter 1994: 11-12).

Moreover, the availability and validity of data varies between advanced industrial states. Generally speaking, the situation in the Netherlands as well as in Germany is much better than in Britain. The 1995 British white paper for waste management, for instance, was still based on figures for 1990 (DoE 1996b). But also in the Netherlands, for example, the situation is far from perfect. A wide range of figures is each describing the same phenomenon. Even quite systematic studies come to different solutions: in the context of the implementation of the Dutch Covenant, the figures of Cooper & Lybrands measured packaging attached to products (input), and the RIVM, measured the packaging at the consumer stage (output), differed for almost all materials by more than 10 per cent and in one case even 100 per cent (Commissie Verpakkingen 1993: 9-11).

The private household is one of the sites where used packaging, mostly sales packaging, end. Together with the packaging waste of small business (e.g. offices, shops), and other sites 'comparable' to domestic consumers (e.g. restaurants, schools, hospitals) this used packaging belongs to the category of municipal waste (OECD 1995b 7,8). According to the OECD definition, municipal waste is that which is collected and treated by and for municipalities (OECD 1995a: 158). Hence there is a distinction between '*Hausmüll*', 'Household Waste/Domestic Waste', '*Huishoudelijk Afval*' on the one side and '*Siedlungsabfall*', 'Municipal Waste' and '*Gemeenteafval*' on the other side. To the second category belongs besides household waste also '*hausmüllähnliche Gewerbeabfälle*' or '*Kantoor, Winkel Diensten (KWD) Afval*', bulky waste, market and garden residues. This does not mean, however, that the definitions within the same category delineate the same scope of packaging. To complicate things further, not all commercial waste can be seen as municipal waste. Industrial packaging waste, mainly transport packaging and industrial sales packaging, is not always collected by local authorities. A number of statistical problems are related to the existence of different categories. The Commission of the European Community stated that some countries measured packaging in domestic waste, while others measured it in all sectors, including industry and commerce (CEC 1990: 16, referred to in Porter 1994: 12,13).

Besides the distinction between household waste and commercial waste there is another important distinction, namely between transport packaging, secondary packaging and sales/primary packaging. These types of categories overlap in to a certain degree. Most of the commercial packaging waste is made up of transport packaging, while almost all household waste consists of sales/primary packaging. While policymakers and policies in Britain and the Netherlands primarily use the household/commercial categories, in Germany and on the EU level the distinction between transport and sales packaging is dominant. Neither analytical distinctions are done *l'art pour l'art*. This study will show that these distinctions are important for the politics of packaging waste, since transport/commercial packaging carries different economic and political logics as compared to sales/household packaging. While the recycling of packaging waste from commercial sites is cheaper, the recycling of packaging waste from households is more visible for the consumer and therefore more stressed by politicians (see for details below in this chapter).

In the absence of a comprehensive national packaging waste policy no systematic nation-wide data for the category 'packaging' was collected in the past. Since most packaging waste, some 60 per cent (OECD 1995b: 8), is generated by households and small businesses, it makes sense to look more closely at the figures provided for municipal and household waste generation. This is also done to highlight further the statistical problems involved. Note that some 20 to 30 per cent of the household/municipal waste is made up of used packaging.

Table 5 Municipal (M) and household (H) waste generation (kilogram per capita)

	1980	**1985**	**1990**	**1994**
Germany (figures up to 1985: West-Germany)	350 (M)	320 (M)	360 (M) (W-Germ. 340)	488 (H) 585 (M)
The Netherlands	390 (H) 500 (M)	360 (H) 440 (M)	410 (H) 500 (M)	468 (H) 566(M)
UK	310 (H)	340 (H)	350 (H)	no data available
OECD	430 (M)	440 (M)	480 (M)	no data available

Source: OECD 1995a: 159, VROM 1997a

These figures are based on surveys by OECD members. A study commissioned by the Dutch Ministry of the Environment concludes that the figures used for the period 1990-1992 are more reliable than those from earlier periods but still not sufficiently reliable to make valid cross-national comparisons. The main reason for this is that definitions used by Member States 'vary considerably' (OECD 1995a: 152) and that data bases are poor. In the case of the United Kingdom, for instance, no definition for municipal waste exists. Also, participation of local authorities in the national survey of household waste was voluntary. In 1992 only half of them took part. Figures were often not measured but estimated by the local authorities (VROM 1997a: 23-24). In general, countries with more sophisticated methods show higher figures. For the period 1993-1994 figures were based on the OECD definition of municipal waste and subsequently the quality of measurement improved. A convergence of figures between Member States took place, which can be related to more sophisticated methods. Still, experts say that a margin of +/- 10 per cent would be realistic. The general trend, however, is an increase in the generation of municipal waste, which may partly reflect more sophisticated measuring methods (VROM 1997a: 43).

As already said, until recently packaging was not used as a category in a systematic way. Therefore, only patchy evidence can be provided. According to the French Environmental and Energy Agency the amount of packaging in municipal waste increased in France from 36 kg per capita in 1960 to 120 kg in 1990, that is 440 per cent, whereas municipal waste in general increased from 220 kg to 360 kg, that is only 63%. A virtually continuous rise in the volume of packaging output has been observed in Germany since 1955, when records first began (OECD 1995b : 5). According to figures of Cooper & Lybrands and the RIVM, the amount of packaging (in weight) increased roughly by 20 to 25 per cent between 1986 and 1991 in the Netherlands (Commissie Verpakkingen 1993: 11).

2.2.2 The composition and fate of packaging

Most of the used packaging, in particular in the United Kingdom, ends in the domestic dustbin. Only a small proportion, bottles in particular, is refillable. The following table provides an overview of the refillable quota of three type of drinks in 1987.

Table 6 Share of refillables in D, NL, UK (1987)

	Germany	The Netherlands	United Kingdom
Carbonated Soft Drinks	72,9	88,8	19
Beer	84,3	93	23,3
Bottled Water	84	79,2	0,0

Source: Friends of the Earth 1992: 9

The packaging waste stream is made up of a variety of materials, combinations of materials and forms. The British market research institute Rowena Mills Association made an estimation for different Member States (Verpakken 2/1996: 45).

Table 7 Material composition of packaging

	UK	D	NL	F	I
glass	6	7	5	18	6
paper/cardboard	35	39	39	34	21
metal	23	22	14	15	17
plastic	28	29	36	25	44
wood	2	3	6	7	12
rest	6	-	-	1	-
total	*100*	*100*	*100*	*100*	*100*

Source: Verpakken 2/96: 45

The differences in the share of materials shown in the table above, reflect differences in consumption and production patterns. The French, for instance, produce and consume a lot of cosmetics, distilled drinks and wines and sell them in glass packaging. The following tables provide an overview of consumption sites, main area of use, composition, usual options for re-use and the status of recycling technology for the most important packaging materials.

Table 8 Glass, paper & cardboard, tinplate, aluminium recycling and recovery

Material	**Glass**	**Paper/Cardboard**	**Tinplate**	**Aluminium**
Consumption sites	private house-holds/small business	private households and commercial sector	see paper/cardboard	see paper/cardboard
Main area of use	semi-liquid and liquid foodstuff (drinks)	transport packaging (corrugated cardboard) sales packaging for „dry' goods (e.g. washing powder)	for private end users and small business: cans (foodstuff, drinks, chemical products) for commercial sector: cans, barrels	aluminium cans (e.g. drinks, aerosol) aluminium tear-off lids, tubes, dishes, foil aluminium-coating in composites
Composition of packaging	glass, different colours, usually clear, green, brown	pulp fibre	steel with tin coating	aluminium partly aluminium combined with paper or plastic
Usual option for re-use	re-usable packaging for drinks and other products	-	reconditioning of barrels and other bulk packs	-
Available recycling/ recovery methods	recycling of material to make new glass	recycling of material to produce paper/cardboard or recycling of material by making compost, energy recovery	recycling of material in steel production	recycling of materials in aluminium production
Status of recycling technology	practised world wide in glass production	recycling of materials: practised world wide as part of paper production	practised world wide	recycling of aluminium materials tested world wide

Source: based on OECDb 1995: 10-14

In the European Union, the most important waste management options are landfill (about 62 %) and incineration (about 25 %). Recycling amounts to some 8 per cent and composting to some 6 per cent (Das Parlament 9.7.93, S. 14). Most of the packaging waste was landfilled and incinerated until the early 1990s. Figures for the recycling of packaging waste are rare for this period. The following table provides an overview of the development of recycling for a number of waste materials, from which most stem from used packaging. As all other waste statistics, statistics about recycling have to be read with care. Most figures on recycling rates are supplied by industry organizations with a vested interest in making the recycling achievement look as good as possible (EC Packaging Report March 1994: 12). The OECD argued '*Even when one can agree on the definitional questions..., there can be differences in estimates because of the method used to gather the data...each participant in the process is free to choose a method which best represents its interests. Even without vested interests at stake, the choices will differ from country to country depending on time, money and tools available, and the degree of access allowed by those who control recovered material*' (OECD 1992, quoted in Porter 1994: 18).

Table 9 Plastic and composite recycling and recovery

	Plastic	Composites
Consumption sites	commercial sector, private end-users	beverage cartons: primarily private end-users, others: sales packaging private and commercial
Main area of use	transport packaging (foil) wide range of flexible and rigid sales packaging: foodstuff other liquid goods, goods sensitive to moisture (bottles, canister, bags, foils etc.)	Beverage cartons: liquid and semi-liquid foodstuffs, others: primarily foodstuff
Composition of packaging	great variety of packaging plastics e.g. polyethylene (PE), polystyrene (PS), polypropylene, polyvinyl chloride (PVS), polyethylene terephtalate (PET)	wood pulp (paper), aluminium, various plastics
Usual option for re-use	refillable beverage bottles and rigid containers for washing powder and cleaning products, re-usable boxes and pallets for transport packaging	-
Available recycling/ recovery methods	recycling or plastic materials (reshaping packaging material in other plastic objects), feed-stock recycling (turning plastic into multi-purpose plastic input stock), energy recovery: incineration of pure plastic not tested world wide, incineration of plastic as part of mixed waste technically feasible	recovery and recycling of paper used in beverage cartons is practised in Germany on a large industrial scale. Partial recycling of other composites using the main material in each case, energy recovery along with mixed waste
Status of recycling technology	recycling of plastic material: conventional methods with a limited market potential, feedstock recycling: new technology on the base of coal oil processes, pilot plants in Germany	recycling of beverage cartons introduced in Germany, recycling of other composite packaging in some cases together with the „main material'

Source: based on OECDb 1995: 10-14

Even the association in Brussels whose main objective is to promote recovery and recycling - the European Recovery and Recycling Association - did not have up to date inclusive figures for material recycling rates across the European Union in the mid 1990s (EC Packaging Report March 1994: 12). The following figures give only a general idea about national developments and cross-national variations.

Table 10 Development recycling quota paper & cardboard and glass for D,NL,UK

Material	Country	1980	1985	1990	1995
Paper/ Cardboard	D	33	40	40	90
	NL	46	50	50	62
	UK	30	28	32	49
Glass	D	23	44	54	82
	NL	47	49	67	74
	UK	5	12	21	28

The figures up to 1990 stem from the OECD (OECD 1995: 171). These figures include not only material from used packaging, but also from other sources. This is particularly important in the case of paper, where the larger part stems from old newspapers rather than packaging. In the UK for example, packaging material represents around 43 weight per cent of the production of paper and board industries (Key Note Market Review 1993a: 10). The figures for 1995 - at least officially - refer only to packaging material (Commissie Verpakkingen 1996: 19, DoE 1996:41, DSD 1996: 19). Until recently, there were no systematic data for the recycling of the other packaging materials available. The figures for 1994 (UK) and 1995 (D, NL) read as follows.

Table 11 Recycling quota plastic, tinplate and aluminium, D,NL,UK

	Plastic	**Tinplate**	**Aluminium**
Germany (1995)	60	64	70
The Netherlands (1995)	11	56	20
United Kingdom (1994)	5	14	16

Source: Commissie Verpakkingen 1996: 19,21; DoE 1996: 41; DSD: 1996: 19; Material Recycling Week August 1995

A number of general conclusions can be drawn. First, paper, cardboard, and glass are the materials most frequently recycled. Second, Germany and the Netherlands achieved substantially higher recycling quota than the United Kingdom. Third, from 1990 to 1995, the German recycling quotas increased rapidly, while the Dutch and the British quotas increased modestly. Finally, Germany recycled a substantial share of all packaging waste materials, while the Netherlands concentrated on the most economically effective materials such as glass, paper, cardboard and tin plate.

2.2.3 Packaging and environmental quality

Used packaging belongs to the artefacts of modern societies which are popularly perceived as an environmental problem. Research among consumers reveals that many of them find packaging to be an unnecessary waste of resources, not least because of its short life span (Milieudefensie 1997: 30). It has to be noted, however, that packaging waste is only a small fraction of the overall waste stream. From the 2000 million tonnes of waste generated in the EU in 1989, used packaging amounted only to 50 million tonnes, which is about 4 per cent. Agricultural waste (1,100 million tonnes), residues from mining and power plants (400 million tonnes), waste water sludge (230 million tonnes) and construction waste (160 million tonnes) make up the lion's share of the overall waste stream (Müllmagazin 4/1992). Still, as an OECD study concludes, packaging in terms of its weight and its volume makes a greater contribution to municipal waste than any other group of manufactured goods (OECDb 1995: 5). And packaging waste is highly visible for the consumer. Up to 50 per cent of the volume of the average household dustbin is made up of plastic bags, tins, bottles etc (OECDb 1995: 5).

Fifty million tonnes of used packaging in the EU means that each inhabitant produces 150 kg of it, annually. For the production of packaging resources and energy are needed. Moreover, '*it is estimated that for every tonne of waste produced by the consumer, five tonnes have been produced at the manufacturing stage and 20 tonnes in extracting raw material*' (WARMER Bulletin 11/94). Inappropriate waste disposal can lead to soil, water and air pollution because packaging may contain harmful ingredients such as heavy metals or chlorine (OECD 1995b: 16). As waste accumulates in landfill sites, suitable locations become scarce. Note that statistical comparisons usually use the weight of packaging as indicator in describing amounts. The volume of packaging waste has increased faster than the weight. This is important for waste collection and waste treatment, for it is the volume of packaging which determines the capacity of garbage bins, transport containers and landfill sites.

The decreasing capacity and the increasing technical requirements of waste disposal facilities result in an increase in disposal costs, which put pressure on the whole system to prevent, re-use, recycle or incinerate waste. Incineration, re-use, and recycling also have negative effects on the environment. Incineration may result in the release of harmful substances such as dioxin and furane. Moreover, incineration does not result in complete disintegration. It is merely a form of volume reduction. The residues, some 10 per cent, have to be landfilled anyway.

Re-use also has its environmental costs. Re-usable packaging is usually heavier than one-way packaging, hence more material is needed to produce it. This can be easily outweighed by a large number of trips of the re-usable packaging, but transport causes another problem. Longer transport distances between the points of filling and sale increase the use of energy and emission of air pollutants, to mention just two negative environmental consequences of re-filling systems.

Recycling also poses environmental problems. Material has to be collected, sorted, upgraded, and recycled. This involves transportation, which costs energy and causes air pollution, particularly, when, as is the case in Germany, the material is transportated to countries like China or North Korea. Moreover, it is estimated that the recycling of a tonne of paper produces half a tonne of sludge, containing heavy metals. The recycling of one tonne of aluminium produces half a tonne of slag containing dioxins, and has to be treated as hazardous waste or has to be upgraded in a costly way. Melting glass and tinplate costs energy. In the case of tinplate it also produces potentially dangerous emissions. The recycling of plastic food packaging is possible only for lower quality products (downcycling) because for reasons of hygiene the recycled material may not come into contact with food. In addition, critics say that the new forms of raw material recycling practised in Germany cost more energy than they save, and are therefore less environmentally beneficial than burning (Der Spiegel 21.6.1993). The OECD observed that '*(i)ronically, the packaging materials that pose the greatest difficulty for recycling are those that make the greatest contribution to waste reduction: plastics and composites made from thin layers of more than one material. Because of the difficulty and cost of separating the layers and resin types, these materials are not recycled. On the other hand, composite packages and plastics are usually lighter and occupy less volume than more traditional packaging they replace*' (OECD 1992: 12, quoted in Porter 1994: 15).

The answer to the question of which packaging system and which waste management option is the most beneficial is contentious and depends on a number of assumptions and conditions. These aspects pointing at the problems of life cycle analysis will be advanced in section 2.5. The next section will deal with the economic interests involved in the issue of packaging and the environment.

2.3 Packaging and business

2.3.1 Introduction

Packaging serves various important functions in advanced industrial society. Transport packaging enables or facilitates the movement of goods over long distances, providing safety, protection and hygiene. It also helps to save space by providing the possibility of stacking during transport and storage. Secondary packaging, like cardboard for tooth paste, is wrapped around sales packaging for market reasons, to allow for self service in supermarkets and hinder theft of goods. Sales packaging also protects goods, is needed for marketing reasons and enables or facilitates self service. Moreover, it provides for conservation and consumer convenience in handling. The marketing function of packaging and the facilitation of handling and turnover of products make packaging a crucial factor in our society. '*Die Verpackung ist somit eine Voraussetzung unserer konsumorientierten und sich schnell wandelnden Lebensweise geworden*' (Bünemann and Rachut 1993: 12).

The central functions of packaging for the whole economy imply that packaging waste policies usually affect large parts of the economy if not the economy as a whole. Comprehensive policies adopted by governments usually apply to tens of thousands of companies or even several hundred thousand. On the most aggregated level, business is principally against packaging waste policy interventions. The peak associations of national trade and industry, the German Peak Association of Regional Boards of Trade and Industry (*Deutscher Industrie und Handelstag*, DIHT) and the Peak Association of German Industry Associations (*Bundesverband Deutscher Industrie*, BDI), the Confederation of British Industry (CBI), the Dutch peak association of employers and industry association (*Verbond van Nederlandse Ondernemingen - Nederlands Christelijk Werkgeververbond* VNO-NCW), the European Chamber of Commerce and the Union of Industrial and Employers' Confederations of Europe (UNICE), generally argue against any measures that intervene into markets, which may have redistributive effects, or serve as technical barriers to trade. The case studies will show, however, that these organizations vary in a) the determination and precision in which they put forward these arguments; b) the role they play in packaging waste policy making and c) the degree to which they can commit their affiliates to these positions.

Any ban on packaging materials or types of packaging, mandatory deposits and take-back obligations, or mandatory recycling quotas are generally opposed, however. The ban of a certain packaging material or type of packaging, probably the most interventionist measure, could immediately threaten the existence of a specialized packaging manufacturer. The shift from one-way packaging to re-use systems would result in the short run in high investment costs, for instance, for filling facilities and logistical systems (Spies 1994: 274). More storage capacities would be needed, and the adaptation to changes in market requirements would be more difficult, given the need for standardization associated with refillable packaging (Koppen 1994: 165). Recycling may require the establishment of collection, separation and upgrading, e.g. cleaning/compressing, and adequate recycling technology needed for high quality recycled material, sufficient recycling capacity, and sufficient markets for the recycled products. Recycling may require technical innovations and adaptation at the production stage in order to make products more easily recyclable (SRU 1991: 252-253). The recycling of plastic and composites is quite difficult compared to paper, cardboard, glass, aluminium and tinplate. Recycling usually involves costs for energy, transport and labour which cannot always be covered by the price of the recycled material. Profits can usually be made with aluminium and steel, and most of the time also with paper/cardboard, though this market is very volatile (Interview British Industrialist 24.6.1996). But there is only a limited market for recycled plastics, partly because these may not come into contact with food for hygienic reasons. The recycling of composites, primarily drink cartons, is only just beginning.

When industry is asked by government to increase its recycling efforts, industry usually refers to recycled waste from commercial sites, such as transport packaging or industrial sales packaging. Recycling of commercial waste is cheaper as compared to household waste since in contrast to the latter a) there is a relatively small number of sites where waste occurs; b) there is a large amount of packaging per site; c) a limited number of different packaging materials occurs at each site; d) each item of packaging is large in size; e) the emptied packaging is usually thoroughly emptied for economic reasons and f) the packaging is mostly treated as industrial waste, therefore the disposal has to be paid for by the company anyway (Michaelis 1995: 231, OECD 1995b: 7,8).

Household waste, however, is more visible to the consumer. Governments therefore tend to focus on this waste stream, which then meets resistance from industry because of the bad economics of recycling. For instance, while the part of the German packaging ordinance dealing with transport waste was implemented in silence, the part dealing with sales packaging was very controversial and dominated the whole political debate. The different economics of recycling played an important role in the political processes as will be shown in this study. To give an example: British retailers, obliged to take responsibility for roughly half the recycling and recovery targets of the UK, fulfilled this obligation by recycling transport waste generated at their backdoors, instead of contributing to the much more expensive recycling of sales packaging. This caused objections by other parts of the packaging chain.

Packaging waste policies affect companies in different ways, depending, among other things, on the position of the company in the packaging chain and the issue at stake. The packaging chain includes producers and deliverers of the raw materials for packaging, packaging manufacturers, filling and packing industry, and retailers and wholesalers. It is important to note that business interests regarding other issues of environmental regulation may be quite homogenous, but in the case of packaging, a complex and dynamic configuration of overlapping and contradictory interests is at stake.

Packaging waste policies do not only affect the packaging chain, though. Policies aiming at waste prevention, re-use, recycling, incineration and disposal create a market for environmental industries and services, most notably waste management companies. They generally have an interest in a strict and comprehensive legal framework. To a large extent, the size of their markets is delineated by the norms and standards set by government. For example, when in the course of the amendment process of the German Packaging Ordinance, the German government planned to lower the minimum recycling targets, the association of the German waste management industry (*Bundesverband der Entsorgungswirtschaft*, BDE) argued strongly against it (Entsorga Magazin 19.6.1996). In the following section the interests concerning packaging and packaging waste of the various parts of the packaging chain (2.3.2) and the waste management industry (2.3.3) will be deduced from the assumption that companies are interested in long term profit maximization.

2.3.2 The packaging chain

Packaging and the environment formed a controversial issue for more than 25 years. As a reaction to the societal and political debate, companies of the packaging chain have founded cross-sectoral organizations in all the three countries investigated and on the European level. All organizations deal with a single issue: packaging and the environment. The organizations are the German Working Group on Packaging and the Environment (*Arbeitsgemeinschaft Verpackung und Umwelt*, AGVU), the Dutch Foundation of Packaging and the Environment (*Stichting Verpakking en Milieu*, SVM), the British Industry Council for Packaging and the Environment (INCPEN), and the European Organization for Packaging and the Environment (EUROPEN). These cross-sectoral and single issue organizations all emphasize the important economic functions of packaging and its small share of the overall waste stream. They deny the need for governmental packaging waste policies for environmental reasons and also claim that market forces are sufficiently able to optimize packaging since these give an incentive to cut costs by saving material and energy.

In line with the peak associations of trade and industry, these organizations argue against the discrimination of certain materials and types of packaging. When government initiatives are planned they argue against legal command and control measures, which are seen as rigid and inflexible and against mandatory economic instruments such as deposits, taxes or charges. As a lesser evil, they favour the voluntary agreements and codes of practice. The line of reasoning, however, varies between organizations and across time, as does the function of the organizations in the policy process and their disciplinary authority *vis-à-vis* their member firms or affiliates.

Packaging material suppliers

Raw material suppliers are directly and negatively affected by all measures aiming at the reduction of packaging, be it packaging prevention, re-use or recycling. Discrimination against one material, for example plastic, however, is beneficial for the supplier of another material, such as glass. The concrete effect therefore depends on the specific character of the measures adopted.

Multinational oil and chemical companies are the main suppliers of raw material for plastic packaging. ICI, for instance, is the principal supplier of PET. Other large companies involved are Shell Chemicals, BP Chemicals and Exxon (Key Note Market Review 1993a: 35). These companies are among the biggest companies in the world. They had an annual turnover of tens of billions of pounds in the early and mid 1990s (The Times 1996: 90-91). But their interest in packaging material supply is only one among others. The markets for plastic packaging material is global and volatile, reflecting changes in the price for oil.

The raw material for paper packaging, wood pulp, is also provided by large companies especially from Canada, Sweden and Finland (Bletz 1996: 39-41, Key Note Market Review 1993a: 15). As in the case of plastic, the market for wood pulp has always been volatile. Moreover, political intervention across Europe has exaggerated and accelerated price movements, according to waste expert David Perchand (ENDS Report Nov. 1994: 11). To give an example, the wood pulp price doubled to 700 USD a tonne in summer and autumn of 1994, partly as a consequence of growing environmental restrictions on forest products and the threat of a strike in British Columbia - one of the world's, biggest suppliers - and changes in recycling rules in Germany (Financial Times 7.10.1994).

As in the case of plastic, paper and cardboard, the supply of steel and aluminium is dominated by a small number of large firms, such as British Steel, though these firms are not as large as the oil and chemical multinationals. But the markets for steel and aluminium are less volatile than those for paper/cardboard and plastic.

The packaging industry

The packaging industry has become an important one in terms of turnover and employees in any advanced industrial economy. In Germany the packaging industry is the sixth largest branch. It had a turnover of 30 billion German marks at the end of the 1980s (BMU press release 10.8.1990). In the Netherlands the packaging industry's turnover was 13 billion guilders in 1994 (Bletz 1996: 39), for the UK this amounted to 9 billion pounds in 1991 (Key Note Market Review 1993a: 7). Packaging manufacturing has always been characterized by quite a large number of relatively specialized companies. This holds especially true for plastic packaging manufacturers (Key Note Market Review 1993a: 34). In recent years, however, a process of concentration has been taking place. In the UK for instance, approximately 90 per cent of metal packaging production is in the hand of five companies (Key Note Market Review 1993a: 51). Among the largest are international companies like Carnaud Metalbox (France), Continental Can (Germany), Alcan Aluminium Canada and Van Leer (The Netherlands). Seven companies control 90 per cent of the market of glass container manufacturing. In Europe, nine companies are responsible for 60% of the glass packaging output (Key Note Market Review 1993a: 79).

Large paper makers have interests in the paper and board-packaging industries. A few groups led by Bowater dominate this sector in the UK (Key Note Market Review 1993a: 15). Concentration processes have been particularly strong in cases of companies with a main interest in plastic and paper packaging. Companies like the British Polythene Industries or Bowater Packaging expand rapidly through strategic acquisitions of well-established packaging companies (Key Note Market Review 1993a: 35).

There is strong price competition between packaging companies. They therefore have an economic incentive to use as little packaging material possible for a certain function. Making ten containers rather than five from the same amount of raw materials halves one important cost factor (INCPEN 1995: 3). There is also competition between the producers of different packaging materials and types of packaging. Therefore the result of the debate on the environmental impact of certain materials is of critical importance for material producers (Mingelen 1995: 31) and policies intending to reduce the amount or the composition of packaging, i.e. substitution of packaging materials, have an immediate and strong effect on the economic performance of specific packaging producing companies. Prohibition of certain types of packaging or materials might immediately result in bankruptcy. More strict environmental standards can have advantages for those packaging manufacturers who are flexible and anticipate such changes.

The substitution of one-way packaging by re-usable packaging is in the interest of only very few sales packaging manufacturers, because only glass and, in recent years, plastic are used in re-use systems. Since the turnover of glass producers depends only for one third on re-usable glass containers, in this branch also economic interests would be threatened by the substitution of one-way packaging by refillables.

The interests at stake with regard to recycling depend on the design of the measures adopted, the burden of costs they induce, the degree to which the costs can be parcelled down the packaging chain, and characteristics of the recycling market. Measures trying to increase the recycling of packaging have divergent impacts on the packaging industry. Given the difficulties and costs involved in the recycling of plastic packaging and composites (see above), the resistance of producers of such packaging against political intervention is the strongest. The need for technical adaptation and the high costs of recycling will weaken their competitive position. Therefore in Germany, plastic producers advocate waste incineration. The aluminium industry as well as the tin plate industry, however, expect economic advantages of political measures raising recycling quotas. The paper packaging producers and the glass packaging manufacturers are generally positive about recycling, but are critical about political intervention in the packaging market (see Spies 1994: 275).

Though packaging manufacturers have increased in size during the last decade they still are economically the weakest part of the packaging chain. They are trapped between the large companies at the material supply side and those at the packaging demand side, where they are confronted with large multinational consumer good manufacturers and retailers (see below). Packaging manufacturers do not have a lot of influence on purchase and sales prices (Bletz 1996: 39). What is more, the large packers and fillers and the retailers determine the shape and the composition of packaging by their demand (Interview Specialized Journalist 23.7.1996, Interview British Industrialist 24.6.1996, Key Note Market Review 1993a: 28, 33).

Packers and fillers (consumer good producers)

The packers and fillers, in particular the producers of fast moving consumer goods, are the clients of the packaging producers. Some 60 per cent of the overall amount of packaging is used in the food and drink sector. The market for consumer products is dominated by a small number of large multinationals which have an annual turnover of tens of billions of pounds, such as Unilever, Nestlé, Procter & Gamble, and Coca Cola, and a larger number have an annual turnover of more than 5 billion pounds such as Grand Metropolitan, Allied Domecq, Guinness, Heineken, Cadbury Schweppes, or Danone (Financial Times 17.12.1996).

Despite the various important functions of packaging for packers and fillers, packaging is also a cost factor to be minimized. The market for fast moving goods is also quite competitive, and competition is often price competition rather then quality competition. '*When bread is offered for 99 pence in a supermarket and the price for the plastic foil in which it is wrapped increases by 2 pence, this price increase will not lead to an increase in the product price. The retailer or producer will try to cut costs elsewhere*' (Interview British Industrialist 24.6.1996). Packers and fillers are therefore very responsive to price changes for certain types of packaging. Reduction of material used, for instance, by light weighting, is used by companies to cut costs, also in the absence of packaging waste policies.

The introduction of new refillable systems would in particular be opposed by producers of brands. Refillable systems imply standardization. This is difficult to reconcile with the brand producers' interest in marketing and identification of the product. Those companies, however, that have re-use systems in place, may have an interest in continuing or increasing their usage because of sunk costs and market protection. Cases in point are the German beer breweries and mineral water companies. The producers of brands are also the packers and fillers that are most responsive to critical consumers, because they depend very much on the positive image of their brands. Some of them are therefore driving forces in initiatives dealing with environmental aspects of packaging (see case studies for details).

Retailers

The retailers have a strategic role in the packaging chain since they link the consumer with the product. There has been a strong process of concentration in the last decades. A large part of the turnover is generated by few companies. It is estimated that 40 per cent of the British food retail trade is under the control of three companies, namely Sainsbury, the Argyll Group (Safeway) and Tesco (Lang and Raven 1994: 124). The situation is no different in Germany and the Netherlands, where a small number of companies dominate the market. Like their British counterparts, the German food retailing groups Karstadt, Asko and Kaufhof and the Dutch Ahold (Albert Heijn) have an annual turnover of more than 10 billion pounds (Financial Times 17.6.1996), although these companies are more active in foreign markets, in comparison with the British ones, and the link between turnover and domestic market dominance is therefore weaker. Note that many retailers are also packers and fillers because they often produce own products.

With regard to packaging waste prevention, an argument in favour of smaller packaging is that less space is needed on the shelves. A case in point is concentrated washing powder. Small packaging, however, is less attractive for marketing reasons. The retailers, especially the small ones in the inner cities, would bear the main burden of an increase in re-use systems, such as the introduction of mandatory deposit schemes. They would be confronted with a huge amount of used packaging, which would occupy a lot of handling space, which cannot be used for exhibiting products. Moreover, additional personnel is needed for handling and to clean packaging from residues. Since especially retailers are exposed to the consumers they have an interest in a green image. They therefore establish or participate in a number of local recycling schemes, in particular for glass packaging and metal cans (Key Note Market Review

1993b: 45). These retailers are, at least in the German and the Dutch case, active members of the cross-sectional trade associations dealing with packaging and the environment.

2.3.3 The environment industry

Packaging waste policies may put costs on companies in the packaging chain. However, what costs are for some companies are benefits for others. Companies who supply products and services to collect, separate, upgrade, transport, recycle and dispose of packaging benefit from policies that increase the amount, frequency and standards of those products and services. These firms generally have an interest in high environmental norms and standards since they define their markets (The Economist 8.1.1994: 65). The companies' interests often coalesce with the interests of the pro-environmental advocates in government agencies. With the increased scope and intensity of environmental regulations these companies have become more important. Nevertheless their relevance varies cross-nationally. They have become particularly important in Germany.

In the course of the privatization of waste disposal functions, this market has become very attractive for private companies and a concentration process has taken place. Several large companies started to play an important role, for instance, in British waste disposal, including Biffa Waste Services, which is considered the largest one, and Shanks and McEwan (Key Note Market Review 1993b: 32). The market for recycling has been traditionally dominated by small and medium sized specialized companies. Concentration processes have also occurred in this sector, however. In the case of plastic recycling in Germany, large utilities such as RWE and oil companies such as VEBA became involved and acquired well-established companies (see German case studies for details). In Britain, water companies started to enter the waste market (Key Note Market Review 1993b: 69). Recycling is often done by companies who are also the supplier of the packaging material. In Britain for instance, Alcan Aluminium is the leading company in aluminium can recycling, while British steel is the biggest in steel can reprocessing, and D.S. Smith in paper and board recycling (Key Note Market Review 1993b: 44,45,51).

2.3.4 Interests cutting across sectors, products and materials

Generally speaking, large companies with a solid financial base are more able to adapt to a changing social, technical and economic environment, than small, specialized and not so well funded enterprises. This also holds true for changes in the political environment. A changing legal context which demands new information, new investments and which places administrative burdens on companies, is more easily tackled by large companies. The latter have the financial resources, and the specialized staff to react to new political demands. Packaging prevention or the switch to packaging which is more easily recyclable is easier for a packaging company which has a technological advantage (Bletz 1996: 39). To give an example: the switch towards hybrid packaging systems, generally associated with a good environmental image, gave Unilever a competitive advantage over smaller companies (Interview Unilever 24.6.1996).

Large multinationals like Unilever, Coca Cola or Procter & Gamble may share with other companies of the packaging chain the objection against government intervention. However, when national governments have intervened, these companies had a vital interest in harmonizing national regulations. It is difficult for them to deal with fifteen different packaging regimes within the EU, which may ask for nationally distinct labelling, national distinct requirements for the compositions of packaging, or the type of packaging, i.e. one-way or refillable. Such an interest is not shared by companies who produce for only one national market.

2.3.5 Conclusion

Packaging serves important functions in our society. Strong and often convergent interests are at stake. But the companies' preferences and strategies as expressed in the political process depends on their vertical (sector) and horizontal (e.g. material) position in the packaging chain, next to their size, financial power, their degree of export orientation and the scope and intensity of existing and planned policies. Politically important are also the degree in interest aggregation in business association, the general attitude towards government, more salient features of government-industry relations and the broader political context. The latter factors in particular vary across time and Member State. They will be discussed in detail in the case studies.

2.4 Packaging and science

'*Life is a success with ignorance and confidence*' Mark Twain

Packaging waste issues are often highly technical, therefore experts like natural scientists, engineers and economists play an important role in the formulation and implementation of packaging waste policies. Moreover, research on topics discussed in the political arena is often in an embryonic stage. Scientific uncertainty exists in respect to many topics and research results are often ambiguous. There is no clear agreement to what extent re-use is better than recycling, from an environmental standpoint, recycling better than incineration, incineration better than disposal, or glass bottles better than plastic bottles. As Fairly argued: '*Countless studies and reports compare the environmental effects of different recycling and waste management schemes for packaging. But the wildly differing criteria used to assess environmental performance...mean that almost any conclusion can be reached*' (quoted in Porter 1994: 15-16).

The complexity increases if one focuses not only on environmental aspects but also on economic ones. The uncertainty and the ambiguity about the alternatives mentioned above is due to the complexity of life-cycle analysis (LCA). This uncertainty and ambiguity plays an important role in the political process. This study will show, however, that the importance of LCA in the policy process varies according to the principles guiding the traditional regulatory style i.e. to what extent political action has to be based on sound empirical evidence.

Uncertainty and ambiguity about the environmental impact of certain packaging materials and types of packaging as well as certain waste management options is linked to the complexity and 'newness' of the life cycle analysis method. Life cycle analysis, also called life cycle assessment, is defined by the German Federal Environment Agency (*Umweltbundesamt*, UBA) as '*an objective process to evaluate the environmental burdens associated with a product, package, process, or activity by identifying and quantifying energy and material uses and resultant environmental releases during the entire life-cycle (...) and evaluating opportunities and implementing changes to affect improvements*' (UBA 1992: 19). In the past life cycle analyses have been done for several products and processes. Packaging is, however, one of the major objects of life cycle analyses. A review of literature carried by the UBA in 1992 found that 42,9 per cent of the 112 sources identified dealt with packaging (UBA 1992: 21). Other terms which are used instead of life cycle analysis are 'Cradle to Grave Analysis', Material Flow Analysis, and 'Ökobilanz' (see UBA 1992:18, WARMER Bulletin August 1995).

LCA in practice is not such an objective process as the definition suggests. Scientific uncertainty and ambiguity create room for debate and influence, which is frequently and strategically used by all those with interests at stake. Economic actors are more successful in shaping the LCA process than politicians, consumer organizations and environmentalists. Often companies or their trade associations carry out these analyses themselves, or they commission the projects to research institutes. Business has the money and - probably more important - controls most of the data needed to carry out a LCA. In fact, the capacity of governments to collect, select and analyse data varies significantly among the three countries. The German government allocates more resources to LCA's than the Netherlands and the United Kingdom. LCA's are conducted in Germany under the supervision of the Federal Environment Agency. Moreover, business organizations have to provide more data in Germany than in the UK. In fact many UK companies and trade associations use the lack of data for strategic purposes. On the one hand business is reluctant to provide data, arguing that data are too expensive to collect and on the other hand, it often rejects governmental plans with the argument that the political action is not based on sophisticated knowledge.

There is no standard method for LCA yet. Choices have to be made which cannot be based on 'objective' scientific criteria. It may be that in the future among the scientific mainstream a convention will be developed on how to make choices in certain respects. The following discussion will show, however, that choices still remain to be made which have to be based on values or interests. In regard to packaging and the environment, LCA is particularly important in deciding whether one type of packaging is more or less environmentally friendly than another.

First, it has to be decided which alternative packaging systems are included in the LCA: only those already established on the relevant market, or those already established in other countries or for other products, or those not yet established, those still being developed. The field of packaging development is quite dynamic. New products are developed all the time; like light weight glass bottles, or refillable PET-bottles (Key Note Market Review 1993a: 48, 84, SVM 1994b). In the course of the LCA's which were to be carried out in the context of the Dutch Packaging Covenant (1991) for example, a conflict arose whether the refillable PC bottle developed by General Electric, which was not yet on the Dutch market, should be included in the LCA for milk bottles (Interview Steering Committee LCA 11.9.1997, see also chapter 5.3.6).

Second, a choice has to be made concerning the environmental aspects to be included in LCA's. The first obvious aspect is 'waste': the quantity and quality of waste generated by a certain type of packaging. In this respect, refillable glass bottles usually have an advantage over one-way plastic bottles. In recent years, however, a second aspect became equally important, namely 'energy consumption'. On this criterion one-way plastic packaging usually performs better than refillable glass bottles, because refillables have to be transported over longer distances in their lifetime and are heavier than one-way plastic bottles. In political debates therefore, producers of re-usable glass bottles will usually concentrate on the waste aspect, while one-way plastic bottle producers will try to focus the attention on the 'energy consumption' aspect. While 'waste' and 'energy consumption' are now routinely included in LCA, there is no widely accepted standard for the inclusion of other aspects like water and air pollution, nuisance (smell, noise), environmental and human toxicity, impact on ozone depletion, global warming, or acidification.

Even when LCA's include the same aspects such as 'energy consumption' studies may still vary in the depth of the inquiry. Do they look at the consumption of energy at the production of the product, or at the disposal or both? Do they look at whether the energy used comes from fossil resources or renewable ones? Do they look at how much and what kind of energy was used to build up and maintain the production plant, or the disposal plant? Does the study take into account whether the employers of the plant travel to work by bike, by public transport, or by car. A specialized journal claims about a typical LCA: '*(S)uch a study would normally ignore second generation impacts, such as the energy required to fire the bricks used to build the kilns used to manufacture the raw material. However, deciding which is the "cradle" and which the "grave" for such studies has been one of the points of contention in the relatively new sciences of LCA*' (WARMER Bulletin August 1995).

Life cycle assessment is based on scenarios. Researchers have to make assumptions and the kind of assumptions they make strongly influences the results of the LCA's. A case in point are LCA's that compare refillables with one-way packaging alternatives. Two assumptions are crucial for such a LCA. First, the rate of return of the refillable, that is the number of re-uses. If a relatively high rate of return is assumed, the environmental effects of refillables in comparison to one-way bottles decrease. Secondly, assumptions about the distance between the location of consumption and the location of refilling are important. Each trip causes emissions into the air, and losses of natural resources. If the distance to the refilling station is long, one-way packaging is in a more favourable position. A mineral water bottle consumed in the Netherlands and refilled in the south of France is probably more harmful to the environment than a one-way French bottle. The crucial importance of the assumption concerning the number of trips of refillables became obvious in a LCA concerning milk and juice packaging. The break-even point for light-weight milk bottles in comparison to one way-cardboard boxes is seen to be 20 trips. The present rate of return is probably 20 to 30. To give an example: Tetrapak, the main producer of cardboard boxes for milk, commissioned a study where it underestimated the trips in order to reach a better result for its own product (Dam and Oevelen 1993: 5). Other assumptions which are important, for instance, whether and to what extent the material of the one-way packaging will be recycled, the quantity of water and the toxicology of chemical used to wash returnable bottles after use, or the type of transport used to bring refillables back to the refilling station.

Even when all interested actors could agree on the scope, depth and assumptions of the LCA, a major matter of judgement remains: the comparison of different kinds of environmental impact. How can one compare high energy demand with high water use? Which imposes a greater environmental burden? How should the use of non-renewable mineral resources like oil or gas, the ingredients of plastic, be compared with the production of softwoods for paper? *'Soll für die abfallverringernde Kunststoff-Verpackung aus Polyethylen weiter auf die Ausbeutung fossiler Rohstoffe gesetzt oder Glas und Papier bevorzugt werden, die auf der Deponie zwar mehr Platz beanspruchen, aber mit Sicherheit keine Schadstoffe emittieren und sich zudem gut wiederverwerten lassen?'* (Ökologische Briefe 12.11.1992: 16).

In short: LCA's are not confined to objective methods. There are choices included which are based on values or interests. Note, that the exercise becomes further complicated when not only environmental aspects are analysed and compared, but also other dimensions, including economic, technological, employment, health and safety aspects etc.

2.5 Conclusion

Given the environmental, economic, symbolic, technical and political dimension of packaging it is not surprising that market interventions by governments cause fierce political battles. The values, interests and scientific uncertainty involved also have the consequence that - as in most other areas or regulation - no easy functional relationships between problems and solutions exist, so that national regulations are shaped by values, interests and institutional rules. Let us now see how three Member States of the EU tackled the issue of packaging and the environment up to the early 1990s.

3. National packaging waste policies: anything goes

The first empirical part of the case study concentrates on the packaging waste policies prior to the process of Europeanization. This part will answer the question of whether, and if so, why national policies concerning packaging waste differed in the early 1990s. It starts by answering the first research question: what were the differences and similarities of the German, Dutch and British packaging waste policies in the early 1990s? This will be done by a synoptic comparison of national packaging waste policies according to the policy elements as developed in chapter one. In order to explain the differences and similarities found (research question two), each country will then be analysed separately. For each country, first the foundation and development of national environmental and waste policies and their shaping ideas, interests and institutions will be discussed. This exercise is necessary because national packaging waste policies emerged against the background of these developments. In the second step I will analyse the policy processes that resulted in the national packaging waste programmes of the early 1990. These country studies are increasingly comparative. The German development will be taken as a starting point (chapter 4). Here the focus will be confined to the country itself. The Netherlands will be more explicitly contrasted with the German development in order to account for variations in policy development (chapter 5). The United Kingdom finally will be contrasted with both Germany and the Netherlands in order to account for the specific nature of the British case (chapter 6). In the final chapter of this first empirical part, the shaping factors of the national variations in packaging waste policies which have already been discussed in the nationally-oriented chapters will be re-examined from a cross-national perspective. This will be done according to the five policy elements developed, and will include references to the specific features of the issue of packaging waste as well as the broader historical and cultural context in which the actors and institutions are embedded (chapter 7).

Table 12 National policies prior to the European packaging directive

Country **Dimension**	**Germany Packaging Ordinance 1991**	**The Netherlands Packaging Covenant 1991**	**United Kingdom White Book, 1990 & Codes of Practice 1992**
FORMALIZATION			
degree of legal codification	Ordinance: generally binding under public law	Covenant: voluntary agreement binding under private law for signatories	government and industry statements, not legally binding
STRICTNESS OF STANDARDS			
packaging reduction targets	no	stabilization below the level of 1986 (i.e. -15%, a.c.t.1990)	no
re-use target	maintenance of share of drinks sold in refillable (72%)	increase of re-use under certain conditions	no
material recycling targets (levels already achieved)	sales packaging: 64/72% depending on material (1989= 1%-58%)	60% (1986 = 25%)	half of recyclable *household waste*, i.e. some 25%, incl. composting (1990=5%)
incineration target for packaging waste	prohibited for collected sales packaging	increase to max. 40%	no explicit targets
landfill for packaging waste	prohibited for collected sales packaging	phase out	no explicit targets
time horizon	short (1.7. 1995)	longer (2000)	longer (2000)
prescription of implementation	detailed prescriptions (targets, types of material, deadlines)	less detailed prescriptions	general vague target, no prescription of implementation
flexibility	possibility for self regulation within tight legal framework	self-regulation within broad framework, change of targets under certain conditions	complete flexibility
ALLOCATION OF RESPONSIBILITIES			
separate collection for recycling	private	public and private, including pilot schemes	public and private pilot schemes
recycling	private	private	public and private pilot schemes
licensing	public: *Länder*	-	-
monitoring/assessment	public: *Länder*	public-private committee	
dispute settlement/ enforcement	*Länder*/administrative courts	arbitration/civil law courts	-
MODES OF INTEREST INTEGRATION			
intensity of integration	medium	high	low
mode of interest integration	corporatist, peak trade associations, and single issue cross sectional foundation	corporatist single issue cross sectional foundation	pluralist, companies and trade associations
ORIENTATION TOWARDS TARGET GROUP			
general orientation	impositional	mediating	laissez -faire
character of incentives	repressive: obligations, mandatory deposits	repressive and stimulating: deposits and subsidies, provision of infrastructure	stimulating: recycling credits, provision of infrastructure

3.1 Overview

Industrialized liberal democracies share a number of developments and trends in environmental policymaking. Against this background the variations in national approaches to packaging waste are striking. This section will compare the differences and similarities in national packaging waste policies, using the conceptualization of "policy" as developed in the theoretical part of the study. The policy elements which refer directly to the content of the policy programme, i.e. formalization, strictness of standards, and allocation of responsibilities, will be discussed in more detail than the elements that refer to the policy process, i.e. modes of interest integration and the orientation towards the target group. The latter characteristics will be fully fleshed out in the analysis of these policy developments in the various countries. Table 12 provides an overview of the national policies preceding the European Directive on Packaging and Packaging Waste.

3.2 Formalization

The degree of legal codification of the **German** packaging waste policy was high in the early 1990s. The German programme was codified as public law in an Ordinance. The Packaging Ordinance (*Verpackungsverordnung,* BGBl. 1991 I, p. 1234) was based on Article 14 of the German Waste Act (Abfallgesetz, BGB 1 III no. 2129-15, 1986). This Article constituted the legal basis for producer responsibility concerning the prevention, reduction and recovery of waste. Ordinances may be adopted by the federal government, not by individual ministers, after consultation with interest groups and with agreement of the Chamber of state (*Länder*) governments (*Bundesrat*). The law enabled the government to mandate a) the marking of packaging; b) the restriction to packaging which is re-usable or easily recyclable; c) the take-back and deposits; d) the separation of used packaging and e) the restriction of usage for certain purposes. The Packaging Ordinance was based on this law and applied to all German and foreign packaging manufacturers (*Hersteller*) and distributors (*Verteiler*) of packaged goods, i.e. packers and fillers and retailers including mail-order firms, who market their products in Germany. The Ordinance dealt with transport packaging, secondary packaging and sales packaging. It did not apply to packaging containing environmentally damaging or unhealthy substances such as packaging contaminated with pesticides or solvents acids. These forms of packaging had to comply with other regulations e.g. regulations on hazardous waste.

Unlike in Germany, the **Dutch** government had not introduced a generally binding mandatory scheme for the control of used packaging. The Ministry of the Environment and the Foundation of Packaging and the Environment (SVM) established a voluntary covenant scheme. This Packaging Covenant (VROM 1991) is legally binding under private law for the signatories. Initially, the covenant was signed by the Minister of the Environment on behalf of the Netherlands and the Foundation for Packaging and the Environment (SVM) representing some 60 to 70 per cent of the turnover of relevant markets. Declarations to join the covenant scheme were signed by some 200 companies, both affiliates of the SVM and other companies from all parts of the packaging chain including recycling companies. These companies represented roughly 80 per cent of the turnover of relevant markets. The Covenant applied to packaging waste from households and commercial sites. This scope was roughly equivalent to the German system with its distinction of transport, secondary and sales packaging. The covenant will expire on 1 January 2001.

The **British** government had not introduced any binding packaging waste regulations for business. The government stated a general non-binding recycling target for municipal waste and a number of requests to industry in its White Paper '*This Common Inheritance*' (HM Government 1990). Trade associations reacted with a non-binding business plan and several codes of practice. Sub-national packaging waste measures are codified in the Environmental Protection Act (EPA, 1990): local authorities have the obligation to develop and submit recycling plans. They are, however, not legally obliged to implement them. The EPA also codified local recycling credit schemes financed by central government.

3.3 Strictness of Standards

The **German** government not only introduced the most formalized and most comprehensive packaging waste programme but it also set the most ambitious standards, in particular for recycling. However, no explicit targets for packaging and packaging waste reduction were set. The German government expected a reduction in the amount of packaging waste as a consequence of the other programme provisions, in particular the provision in regard to recycling. As far as *re-use* was concerned, the Packaging Ordinance aimed at maintaining the aggregate market share of beer, mineral water, soft drinks, juice and wine sold in refillables at the 1991 level in the various German regions and set an overall national quota of 72 per cent. The quota for pasteurized milk packaging was 17 per cent. These quotas were meant to stop the trend towards one-way drink packaging. The Ordinance also stated that the government would decide three years after the enactment of the Ordinance whether an increase and differentiation of the re-use quota was necessary. The government set ambitious targets for *recycling,* meaning material recycling. Transport and secondary packaging had to be taken back and had to be re-used or recycled independently of the public waste disposal system. This did not mean, however, that re-use/recycling quota is 100 per cent. The provision has to be read against the background of the 1986 Waste Act, which stated that one could not demand action of industry that is not technically possible or economically feasible (Interview BDI 6.12.1996). The Ordinance also stated that sales packaging had to be taken back and had to be re-used or recycled[10]. To stimulate take-back, the Ordinance introduced the obligation to charge a deposit on one-way packaging for washing powder and cleansing agents (DM 0,50; <

[10] If consumers require the goods with the transport or secondary packaging, the packaging is treated as sales packaging.

1,5 liter > DM 1) as well as for emulsion paints (DM 2,00). In the case of sales packaging, however, an escape clause was included in the Ordinance. The individual take-back responsibilities and the mandatory deposits could be replaced by participation in a collective scheme. This collective scheme had to meet quantified collection and sorting targets. The targets were as follows:

Table 13 Quota set for sales packaging in Germany

	Deadline 1.1.1993			**Deadline 1.1.1995**		
	collection	**sorting**	**recycling**	**collection**	**sorting**	**recycling**
glass	60	70	42	80	80	64
tinplate	40	65	26	80	90	72
aluminium	30	60	18	80	90	72
paper/cardboard	30	60	18	80	80	64
plastic	30	30	9	80	80	64
composites	20	30	6	80	80	64

The quotas for collection referred to the minimum share of collected sales packaging of the overall amount of sales packaging that entered the market, while the sorting quotas defined the share of material to be sorted out of the collected sales packaging for recycling. The recycling quotas were the result of a multiplication of the respective collection quota with the respective sorting quota.

The Ordinance also stated that *landfill* and *incineration* were not permitted as viable disposal routes for packaging waste that has been sorted. These management options were therefore confined to not re-usable/recyclable packaging, and to sales packaging that had not been sorted.

The time horizon for meeting the obligations was quite close at hand. For transport packaging, the obligations came into force in December 1991, for secondary packaging in April 1992, and for sales packaging in January 1993. Unlike the cases of the Netherlands and the United Kingdom, the deadlines for the recycling targets had to be met within a rather short time. Intermediary targets had to be met in 1993 and the final targets in the second half of 1995. The degree of detail in which the implementation of the Ordinance had been prescribed also contributed to the strictness of the German standards. The quotas for refillables and the collection and sorting targets were quantified. The collection and sorting requirements were specified for separate packaging material streams. This prevented industry from switching between materials to be recycled.

Although the German government adopted a rigid and tight legal framework at least some flexibility was left for industry. The 72 per cent refillable quota does not apply to separate drinks but to the aggregate market share of a number of drinks. In general, the industry was free to determine the way in which to achieve the targets and fulfil their obligations. Industry used this room for manoeuvre for self regulation which ensured a certain degree of flexibility. Intra-industrial co-ordination took place by private law, through contracts between business partners, and economic instruments, i.e. charges. The provision of self-regulation, specifically in the form of a collective scheme in the case of sales packaging, allowed for a certain flexibility of financing and organising of waste collection, and the sorting and recycling processes, but the detailed definitions, the fixed and demanding targets and the strict time frames made the German approach less flexible than the Dutch and the British approaches.

The **Netherlands** is the only country included in this study that specified a quantified ceiling on the generation of packaging waste. VROM and SVM agreed that the packing industry would reduce - by the year 2000 at the latest - the amount of packaging newly introduced into the market to a figure below the quantity of 1986. Moreover, beside this mandatory target, the packing industry declared the intent '*to do everything within its power and to make great sacrifices*' to achieve an absolute reduction of 10 per cent by the 2000[11].

Table 14 Waste management obligations set in the Dutch Covenant

	intermediary targets	**2000**
prevention	1997, 3% less than 1991	less than 1986
take-back		min. 90% of used one-way packaging if it is collected separately
collection	min. 50% glass collection, separated according to colour (1994), min 50% paper and cardboard	min. 75% paper and cardboard
recycling	min. 40% of used one-way packaging, (before 31.12.1995) [50%]	min. 60 % of used one-way packaging
incineration		max. 40 % of used one-way packaging (940 kilotonnes)
landfill	max. 40 %, of waste offered for landfill in 1986 (before1.1.1996) (940 kilotonnes)	

In contrast to the German Packaging Ordinance, the Dutch Covenant did not include obligations to safeguard the current level of refillables used for drinks. But the signatories bound themselves to introduce a deposit of Dfl 1,- on all carbonated mineral waters and soft drinks in one-way bottles made of plastic materials per 1 October 1991. This provision discouraged the use of one-way plastic bottles and therefore indirectly stimulated re-use. Moreover, while the German government merely postponed the decision on the increase of refillables, the Dutch government announced that *'If research shows that replacement of one-way packaging by re-usable packaging would cause clearly less damage to the environment and that there are no preponderant objections to such a change-over on market economic grounds then the packing industry undertakes to switch over to using re-usable packaging rather than one-way packaging'(art.8).*[12] The research, namely life cycle analysis and market

[11] The Covenant contains two type of provisions: Obligatory results, and obligatory efforts. The latter kind of provision means a declarations of intent to undertake best efforts.

[12] Art 8 reads as follows 'Indien uit onderzoek blijkt dat het vervangen van eenmalige verpakkingen door meermalige verpakkingen duidelijk minder schade voor het milieu oplevert en hiertegen op markt-economische

research had, to be paid for by industry and had to be carried out by independent research institutes for a number of products, listed in a supplement, by 1 December 1992.

As in Germany, ambitious targets were set for material recycling. The covenant stipulated that the quantity collected should be at least 90 per cent of the overall packaging waste, given that the material is collected separately. The quantity recycled should be at least 60 per cent by the year 2000. The packaging chain undertakes '*maximum efforts*' to recycle a minimum of 50 per cent used packaging before 31 December 1995. No material specific obligatory targets were set however. The packaging chain should '*aim*' at the following results: 80 per cent glass, 60 per cent paper/cardboard, 50 per cent high grade plastic bottles and flasks, 75 per cent metal. The signatories were also bound to reduce the quantity of packaging waste to be landfilled to 40% of the packaging offered for landfill in 1986. Landfill had to be stopped by the year 2000.

The covenant contained detailed and precise provisions as far as the objectives and targets are concerned. The objectives for prevention, recycling and incineration were set in percentages. In contrast to Germany, these quotas referred to an absolute amount of packaging waste (namely the level of 1986). Given the trend towards more used packaging, the Dutch targets were not as modest as they look at first glance, since the German targets were '*relative*', they could be achieved even when the absolute amount of used packaging doubled or tripled. However, Dutch industry was not forced to recycle sales packaging as in Germany. It could maximize the recycling of the economically advantageous transport packaging made from cardboard and plastic.

The Supplement of the Covenant listed dozens of rather specific measures through which industry wanted to achieve the envisaged general prevention and recycling goals. The following examples give a taste of the kind of measures announced. *'The transition to compact washing powder will result during the next 12 months in the replacement of conventional washing powders up to 2/3 of their turnover (savings on packaging materials: 1 mln kg compared to 1986)'* or *'Reduction, before the end of 1992, in the weight of tea boxes by 8 per cent compared to 1986'*.

The link between these concrete measures and the general prevention and recycling goals remained unclear, however. Moreover, a report of the Law Faculty of the University of Amsterdam commissioned by Dutch environmental organizations claimed that formulations as *'to do everything within its power and to make great sacrifices'* (the obligatory efforts) was not enforceable (Centruum voor Milieurecht 1991: 6). The same held true for the quality standards for recycling. Recycling had to take place to the 'highest grade possible' ('*zo hoogwaardig mogelijk*'). This provision was also seen as unenforceable. More in general, it was questioned, whether the SVM and/or individual companies could be held legally responsible for the global targets.

In order to achieve the targets, and to monitor the implementation, the industry had to draw up one or more implementation plans for carrying out the objectives listed above. The

---

gronden geen overwegend bezwaar bestaat, verplicht de verpakkende zich over te gaan op het gebruik van meermalige verpakkingen ten koste van eenmalige'

implementation plan(s) has/have to consist of a) a summary of measures to be taken (see example above), their significance for the objectives, and the way they will be carried out; b) a summary of the results achieved during the previous year and c) a test of the results achieved against the targets stipulated.

The covenant comprises devices for policy-oriented learning as well as for adaptations necessary to adapt to external developments. Devices for policy-oriented learning are provided by the life cycle analysis, the market research and pilot projects for the collection and recycling of waste. The possibilities for corrections are as follows. In the case of unforeseeable and substantial socio-economic, international or technological developments, changes of the content of the covenant were possible by mutual agreement. Objectives could be cancelled, and obligatory efforts could be transformed into obligatory results.

In its 1990 White Paper on the Environment the government of the **United Kingdom** stated that it would encourage the minimization of waste, and promote the recycling of as much waste as possible, including the recovery of materials and energy (HM Government 1990: 186). In addition, the government adopted the Environmental Protection Act (EPA) which, according to the White Paper, would by means of increased waste disposal standards provide a '*strong*' incentive for industry to cut down the volume of waste (HM Government 1990: 187). The UK government did not adopt a comprehensive packaging waste programme. There are no general waste prevention and re-use targets. There are also no recycling targets that deal explicitly with packaging waste. The government stated, however, that half of the recyclable household waste would have to be recycled by the year 2000. Since the government wrote in its White Paper that glass, cans, plastic bottles and paper would be readily recyclable (HM Government 1990: 190) it could be concluded that a large proportion of the recycled household waste would have to consist of used packaging. The government estimated that 40 per cent of the household waste would be recyclable by material recycling and 10 per cent could be composted. Therefore it estimated that the recycling target would have to be 25 per cent, including composting (HM Government 1990: 190). There were for a long time, however, no clear answers to the question, how much of the household waste the government regarded as recyclable. Hence, rather than setting a quantified target, the government stuck to broad estimations. The White Paper requested industry to reduce unnecessary packaging, but there was no circumscribed context for industry action. Therefore, in contrast to Germany and the Netherlands no clear targets deadlines, or provisions for sanctions were adopted in the UK.

The Environmental Protection Act (1990) included more concrete provisions which are, however, primarily aimed at local authorities. Local waste collection authorities have to set local targets and develop recycling plans which must be submitted to the Minister of the Environment, in draft, by 1992. A DoE waste management paper gives guidelines for the content of these plans, but these guidelines are only suggestions. Moreover, local authorities are not required to meet any commitments made in the plans. Their implementation is not obligatory.

The government employed a minimalist but very flexible approach. Rather than regulating specific packaging waste materials by setting mandatory and quantified targets as in Germany and the Netherlands, the British government set a very general not binding household waste recycling target. This approach leaves a great leeway to sub-national and private actors.

3.4 Allocation of responsibilities

The **German** Packaging Ordinance placed the responsibilities for the collection, sorting and recycling of used packaging with the private sector. Hence it introduced a comprehensive producer responsibility for used packaging. For transport packaging the last owner of the packaged goods had the take-back obligation. For secondary packaging, the regulation applied to the distributor who delivered the packaging to the final customer regardless of whether this was a private or commercial customer. For sales packaging the retailers and the mail-order firms were responsible for taking back the used packaging of the consumers and the rest of the packaging chain was responsible for taking it back from the retailers and to re-use or recycle it. Companies could delegate the implementation of the tasks to third parties, but they remained legally responsible.

In the case of sales packaging companies could escape of their individual responsibilities if they joined a collective scheme. In Germany one collective scheme was established: the Duales System Deutschland GmbH (DSD). The DSD ran a nation-wide packaging waste collection, sorting and recycling system. Since this collective system worked parallel to the traditional municipal waste management system, it was given the name *Dual* System. The DSD was set up under the auspices of the German peak associations of trade and industry BDI and DIHT. Some 600 companies from all parts of the packaging chain were shareholders. The DSD established the Dual System which had to cover the whole territory of Germany. It established an intricate system of contracts with more then 16,000 companies.

Packers and *fillers* who wanted to join the DSD had to pay a charge, "the green dot" (Grüner Punkt), which had initially been calculated by the volume of packaging, and later according to the amount of packaging used in terms of weight and the kind of material. The DSD provided each *household* with a special yellow bin or bag for sales packaging, except for glass and paper for which neighbourhood containers were used. DSD set up contracts with regional waste disposal firms, either from the public or private sector or a mixture of the two, to undertake the collection. The actual operating systems differed between *municipalities,* and in the cases when satisfactory collection systems already existed they were incorporated into the framework of the Dual System. To fulfil the collection quotas of the Ordinance, the DSD relied on the support of the *consumer*. Therefore, the DSD organized nation-wide publicity campaigns and information programmes for individual administrative districts and cities. Once the packaging waste was collected, DSD arranged for sorting into the different material fractions. The DSD entered into contracts with *public or private waste management organizations* to carry out this function. The DSD had approximately 400 sorting plants in operation.

Figure 5 Interactions within the Dual System

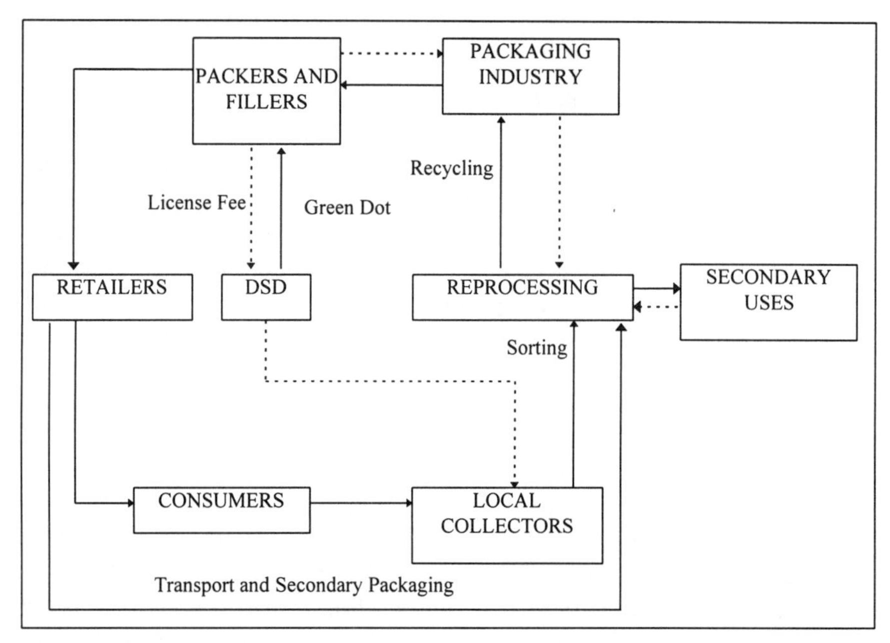

Source: Based on Michaelis 1995: 236

DSD forwarded the sorted materials to the *'guarantor'* who accepted and recycled packaging materials collected by the DSD. The 'Guarantor' was responsible for ensuring that collected and sorted material was recycled. Guarantors were usually the raw material producers of packaging. Note that the collection and recycling guarantees included not only domestically manufactured products but also imported goods (IW 1992: 2).

These systems have to be organized in such a way that the given quotas for the collection, sorting and recycling of packaging materials are met either by a) establishing return systems such as door-to-door collection systems; b) installing containers easily accessible to the consumers, or c) a combination thereof. The return and management systems have to be co-ordinated with the existing local waste management systems. The systems also have to ensure that the collected packaging will be effectively re-used.

The German *Länder* are responsible for the licensing, monitoring and enforcement of industry activities. The Packaging Ordinance did not pay very much attention to the monitoring and enforcement of individual responsibilities. There are no provisions that force individual companies to prove that they fulfil their requirement. The Ordinance was designed under the assumptions that a) the take-back and re-use/recycling of transport packaging would not cause any problems and (b) that all packers and fillers would join a collective scheme.

The collective scheme had to prove for each Land separately that their collecting and recycling arrangements were sufficient to meet the requirements. When this was the case, the respective Land declared the exemptions to the individual take-back requirements (Freistellungserklärung). The *Länder* had the right to revoke the declaration if the collection and recycling obligations were not met. They could restrict the revocation to specific materials. Thereby, the *Länder* had refined instruments which could be used as the big stick. Each year the DSD had to prove for each Land, that the collection and recycling targets were met.

The **Dutch** covenant did not allocate the recycling responsibilities in a clear and comprehensive way. The covenant obliged the Minister of the Environment to set up the legal framework for the collection of packaging waste, the increase of waste incineration and the phase out of landfill. The industry took responsibility for the collection of glass and paper and cardboard packaging to a certain extent (see table 19). In addition, the packaging chain was obliged to commence several trial projects regarding separate collection systems and to evaluate with VROM whether they were successful. These pilot projects included kerbside collection for plastic and metal packaging as well as drinking cartons (Tetra-pak). It was concluded that the separate collection of these materials was - in contrast to the separate collection of glass and paper and cardboard - not sensible, given the low response, the bad quality and the high costs (Mingelen 1995: 21-22). Hence in contrast to Germany, Dutch industry, was not responsible for the separate collection of plastic, composites, tinplate and aluminium stemming from sales packaging, at least as long as the overall recycling targets were met. The precise allocation of responsibilities between industry and local authorities, in particular in regard to financing, remained vague however. The Dutch packaging chain was obliged to create recycling capacity and to use recycled material in new packaging.

The Covenant spelled out a complex process of monitoring, assessment and dispute settlement. The SVM had the task of analysing the implementation plans annually and of writing a report. The report had to be sent to the Packaging Committee. The Packaging Committee (‘*Commissie Verpakkingen*’) consisted of two members appointed by the packaging chain, two members appointed by the ministry and one member appointed by both parties together. Two representatives of environmental- and consumer-organizations were allowed as observers at the meetings at which the report of the SVM and the report of Packaging Committee were discussed. The Committee analysed whether a) the implementation of the measures agreed on proceeded as announced in the covenant; b) the targets set out in the covenant could be met through these measures and c) the targets were met. The Committee’s findings had to be laid down in annual reports which were open to the public. In the case of disagreements between the parties of the covenant an arbitration committee had to find out whether circumstances beyond one's control were responsible for non-compliance with the covenant.

In the **United Kingdom** the local authorities carried the main responsibility for implementing the government’s recycling policy. The role of government was confined to the encouragement of local pilot projects for the separate collection of waste materials (HM

Government 1990: 190-191). The local authorities had to adopt implementation plans which specified systems for the collection and processing of recyclable materials. In addition there were a number of local public-private partnerships (see chapter 5 for details). The performance of the local authorities was monitored by the Department of the Environment (DoE). But since there were no mandatory targets or obligations, no formal sanctions were possible.

As a response to the government's request in the 1990 White Paper, industry developed several initiatives to address the issue of packaging and the environment. The three most important ones were announced and established in 1992. The Consortium of the Packaging Chain (COPAC) - made up of six trade associations representing mainly the manufacturers of packaging - announced that it would recover 50% of all packaging waste. Recovering included material recycling as well as incineration. COPAC would only recover used packaging which had already been collected and appropriately sorted. In the same year, the Packaging Standards Council was set up to handle complaints on excessive and unsafe packaging against a one-page code of practice. This was an initiative of the Industry Council for Packaging and the Environment (INCPEN). The third initiative stemmed from the British Retail Consortium (BRC), the retailers' trade association. The BRC issued notes to '*offer guidance to retailers and their suppliers in determining their own company policies on packaging with the aim of achieving the optimum balance between best environmental practice and these vital functions' (safety of consumers, M.H.)(*BRC 1993: 1, see chapter 6 for more details*)*.

3.5 Modes of interest integration

The **German** government had frequent contact with the target group, but no large and enduring network was created for the preparation of the German packaging waste policy as was the case in the Netherlands. The interaction concentrated on a small and selective group of government officials and representatives of the two major peak associations of German industry (BDI and DIHT), and a cross-sectional single issue organization, the Working Group on Packaging and the Environment (*Arbeitsgemeinschaft Verpackung und Umwelt*, AGVU). In this small and informal network the subtle bargaining took place that resulted in the outline of the German Packaging Waste Ordinance and the Duales System Deutschland. Other societal interests played a minor role in the policy process. The official hearing on the draft Ordinance at which more than 130 actors participated had no impact on the shape of the German approach. The German bargaining system lost its exclusivity somewhat when the draft came into the arena of the *Bundesrat* Here more interests were able to voice their preferences in the political process.

The **Dutch** government had the most intense contact with the target group. During the programme formation the government first consulted the industry parties and then negotiated on the programme. In the consultation phase a large network was created including a broad range of actors, such as environmental and consumer groups, research institutions and a process mediator. All actors met regularly and frequently. The function of this network was first to generate information about packaging waste problems and feasible solutions. The second reason for creating the network was to achieve a common understanding of the problem and to create support and legitimacy for a packaging waste policy programme. In the second phase the government negotiated on the concrete provisions of the programme with industry. The network created for this purpose was much smaller and restricted to the government and a few representatives of the packaging chain, in particular the Foundation of Packaging and the Environment (SVM). The SVM is a cross-sectional single-issue trade association which represents large parts of all sections of the packaging chain. Moreover, it has some disciplinary authority to bind its affiliates to agreements made with the government. One can therefore speak of a negotiation network that had a corporatist character.

The **British** government had the least intense contact with the target group. Since packaging waste was not very high on the governments agenda, no close networks were created for the development of a packaging waste policy. The Recycling Advisory Group was set up in 1989, with members drawn mostly from companies and trade associations and a few representatives from local government and environmental groups, which put forward a number of reports to the government (ENDS Report April 1990: 14). There was also *ad hoc* contact with trade and industry, but no detectable network was created with which the government negotiated packaging waste policy measures on a regular basis.

3.6 Orientation towards target group

The **German** government pursued the most impositional approach towards industry. The Packaging Ordinance introduced the principle of product responsibility for the packaging chain. Thus, whereas in the past the industry had no responsibilities concerning the fate of used packaging, now it became legally responsible for the collection, sorting, and the re-use or recycling of packaging. This obligation did not cause many problems in the case of transport and secondary packaging. For transport packaging the existing logistic infrastructure could be used, and transport packaging can be usually quite easily recycled or re-used. Secondary packaging has always been a small amount of the overall amount of packaging: in fact less than 0,3 per cent. Secondary packaging virtually disappeared, or was not left in the retail shop and had therefore to be treated as sales packaging.

The real problems were caused by the obligations for sales packaging. Most representatives of the packaging chain resisted this responsibility. Infrastructure had either to be established or to be expanded. While the minimum collection and recycling quotas for 1993 (after one year) were not very demanding the quotas required for 1995 were quite tough, especially those for plastics, where virtually no material recycling had take place before. Assessments calculated that 7 billion German marks were needed to build up collecting, sorting and recycling facilities. The running costs were estimated to amount to 2 billion each year but turned out to be higher (DM 3 billion, 1993 and 1994; DM 4 billion, 1995; DSD 1994, 1995, 1996). The following figures show that especially in regard to plastic waste recycling there was a long way to go.

Table 15 Recycling quotas of the German Packaging Ordinance

	Target 1 Jan. 1993	**Target 1 July 1995**
glass	42	64
tinplate	26	72
aluminium	18	72
paper/cardboard	18	64
plastic	9	64
composites	6	64

The Ordinance did not explicitly discriminate between specific packaging materials. The system of the green dot, however, made, for instance, plastic relatively more expensive than paper. Thus the overall market for packaging was likely to decrease, and the market share of these materials was also about to change.

If industrial obligations were not fulfilled an administrative offence would be deemed to have been committed. A fine of up to 100,000 German marks could be imposed for each offence. More serious was probably the sanction in the event that industry did not meet the quota for refillables and the recycling targets for the collective sales packaging management system. In case of a failure to meet the refillable quota, mandatory deposits for liquid foodstuffs can be introduced. If the recycling targets for sales packaging were not met, take-back obligations and mandatory deposits for washing powder, detergents and emulsion paint would be introduced. The take-back obligations and the mandatory deposits were a powerful threat by the government to urge the industry to establish the DSD.

The orientation of the **Dutch** government towards the target group was less impositional than in Germany. The approach was consensus-oriented. The government did not impose regulations. The Packaging Covenant was a voluntary agreement. It made industry a partner of government. No company was forced to join. Even if the Covenant was binding under private law, the concrete possibilities for sanctioning non-compliance remained unclear. Given the consensus-oriented attitude of the government, potential sanctions were hardly discussed during the programme formation process. Moreover, the government also employed a great deal of persuasion, advice and stimulation. Still the standards were relatively strict. The amount of packaging had increased by some 15 per cent between 1986 and 1990 (Commissie Verpakkingen 1996: 18). Therefore, the ceiling on the absolute amount of packaging which could enter the market put pressure on the packaging chain to engage in the reduction of packaging usage. Since the amount of packaging used is closely related to the rate of economic growth, the figure for packaging would increase even further if no concerted efforts were made to use less or lighter materials or to carry out other prevention measures.

The recycling target was more modest however. Recycling had to increase from 25 per cent (1986) to 60 per cent (2000). Recycling in 1990, when the covenant was concluded, was, however, already substantially higher than the 25 per cent achieved in 1986. The intervention intensity was further weakened by the fact that the time scales for achieving the recycling targets were more generous than in Germany. Moreover, industry could concentrate on transport packaging and rather easily recyclable materials, such as glass. Therefore, the costs of the Dutch covenant for industry were considerably lower than in the German case.

The **British** government adopted an ambitious recycling target for municipal waste, if compared to its former performance rather than to the levels achieved in other countries. The government estimated that 5 per cent of household waste was recycled in the UK in the late 1980s (HM Government 1990: 190). The target of 25 per cent would therefore mean an increase of 500 per cent. Note that the target refers to household waste. That means, in contrast to the Netherlands, that the recycling of commercial waste did not fall under this target. This made the target even more ambitious. But this also suggests that the target belonged to a strategy of symbolic politics, since household waste collection and recycling is more visible to the voter than commercial waste recycling. Moreover, the recycling target was embedded in a minimalist approach to the problem of household waste. Neither a binding regulation was imposed on industry nor an agreement reached between government and industry. Since no statutory targets and obligations were enacted, no sanctions were possible. The government employed stimulation rather than repressive economic instruments, such as recycling credits. It was very much a *laissez-faire* approach. However, the legal changes in the general waste management system, that is, the contracting out of former local authorities' waste disposal operations and the tougher environmental standards, might have the effect of increasing waste disposal costs and serving as an indirect measure to stimulate recycling.

3.7 Conclusion

The three governments adopted different policies for coping with the problem of packaging waste. The different policy elements have been discussed separately. Though this is heuristically useful in order to provide a comprehensive picture, it is also clear that the policy elements are strongly interdependent. Therefore the elements build up to a rather consistent image. The German approach was the most formal and impositional. Strict standards `were to be achieved within a short time frame. The producer responsibility included even the collection of sales packaging. Heavy sanctions could be adopted if the provisions of the Ordinance were not met. The Dutch system was more moderate. Although a ceiling on the overall amount of packaging which can enter the Dutch market was set, the time scales were more generous and the degree of flexibility much higher than in Germany. Accordingly, the Dutch covenant approach was more consensus-oriented and the interaction with the target group was more frequent. The British system can be characterized by a minimalist and symbolic approach. No provisions were adopted which were sufficient to meet the target for household waste recycling. Accordingly, there was no frequent and regular interaction with the target group. In order to understand the differences between the three countries, I will reconstruct the paths the different countries took. As already said, these developments will first be analysed for each country separately in order to understand the inherently historical logic of policy development.

4. Germany sets the pace

There are good reasons for starting with the German case. With the Packaging Ordinance (*Verpackungsverordnung* 1991), the German government adopted the most ambitious and strictest packaging waste policy in Europe. The German policy process paralleled the development of the Dutch Packaging Covenant and may have influenced it. For most of the other countries, the impact of the German Ordinance was felt during its implementation. The Ordinance has been effective in terms of its own objectives, but had tremendous unintended consequences not only for domestic markets but also for other countries such as the United Kingdom. Germany was setting the pace and the result has been a Europeanization of packaging waste policy.

4.1 The foundations of environmental policy in Germany

In Germany, the creation of environmental protection as a new policy field was associated with a change in the federal government in 1969. The coalition of the two largest German parties the Social Democratic Party (SPD) and the Christian Democratic Party (CDU) was replaced by the social-liberal coalition (SPD-FDP). The development of environmental protection policy in Germany confirms the neo-institutionalist claim that not only society affects government, but that government affects society too. The initiative to move towards a comprehensive environmental policy came from forces inside the government machinery, rather than from demands of organized societal interests or diffused public pressure. Neither election campaigns nor party programmes had placed emphasis on the environment. Analyses of press releases revealed that environmental protection became a public topic after the first activities of government (Margedant 1987: 21-23, Weßels 1989: 276). The new policy field emerged against the background of a general climate of reform and modernization which had already developed under the former coalition, and which was supported by favourable economic circumstances, such as almost full employment and high economic growth. Civil servants in the Ministry of Internal Affairs acted as policy entrepreneurs and constructed the intellectual and institutional framework for environmental protection policy. They were supported by members of the Chancellor's office. Chancellor Brandt was open minded to environmental issues and the Minister of the Chancellor's office was impressed by the environmental protection activities of the United States (Weßels 1989: 276). Early environmental legislation in the US, in particular the National Environmental Protection Act (1969) provided a stimulus, and shaped the intellectual background of environmental policy and to some extent guided its institutionalization. The government's policy was also influenced by international events such as the UN Conference on the Environment in Stockholm (1972).

State agencies as policy entrepreneurs

In Germany, no distinct ministry for environmental protection was created. The Ministry of Internal Affairs (*Bundesministerium des Innern*) assumed a number of competencies from the Ministry of Public Health. The emergence of a comprehensive environmental policy in Germany was constrained by the German federal structure. The main legislative power in regard to environmental problems lay with the German *Länder*, and there were conflicts between the *Länder* and the Federation about the proper allocation of responsibilities (Margedant 1987: 22). In 1971 and 1972, however, the German constitution was amended granting the Federation concurrent legislative power in areas such as waste disposal and air pollution. Concurrent legislative power means that the Federation assumed the right to issue detailed regulations which supersede state law. In other areas, like water pollution control, the federal legislative power is restricted to so-called framework laws, leaving detailed and specific legislation to the *Länder* (Storm 1989: 11).

The German government adopted an environmental programme in 1971, despite conflicts with the German *Länder* about the redistribution of competencies. This programme was the centrepiece of German environmental policy (Umweltprogramm der Bundesregierung 1971). It was prepared by civil servants but representatives from industry associations as well as scientists and other experts participated (Margedant 1987: 22). The programme was based on three main principles: the precautionary principle (*Vorsorgeprinzip*), the polluters pay principle (*Verursacherprinzip*) and the cooperation principle (*Kooperationsprinzip*). The most important guiding principle was the precautionary principle stating that harmful environmental impacts were to be prevented as early as at the construction and production level. The principle legitimize measures despite scientific uncertainty about causes and effects (see Héritier *et al.* 1996: 40). This focus on abstract principles has its roots in the German tradition of codified/Roman law. In contrast to British case law, the German tradition, and in fact the tradition of most West European countries, starts with general principles and then deduces the concrete judgement of cases in the light of these abstract notions (Weber 1994: 178-179). Hence, in Germany, environmental ideas are institutionally embedded in environmental law and serve as guidelines for Court decisions (see Weale 1992b: 83). The principles of the first environmental programme have remained basically unchanged ever since. The principles continued to provide the very terms in which environmental policy is discussed in Germany.

The federal government machinery is relatively small due to the system of federalism which allocates most implementation functions to subnational actors. The government, therefore, started to reduce its resource-dependency and to enhance its problem-solving capacity in this new policy field by creating organizations and networks. The reason for this was not only to increase knowledge but also to increase the legitimacy of its policies. In 1971 the German Council of Environmental Advisors (*Sachverständigenrat für Umweltfragen*,

SRU) was founded. It was modelled after the U.S. Council of Environmental Quality and was made up of independent social and natural scientists. Its reports, the first, comprehensive one, issued in 1974, have always been major contributions to the environmental policy agenda. Three years later the German Federal Environmental Agency (*Umweltbundesamt*, UBA) followed. Again an American organization, the U.S. Environmental Protection Agency, served as a model. But in contrast to the U.S. Agency, the functions of UBA were restricted to technical, scientific and administrative support. It has no regulatory functions and has less autonomy than its U.S. counterpart (Jänicke and Weidner, 1997: 136). Hence, the SRU and the UBA provided scientific support to the Ministry of Internal Affairs, and constituted a crucial resource in cabinet bargaining in an area such as environmental policy with its technical and scientific character of problems and solutions. Both organizations strengthened the position of this department towards other government departments, like the Ministry of Economic Affairs (*Bundesministerium für Wirtschaft*, BMWi) and towards strong economic interest groups. Moreover, these organizations helped to increase the environmental awareness of the public, and therewith legitimate environmental action by the government.

Organising civil society
Organization building went beyond the core of the political system. Legitimated by the government's objective to increase the environmental awareness of the citizens, the Ministry of Internal Affairs started to create networks in society. One of the '*Brückenköpfe*' (Czada 1991: 155) in the society was the Working Group on the Environment (Arbeitsgemeinschaft für Umweltfragen), founded in 1971, a forum where bureaucrats, politicians, industrialists and scientists discussed environmental problems. Civil servants of the Ministry of Internal Affairs also organized societal support by helping to institutionalize a national environmental movement as a counterforce to economic interests. Environmental protection groups were mainly locally-oriented in the early seventies in Germany. In 1972, a central association of the numerous local environmental groups was founded in Frankfurt (*Bundesverband Bürgerinitiativen Umweltschutz*, BBU). The BBU was a peak association of some thousand groups with a total of 500,000 members (Margedant, 1987: 24). Civil servants were personally engaged in the birth of this association. The travel costs of these 'private persons' were paid by the department (Hartkopf, 1986). The BBU foundation is an example of sponsored pluralism. The state nurtured a plurality of interests to maintain its autonomy towards particularistic interests, in this case economic interests (see Czada, 1991, van Waarden, 1992: 48-49). The assistance of civil servants of the Ministry of Internal Affairs in this case does not mean, however, that government and environmentalists had a cosy relationship. In fact there has been rather a distance between environmental groups and the government in the years following the early 1970s (Hey and Brendle 1994). Other important environmental groups in the 1970s were the '*Deutsche Naturschutzring*', founded in 1950, a federation of some ninety nature conservation organizations, the '*Bund Umwelt und Naturschutz Deutschland* (BUND)' founded in 1975, with then 100,000 members, and the '*Deutsche Bund für Vogelschutz*' already founded in 1899, with 120,000 members (Hey and Brendle, 1994: 139-154, Margedant, 1987: 24).

Policy

From a cross-national perspective, environmental policy was rapidly institutionalized in Germany. The first environmental programme in particular was progressive and demanding in its time. Müller attributes this development to the unpoliticized character of the exercise: the civil servants perceived the development of the programme as an intellectual challenge. '*Sie hatten die amerikanischen Programmansätze studiert und formulierten die allgemeinen Teile des Programms weitgehend unbeinflußt von taktischen Durchsetzungsüberlegungen und eigenen Zuständigkeits- und Handlungsschranken. Trotz vieler Abstimmungszwänge im Detail konnten sie in einem 'intellektuellen Freiraum' operieren, die ihnen die Unterstützung des Bundesinnenministers Genscher sicherte'* (Müller, 1989: 5). Additional facilitating factors for the programme were, according to Müller, a) the backing of the Chancellor's Office; b) the general reform climate associated with the government; c) the wait-and-see attitude of the defenders of established policy areas and interests inside the 'state apparatus' and d) the participation of industry representatives in the development of the environmental programme (Müller, 1989: 5/6).

There was broad consensus on the need to increase environmental protection at this time (Weßels 1989: 279). The BDI stated in its annual report of 1969/1970: *'Unsere Umweltpolitik kann sich - soll sie ihr Ziel erreichen - nicht in der Durchsetzung der bislang versäumten Maßnahmen,...erschöpfen; sie muß darüber hinausgehend gleichzeitig sicherstellen, daß bei allem was wir in Produktion und Konsum heute und morgen unternehmen, auch die sich hieraus ergebenden Folgen im Interesse einer weitmöglichen Verhinderung künftig eintretender Nacheile schon jetzt berücksichtigt werden'* (quoted in Müller, 1989: 8).'

In the early 1970s, environmental issues were not politicised in parliament. Müller (1989) argues that '*(maßgebend) hierfür mag gewesen sein, daß die hohe Konsensfähigkeit und Überparteilichkeit des Umweltthemas es für Parlamentarier zunehmend uninteressant werden ließ, hier ein Betätigungsfeld zu suchen*' (6). In this depoliticized environment governmental policy entrepreneurs were instrumental. Their preferences and capabilities were shaped by a number of institutions. The German selection procedures for ministerial civil servants favour specialists rather than generalists. Therefore experts like environmental lawyers, natural scientists, and engineers are selected rather than generalists as is the case in the UK. Due to this selection principle and due to low fluctuation within their jobs, civil servants can accumulate knowledge and experience and often become identified with their task and 'go native' through a socialization process. The foundation of environmental policy in Germany shows that in new policy fields policy actors are willing to learn and capable of drawing lessons from abroad.

4.2 The history of (packaging) waste policy

4.2.1 The Waste Disposal Act (1972): controlled dumping

Since the 1950s, the amount of waste grew rapidly and parallel to economic growth. As early as in the 1960s, the need was seen to reform the fragmented law concerned with waste disposal (Donner and Meyerholt, 1995: 82). In this period waste disposal regulation was the domain of local authorities. The legal framework was provided by *Länder*-regulations concerning local authority activities (*Landeskommunalrecht*). In addition, single federal provisions concerning public health and hygiene, trading regulations (*Gewerbeordnung)* as well as water protection law had to be applied (Bartlsperger, 1995: 34-53, Brückner and Wiechers, 1985: 157). In 1971 and 1972, several *Länder* adopted waste laws. Shortly afterwards, however, the federal government used its recently assumed legal competence for concurrent legislation in this field and adopted the Waste Disposal Act (Abfallbeseitigungsgesetz, BGBL I, pp. 873). The national political discussion preceding the Waste Act was very much constructed by a 'public health and hygiene' frame. This can be partly explained by the problem perception at that time. Institutional reasons were also important, however. In the late 1960s and the beginning of the 1970s the national government did not have legal competencies in the field of waste disposal as such, but it did in the related field of public health. Framing the waste problem as a public health problem helped to legitimize political activities on the national level in the period preceding the changes in the German constitution.

The Act focused on municipal waste. As the name of the Act indicates, it dealt with waste disposal, which was defined as the collection, transport, treatment, storage and dumping of waste. The Act did not provide a general legal basis for waste prevention and waste recycling (Bartlsperger 1995: 35). It was a reactive and legalistic approach aimed at channelling the consequences of increasing production and consumption (Donner and Meyerholt 1995: 83). Its focus on the end of the production-consumption-disposal chain can be explained by the urgent need to upgrade the existing waste disposal sites. Even the 1971 Environmental Programme with its focus on precaution, which would basically mean waste prevention in this context, emphasized orderly disposal of waste rather than prevention. The diverse and disordered ways of waste disposal had to be tackled by planning and organizational means and had to be designed in a more technical and rational manner (Brückner and Wiechers 1985: 159-157). This emphasis on technical measures reflects the German perception of environmental protection, namely as primarily technical protection of the environment, rather than 'nature conservation' as in the United Kingdom (Héritier *et al.* 1996: 40). An important objective was the reduction and upgrading of waste disposal sites in order to avoid nuisance from smell and dust, to prevent the danger of fires, to ensure hygiene by preventing water pollution, and to serve aesthetic needs (SRU 1991: 39).

The government employed a formalized, regulatory command-and-control approach to the problem of waste disposal, an approach that has also been used for other areas of environmental protection. This approach consisted mainly of licensing and laying down standards. In line with the idea of precaution, the standards were based on the latest available technologies. The administrative procedures guaranteed public participation. This approach forced the authorities responsible to formulate detailed technical requirements, and to prove that they kept up with the state of the art in technology and that the requirements were feasible. Implementation and enforcement of these requirements were often time-consuming, controversial and accompanied by laborious administrative efforts at control in the form of monitoring and sanctioning (Jänicke and Weidner 1997: 139).

The only provision that enabled the federal government to adopt measures to prevent waste was contained in Article 14 of the Waste Act. This article should not be too strongly emphasized, however, since the prevailing ideology was end-of-pipe-oriented. Still, the provision is important in the context of the study, because it dealt explicitly with packaging. Article 14 enabled the government to introduce ordinances that deal with packaging and containers whose disposal required a major effort. The government can rule that they can enter the market only ; a) with a distinct sign; b) for distinctive purposes; c) in a limited amount or d) not at all. This provision indicates that packaging was considered an issue already in the early 1970s. It will be shown, however, that the government did not use the authorization for the adoption of statutory packaging waste policy until the late 1980s.

Table 16 Overview German (Packaging) Waste Policy Developments

Year	Policy measure	Objective of standards	Statutory	Effective
1972	Waste Disposal Act	controlled dumping	yes	yes
1975	Waste Management Programme	prevention and recycling of waste	no	too general for evaluation
1977	Voluntary Agreement on Drink Packaging	increase in re-use and recycling	no	re-use: no recycling: yes
1986	Waste Prevention and Disposal Act	towards resource and energy saving	yes	too general for evaluation
1988	Ordinance on Drink Packaging from plastic	increase re-use, recycling	yes	yes
1989/1990	Fixing of Targets	increase re-use and recycling of drink and plastic packaging	no	replaced by Ordinance
1991	Packaging Ordinance	re-use and recycling of all packaging waste	yes	yes

4.2.2 The Waste Management Programme (1975): recycle, please!

The waste laws of the *Länder* and the Federal Waste Disposal Act aimed at controlling waste disposal by command and control measures to avoid harmful effects on human health. This approach prevailed in the 1970s and the 1980s. In the field of packaging waste and some other areas (waste from industrial processes), however, a cautious shift towards waste prevention and recycling measures could be observed, supplementing command and control measures by voluntary agreements and the recognition of economic aspects of waste. This was a small and gradual shift from waste disposal to waste management. The idea of recycling was particularly motivated by the oil crisis of 1973. This event changed somewhat the perception of the nature of waste. Waste was seen from then on not only seen as something which had to be disposed of, but increasingly also as a secondary raw material. Recycling was a means of saving resources and decreasing dependence on raw material imports (SRU 1991: 39). The priority of recycling *vis-à-vis* disposal of waste was first included in the Federal Immission Act (Bundes-Immissionsschutzgesetz) that applied to industrial processes (Bartlsperger 1995: 36, SRU 1991: 39).

In regard to municipal waste, the idea of recycling was formulated in the government's Waste Management Programme (Abfallwirtschaftsprogramm 1975). Similar to the 1971 Environmental Programme, this programme was informed by the precautionary principle, the polluter-pays principle and the cooperation principle. Unlike the 1972 Waste Act, the Waste Programme was not end-of-pipe-oriented, but emphasized the prevention and reduction of waste. Next to the oil crisis, this can be related to the fact that good progress had been made in achieving the 1972 Waste Act objectives of orderly disposal and reduction of disposal sites.

The Waste Management Programme aimed at the reduction of waste by looking at the input to the production process, the application of environmentally friendly production processes, the reduction of production waste, an increase in the durability of products and an increased re-use of products. Also, emphasis was put on material and energy recycling throughout the production process. The primary responsibility for achieving the reduction objective lay with the economic actors, who were asked to take initiatives for an environmentally sound waste management system. The waste programme was not legally binding. Measures proposed in the programme aimed at increasing the informational base, improving the markets for secondary materials and most generally changing attitudes away from the 'throw away' society. The Waste Management Programme emphasized that economic and environmental interests do not contradict each other, and appealed to the cooperation of all actors concerned (Brückner and Wiechers 1985: 158-160).

4.2.3 The Voluntary Agreement on Drink Packaging (1977): refill, please

The idea of waste prevention and recycling as stipulated in the 1975 Waste Management Programme was not codified in public law, however. Why did the government not use the legal provisions from the 1972 Waste Act and adopt a statutory approach to waste management as it had done in other areas of environmental protection? The answer to the question has to be given against the background of a number of more general economic, social and political developments in Germany. These developments resulted in an 'Eiszeit' (Müller 1989: 8), a period of stagnation (Weidner 1995: 7) that lasted from 1974 to 1978. The view of the 1975 Waste Management Programme that economic and ecological interests do not contradict each other reflected, in particular, the perceptions of the civil servants of the Ministry of Internal Affairs. This view was not shared by all actors. The economic recession following the oil crisis of the winter of 1973/74 resulted in increasing criticism of environmental protection measures. In a decisive meeting between government, industry and trade unions in Gymnich 1975 criticism was voiced that a) investments in the amount of 50 billion German marks would be hindered by too strict licensing, the so called *'Investitionsstau'*, and b) that jobs were threatened because of an excessive burden on industry (Weßels 1989: 281). Already in 1974 the charismatic Willy Brandt was replaced by the more pragmatic Helmut Schmidt as Chancellor. This shift '*markiert gleichzeitig das Ende der Reformphase und den Beginn des Krisenmanagement der sozial-liberalen Regierungspolitik*' (Müller 1989: 9). The pro-environment advocates of the Ministry of Internal Affairs lost the support of the Chancellor's Office. Hence, economic interests found their access not only via the Minister of Economic Affairs but also via the Chancellor's Office. Moreover, the CDU/CSU majority in the *Bundesrat* provided an effective veto against prevention-oriented statutory measures, because ordinances based on Article 14 needed the assent of this Chamber.

The political climate was shaped not only by the economic recession but also by developments in society. Nuclear power became a hot issue in Germany. The anti-nuclear movement was relatively radical and there were a number of militant demonstrations against nuclear energy. The Minister of Internal Affairs was responsible not only for environmental protection but also for public order. The militant demonstrations therefore '*schadeten dem Umweltschutz in doppelter Weise. Sie absorbierten zum einen die Aufmerksamkeit und Konfliktbereitschaft des für die innere Sicherheit zuständigen Bundesinnenministers. Sie schufen zum anderen im politischen Raum eine negative Stimmung gegen 'Basisbewegungen', zu denen auch die Bürgerinitiativbewegung für den Umweltschutz gezählt wurde*' (Müller 1989: 9). Had the environmental competencies belonged to a department of the environment as in the UK, the effect of generally radical and sometimes militant societal process might have been less harmful for the dynamics of environmental protection. The economic and social developments had as a consequence that in all parties - in particular the pro-environment FDP - support for environmental protection became weaker and the representatives of economic interests stronger. A prevention-oriented waste policy by statutory measures could not be introduced by the weakened Ministry of Internal Affairs against these prevailing economic and political forces.

Against this background it is not surprising that even though the government had the legal possibility to adopt an ordinance to tackle the problem of packaging waste, the Ministry of Internal Affairs concluded a voluntary agreement with the industrial branches instead. The industry was represented by the peak organizations of the beverage industry, the packaging manufacturers, the glass and steel industry, retailers and scrap firms, and a number of big companies. The objective was to stabilize the share of refillable drink containers, avoide the introduction of plastic bottles, and increase glass and tin plate recycling (Spies 1994: 278/279, Bohne 1982). The voluntary agreement was the result of bargaining, during which the government dropped its plan to adopt an ordinance based on Art 14. The voluntary agreement, however, was not perceived by the government as an effective instrument superior to the command and control approach. It was more as a typical sign of the '*Ausweichstrategie der Ministerialverwaltung, wegen der Blockade im gesetzgeberischen Bereich*' (Müller 1989: 9).

4.2.4 The new dynamic of the 1980s

The 1980s were characterized by a new dynamic which ultimately resulted in a statutory-based and prevention-oriented packaging waste policy. At the end of the 1970s environmental problems re-entered the political agenda. The environmental movement became stronger. Environmental organizations dramatically increased their membership: the BUND, for example, more than doubled in size from 80,000 (1983) to 179,000 (1990) (Hey and Brendle 1994: 149), whereas *Greenpeace Deutschland*, founded in 1980, grew from 80,000 supporters in 1985, to 500,000 in 1989, and 700,000 in 1991 (Hey and Brendle 1994: 151). The federation '*Deutscher Naturschutzring*', which BUND also belongs, had a membership of 3,3 million members in 1985.

In addition, environmentalists formed the Green Party together with activists of the peace movement and the women's emancipation movement. While in England the majority voting system would give a green party almost no chance to enter the parliamentary arena, in Germany the modified proportional voting system enabled 'the greens' to be represented in a *Land* parliament in 1979. In 1983, the Green Party was represented in six *Länder*-governments, and in the same year it managed the 5 per cent threshold of the federal parliament (1983: 5,6%). The successes of the Green Party reflected the electoral potential for green issues and resulted directly and indirectly in the 'greening' of the political agenda. The direct effect was that the green movement could shape the parliamentary agenda. Indirectly, a greening of the agenda was the result of the greening of the other political parties. Between 1979 and 1981, the CDU, the CSU, the FDP and the SPD came up with new environmental programmes (Müller 1989: 11). Hence, while in the early 1970s the polity effected society, in the mid 1980s society effected the polity.

The greening of the political discourse was accompanied by increasingly good economic performance in 1978 and 1979. In the following years the economic situation worsened. The GDP decreased and the number of unemployed approached the figure of 2 million (7,6%). Still, the environmental protection policy remained dynamic. This had to do with a change in the prevailing policy paradigm. The idea of ecological modernization gained in importance. Instead of seeing environmental protection as a burden upon the economy, ecological modernists see it as a potential source for further growth. The idea became appealing to many members of the policy elite in European countries and international organizations (OECD, WCED) during the 1980s (Weale 1992b: 75-79, see also chapter 1.2.5.). Hence, the idea of ecological modernization provided a new framework for the relationship between environmental protection and economic growth. Industry representatives reacted in a more differentiated manner to environmental protection demands. Rather than arguing that environmental protection had general disadvantages, the sectoral effects came to the centre of the discussion. The sharp increase of 'green industries' (environment industry) in terms of turnover, employment, and export orientation, made the general claim that environmental protection had negative consequences for industry less credible. The upsurge of the environment industry was promoted by high standards and technological pushing principles like '*best available technology*' (see Lenschow 1996). Hence, the economic crisis in the 1980s had fewer negative effects on environmental policy than the recession in 1973/1974. The pro-environment forces remained relatively strong in this period.

'The motor of change (toward ecological modernization, M.H.) was now not so much government regulation, but a broad coalition of modernizers stimulating the self-interest of enterprises in respect of costs (energy, materials, waste, transport etc.) and new 'green markets'. The environmental awareness of consumers and the threat of government intervention, ..., were important facilitators in this process' (Jänicke and Weidner 1997: 148). In addition, the participation of the Green Party in a number of coalition governments in the *Länder*, the first one coming to power in Hessia in 1985, contributed to this development. It is worth noting, that in those days two Ministers for the Environment were former environmentalists. Via the institutional channel provided by the *Bundesrat,* green ideas also entered the national policymaking process. The dynamic of the 1980s resulted on the national level in the creation of a separate Ministry of the Environment (1986). The Chernobyl disaster, which preceded this creation, was the occasion rather than the cause of this organizational innovation.

4.2.5 The Waste Prevention and Disposal Act (1986)

Due to the increasing amount of waste and the decreased capacity for landfill and the general environmental dynamic, policymakers began to place greater emphasis on waste prevention and recycling. In 1986, the 1972 Waste Act was replaced by the Waste Prevention and Disposal Act (Gesetz über die Vermeidung und Entsorgung von Abfällen, 1986, BGBl I, p. 1401). As indicated by the name of the Act, waste prevention assumed an important position *vis-à-vis* waste disposal, at least rhetorically (Bartlsperger 1995: 38-41, Donner and Meyerholt 1995: 85). Though waste prevention had a prominent programmatic position in Article 1 of the Act, the only concrete provision that allowed for directly binding measures was provided by Article 14 of the Act. The new Article 14 had a broader scope of application than the old one. It could be applied not only to packaging but to all products (Klages 1988: 481-482). The measures that the federal government could adopt were quite diverse and more broadly defined than those under the old provision. The government could mandate the marking and the banning of products. It could require separate disposal. It could also adopt take-back obligations, mandatory deposits and the requirement to distribute products only if they are re-usable or easy to recycle. Hence, the industry could be obliged to step over to re-use systems. Products could even be banned. The objectives, the conditions for application, and the shape of the measures were rather broad (SRU 1991: 55). This provided the government - in comparison with other German environmental laws - considerable room for manoeuvre. Moreover, the legal competencies that the federal government assumed with this provision meant, according to the logic of the concurrent legislation, that the *Länder* had almost no leeway for the use of legal instruments for the prevention of waste (Klages 1988: 483).

From a cross-national perspective the legal potential for market intervention was quite remarkable. Industry opposition towards this regulation was rather low, however. Industry was still used to a disposal-oriented approach to waste, and to voluntary measures when it came to waste prevention. Moreover, as said above, the provision was not a concrete intervention but simply provided the possibility for market intervention. As a civil servant from the Ministry of Environment explained: '*Ich weiß nicht ob man sich damals in der Wirtschaft schon vorgestellt hat, was man später so alles auf den Artikel 14 stützen wird*' (Interview BMU 12.11.1996).

4.2.6 The Ordinance on Drink Containers made from Plastic (1988)

The 1977 voluntary agreement on drink packaging met its objectives only partially. The share of refillable containers could not be stabilized. The share of one-way drink packaging rose between 1977 and the mid 1980s from 7 per cent to some 25 per cent (Wicke 1993: 275). The recycling objectives for glass and tin plate could be achieved. This was supported by the favourable economics of recycling in these cases (Hartkopf and Bohne 1983: 451). The failure of the voluntary agreement became obvious in 1988, when Coca Cola announced that it wanted to introduce a one-way plastic bottle made from PET. The new Minister of the Environment, Töpfer, an economist and an environmental expert, was determined to prevent this. The government reacted with an ordinance that stipulated mandatory take-back requirements and obligatory deposits on drink containers made from plastic (BGBl. I S. 2455). The Ordinance became also became known as '*Lex Coca Cola*'. Plastic bottles were to carry a mandatory deposit of 0.5 German marks by 1 December 1989, and retailers and producers were obliged to take the bottles back. The Ordinance met resistance from the EC Commission which claimed that the Ordinance was an illegal barrier to trade. The EC Commission asked the government to delay the enactment of the Ordinance until 1 December 1990, in order to allow industry to create the necessary logistics. The Commission also demanded that the Ordinance had to be expanded to all drink packaging in order to reduce its discriminatory effects. The Ordinance had a negative impact on importers, in particular on French and Belgian producers and distributors of mineral water. They exported some 200 million plastic bottles a year and feared losses in turnover, if they were to be obliged to provide for a costly transport and refill infrastructure, while the German producers of mineral water were connected to a dense re-use network (*Mehrwegpool*). The German government argued in defence of the Ordinance. It said that the EC Framework Directive on Waste asked for the reduction of waste by re-use and recycling. Problems with competition would be caused only because other Member States did not adhere to this Directive.

The Ordinance was a success in the sense that before it came into force, Coca Cola announced that it would not introduce the PET one-way bottle because of the mandatory deposit. In the following years they relied on refillable PET-bottles (Feess-Dörr, Steger, and Weihrauch 1991: 352).

4.3 Towards the Packaging Ordinance (1991)

4.3.1 German "Angst" of the waste avalanche

The 1988 Ordinance has to be seen against the background of the widespread fear that waste disposal capacities would not be sufficient for the increasing waste mountains (*Abfallberge*), and that Germany would be buried under a waste avalanche (*Müllawine*). This problem perception was carried extensively by the media and was widely spread among the federal government, the *Länder*, local authorities, industry and the public. Disposal and incineration capacities were becoming scarce. It was expected that half of the 332 landfill sites would be filled in 5 years, and it was feared that millions of tonnes of waste could not be disposed of. The fall of the Berlin Wall (1989) increased the problem. The closure of cheap East German landfill sites decreased the capacities (Michaelis 1995: 23). The reason for the capacity problem was not only a geographic one. It was costly to run a disposal site due to strict environmental standards. More important, however, was the protest against new sites by local authorities, companies and citizens (SRU 1991: 74/75, Spies 1994: 269/270). Long, costly and uncertain planning procedures made it difficult to develop new capacities. In addition, the escape route to more incineration was blocked by public protest against new municipal waste incineration facilities. The problematic character of the waste problem led the Council of Environmental Advisers (SRU) to develop a specific report on this issue. In its report of more than 700 pages (published in 1991), the Council concluded that the *'Abfallproblematik [sich] in den kommenden Jahren noch verschärfen wird, auch dann, wenn Vermeidungs- und Verwertungsanstrengungen zu greifen beginnen'* (SRU 1991: 3, see also BMU 1990: 162).

The critique was also voiced by citizen's groups. The protest of citizen's groups was facilitated in Bavaria by the possibility of plebiscites (*Volksentscheid*), an institutional provision which also exists in some other *Länder*. Art. 74 of the Bavarian constitution says that a plebiscite has to be held when one tenth of the electorate wishes to bring forth an amendment to a law. The expression of such a wish is called *Volksbegehren*. To initiate a *Volksbegehren* 25,000 signatures are necessary. The citizen's group *Das Bessere Müllkonzept* (BMK), established in 1988, managed to collect the necessary signatures to initiate a *Volksbegehren*. The group aimed at a more prevention and reduction-oriented waste policy in Bavaria. The *Volksbegehren* was held in the second half of June 1990 and was signed by 13 per cent of the electorate. This was sufficient to initiate the *Volksentscheid* in Bavaria. This plebiscite took place on 17 February 1991. The Bavarian party in power, the CSU, drew up an alternative bill to the proposal of the BMK (see Eberg 1997: 136/137). Although the CSU bill won by 51 per cent with 47 per cent for the BMK bill, the CSU included some of the elements of the BMK proposal. Moreover, the CSU was forced to take a more waste prevention-oriented position in the *Bundesrat* This ultimately also influenced the shape of the 1991 Packaging Ordinance.

The local authorities were also among the actors who put pressure on the federal government. The local authorities have the legal responsibility for waste disposal. They faced a dilemma where, on the one hand, they were directly confronted with the problem resulting from the growing waste streams and the decreasing capacities and, on the other hand, they did not have the legal and other resources to control the amount of waste by a product-oriented waste policy. Their action capacity was more or less restricted to the end of the product chain, that is waste collection and disposal. As a reaction to this dilemma, the peak associations of the German local authorities asked the federal Minister of the Environment to adopt waste prevention and recycling measures based on Article 14 of the Waste Prevention and Disposal Act (Spies 1994: 271). The solution of the local waste problems was seen as the '*Überlebensfrage der Stadt*' (Handelsblatt 11.1.1990). In addition, the coming federal elections and the election in Bavaria put pressure on government.

4.3.2 The government's approach: internalizing environmental costs

The developments elaborated on above exerted an enormous pressure on government. However, the fact that a lot of government effort to reduce the waste problem concentrated on packaging waste cannot be explained simply by the general capacity problem. Even though almost half of the volume of household waste consisted of packaging waste, there were other waste streams which were much larger and filled more space in the scarce landfill sites than packaging. The government chose packaging for political and strategic reasons as a centrepiece its waste policy. Packaging waste was a highly visible problem to the consumers and therefore to the voters. Related to this, packaging was seen as a useful vehicle to realize the idea of a more prevention-oriented approach to waste policy namely the concept of producer responsibility that has been strongly supported by the Ministry of the Environment. The concept of producer responsibility, sometimes called the concept of product stewardship (Michaelis 1995: 232), was seen by the Ministry as an application of the polluter-pays principle (*Verursacherprinzip*) in the area of waste policy.

'Das Verursacherprinzip im Umweltschutz ist Kostenzurechnungsprinzip und ökonomisches Effizienkriterium. Danach müssen grundsätzlich demjenigen die Kosten einer Umweltbelastung angelastet werden, der für ihre Entstehung verantwortlich ist .... Die Anwendung des Veruracherprinzips ist der entscheidende Anstoß für ökologische wirksame und zugleich ökonomisch effiziente Maßnahmen' (BMU 1990: 27). Packaging was seen as the first case in point to introduce comprehensive producer responsibility, to be expanded later to other products such as consumer electronics, cars etc (Interview BMU 12.11.1990).

As already indicated with the Ordinance on plastic bottles, the government was no longer interested in following the voluntary road. It wanted statutory measures for a number of reasons, primarily, because the government no longer trusted the industry. The Ministry of the Environment did not believe that business would be capable of achieving the necessary reduction of the waste mountains in such a controversial and complex area as packaging waste. The Ministry drew its lessons from the 1977 voluntary agreements which turned out to be rather ineffective: *'[A]uf die freiwillige Selbstverpflichtung hat sich Bonn wohl lange genug verlassen. Inzwischen erkannte auch der Umweltminister, daß bestimmte Ziele von der*

Wirtschaft auf der freiwilliger Basis allein nicht in dem Umfang erreicht werden, wie dies im Hinblick auf kommende Entsorgungsengpässe geboten ist' (Die Zeit 25.5.1990).

The statutory route was facilitated by the EC Commission's approval of the Ordinance on plastic bottles. Moreover, the EC Commission's objections against the discrimination against certain types of drink packaging were very helpful. This was very much in line with the government's desire to adopt a more comprehensive approach including not only types of drink packaging but generally all transport, secondary, and sales packaging. Note that the German government was not forced to adopt broad regulations, instead it used the arguments of the EC Commission to legitimize its own comprehensive approach to the problem of packaging waste.

Not only the Environment Ministry saw the need for regulation. The Minister of Economic Affairs also recognized that '*wenn man sich auf die freiwilligen Massnahmen der Industrie verläßt, kann man schnell verlassen sein'* (Interview BMWi 10.12.1996). The basic agreement of the BMU and the BMWi to pursue a more formalized approach was very important, since the latter Ministry usually resists interventionist approaches. The inter-ministerial agreement in principle on statutory measures was supported by the match of their different policy paradigms or frames to the solution of the waste problem. These frames share a number of elements. Both departments were in favour of the idea of the internalization of external environmental costs. While the Ministry of the Environment expressed this idea with the concept of producer responsibility, the Ministry of Economic Affairs included the internalization argument in its 'privatization of public tasks' frame, in this case the privatization of waste management (Interview BMWi 10.12.1996). Hence both ministries, which usually have different ideas about the level of environmental protection and adequate instruments to achieve them, share a belief in the idea of the polluter-pays principle and the internalization of costs. They used different policy paradigms, 'producer responsibility' (BMU) vs. 'privatization of public tasks' (BMWi), but both came down to the idea of self-regulation of industry within a legal framework. It turned out, however, that the BMU was in favour of a tight and ambitious legal framework, including quotas etc., while the Ministry of Economic Affairs wanted a lean regulatory framework and as little government intervention as possible.

4.3.3 The fixing of targets

Public pressure and matching frames within government resulted in a determined government's approach to introduce a comprehensive packaging waste policy and to adopt strict standards by means of statutory instruments, based on Article 14 of the Waste Prevention and Disposal Act.

The process started with the government's fixing of re-use and recycling targets. The fixing of targets is foreseen by Article 14 of the Waste Act as an instrument preceding statutory action. The rationale is to give a context to industry and to see whether industry is able to meet the targets without state intervention. Target fixing can be seen as formalization of the cooperation principle as laid down in the first environmental programme (Spies 1994: 279). On 26 April 1989 targets were set for the re-use and recycling of waste from drink packaging.

The targets aimed at an increase in both the share of refillables and the recycling of drink packaging that was not refillable.

Table 17 German refill and recycling targets to be achieved by 30 June 1991

Type of packaging	**Targets**	**Level 1987**
	re-use	
beer	90%	86,40%
mineral water	90%	88,37%
carbonated soft drinks	80%	75,94
other soft drinks	35%	29,45%
wine	50%	43,18
	recycling	
glass	1,55 mio tonnes	1,14 mio tonnes (1986)
tin plate	0,3 mio tonnes	not known

Source: Gesellschaft für Verpackungsmarktforschung, quoted in AGVU (1990)

The government argued that it wanted to reverse the trend of the decreasing share of refillables used as drink packaging (from 90% in 1970 to less than 75% in 1987). Moreover, the fixing of targets was seen as the implementation of the European directive for containers filled with liquid food products (85/339/EEC, 6.7.1985). According to Art. 3.1 of the Directive, Member States had to develop programmes for the reduction of the weight and/or the volume of the containers of liquid food products. These programmes can be either statutory or voluntary (see chapter 8 for more details). It is important to note that the fixing of targets was induced by the required implementation of the European Directive but was part of the process towards a policy that was broader and stricter than required by the Directive. The Directive facilitated the government's policy by increasing the legitimization of action, but did not determine the German approach.

In addition to the re-use and refillable targets for drink packaging, the government stipulated objectives for sales packaging made from plastics (*Zielfestlegung Verkaufsverpackungen aus Kunststoff*, 17.1.1990). There were no quantified targets adopted however. The objectives included the identification of the type of plastic used (by 31.12.1990) and the restriction of the number of plastic types used (by 1.3.1990). One of the objectives was the obligation for trade and industry to make proposals about the establishment of a take-back system outside the public disposal route (by 31.7.1990). With this the idea of an industry-financed system for used packaging parallel to the public system for the remaining household waste was officially posed by the government. In this context the role of the FDP leader and former Minister of Economic Affairs was important. He was the first to officially propose a dual system for the management of household waste (Handelslatt 5/6.1.1990).

The legal provision of fixing targets provided the opportunity for voluntary action. But it was clear to government and to representatives of trade and industry associations, that voluntary action would not be sufficient to reverse the trend towards more one-way packaging and to set up an industry-led plastic waste management system in the envisaged time frame. '*Die 1989 formulierten Zielfestlegungen sind mit deutlichen Interventionsdrohungen bis hin zu Verboten flankiert. Und bei den 1990 ... definierten Zielen, deuten die sehr kurzen Umsetzungsfristen darauf hin, daß das Scheitern freiwilliger Initiativen der Wirtschaft bereits einkalkuliert war oder gar provoziert wurde und das Kooperationsangebot der Legitimationssicherung für kurze Zeit später ergriffene ordnungsrechtliche Maßnahmen diente'* (Spies 1994: 280).

What followed was a political process that took place in two loosely coupled arenas. The BDI, and to a lesser extent the DIHT, together with large retailers developed the concept of an industry-led packaging waste, collection, sorting and recycling scheme, and the government developed the first drafts of the Packaging Ordinance. This was, however, not the end of the story. The Packaging Ordinance needed the assent of the *Bundesrat,* and the concept of the Dual System needed the commitment of the whole of trade and industry. This institution of federalism, was reflected in the need to get the assent of the *Bundesrat.* It made the negotiation process between the public and the private actors rather intricate, since it opened a second arena for negotiations. It allowed the *Länder* to influence the shape of the national packaging waste policy and opened another point of access for private interests.

Before discussing the development of the Packaging Ordinance and the Dual System in more detail, I need to emphasize that the whole negotiation process was intricate not only because of the presence and interconnection of the different arenas. What is more, public pressure and the coming elections resulted in serious time constraints, at least from the perspective of the policymakers. Therefore, the whole process was sometimes quite chaotic, and several provisions would probably not have been made, or would have taken another form, if more time had been spent.

4.3.4 The draft Packaging Ordinance: threatening industry

The first draft of the Packaging Ordinance (23.4.1990) made clear that the central instruments the government wanted to use to put pressure on industry were take-back obligations for used packaging and mandatory deposits. The government drew lessons not only from the failure of the voluntary agreement of 1977 but also from the success of the 1988 Ordinance on PET-bottles. Statutory take-back obligations and mandatory deposits had worked in this case as Coca Cola had not introduced the one-way PET bottle. *'Nach dem öffentlichkeitswirksamen Erfolg der Kunststoffverordnung setzte das BMU auf Pfand- und Rücknahmeregelungen'* (Feess-Dörr, Steger, Weihrauch 1991: 352). The take-back obligations for sales packaging were a threat to industry in particular. While for transport packaging, such as pallets, infrastructure was available to pass used packaging up the packaging chain, no logistic was in place to take back used yoghurt containers and such like from consumers. Sales packaging arises at many points in very different forms and composition. This makes a take-back system that has to sent the packaging back to the packaging producers extremely difficult to manage. Yet, it was strategically a very important step to allow consumers to bring back their sales packaging to the point of sale, i.e. the retailers. Minister Töpfer took advantage of the power relations within the packaging chain. With the idea of take-back obligations for sales packaging, he forced the powerful retailers to support an alternative and industry-led solution. Many retailers feared that they would become the national dumping site, *'Müllkippe der Nation'* (IW 1992: 2). They put the pressure on the rest of the packaging chain. Many packers and fillers, fearing that the retailers would send the used packaging back to them, also recognized that the take-back of sales packaging at the retailer point would not be a workable solution.

The Ministry of the Environment did not really intend the retailers to take back the sales packaging. This is also confirmed by the interviews with government officials. The take-back obligation was from the very beginning a stick to force industry to established its own comprehensive waste management collection, sorting and recycling system. This was also in line with another commtent of the EC Commission on the Ordinance on plastic bottles. *'Offensichtlich hat der Minister aus seinem Kampf gegen die Plastikflaschen gelernt. In seinem neuen Vorschlag zum Thema Verpackungsmüll läßt Töpfer nun Alternativen zur Rücknahmepflicht der Geschäfte zu - ganz im Sinne der EG-Kommission*" (Die Zeit 25.5.1990). A civil servant of the Ministry of Economic Affairs summarized the strategy of the government as follows:

'Die VerpVO war so konzipiert, daß das Duale System nach dem Par 6. Abs 3. gewissermassen Voraussetzung war für die Umsetzung des Ganzen. Wir waren uns alle einig, daß es bei den originären Grundpflichten, nämlich jede Art von Verpackungen an der Ladentheke zurückzugeben nicht bleiben kann. Dies wäre ein allgemeines Chaos gewesen. Man hat also gewissermassen über die VerpVO, das Duale System inititiert und die Rahmenbedingungen für dieses Duale System geschaffen' (BMWi Interview 10.12. 1996).

4.3.5 The Dual System: choosing the lesser evil

The development of the first draft of the Packaging Ordinance went hand in hand with the development of a concept for the Dual System. A pivotal role in this was played by the German liberal party (FDP), which was a strong proponent of the privatization of waste industry. The party traditionally had good contacts with industry and held the post of the Minister of Economic Affairs. As noted above, back in January 1990, the party leader Lambsdorff had already proposed an industry-led system, and the Minister for Economic Affairs Hausmann supported such a concept. The idea of self-regulation within a legal framework was supported by the peak association of the German trade and industry associations BDI and DIHT. The reason for this was its wish to prevent more harmful measures such as take-back responsibilities (Interview BDI 6.12.1996). The BDI and the DIHT commissioned the Working Group on Packaging and the Environment (AGVU) to develop a concept for the establishment of a dual household waste management system. The AGVU concept was based on a research project by the Institute of Public Finance at the University of Cologne. This project was carried out in 1989/1990. It was initially commissioned as an advisory project concerning a comprehensive pilot project in the German *Land* Saarland. The report of the University of Cologne dealt with the tasks, the organization, and the financing of an industry-based system of waste reduction by recycling of household waste. Based on this report, the AGVU presented its concept at 24.8.1990 in Bonn (AGVU 1990: 97pp.) This concept already included the central elements of the Dual System, which was officially set up one year later, namely (a) an industry-financed system for the collection of packaging waste near households (*haushaltsnahe Erfassung*); (b) the take-back and recycling guarantee by packaging producers and packaging material manufacturers of the collected material and (c) a finance system through charges (AGVU 1990: 97pp).

The BDI, the DIHT and large retailers accepted the concept. They said that it was a '*tragfähige Grundlage für die Einführung des Dualen Systems*' (AGVU 1990: 1). The decision for a Dual System was, therefore, already met ten months before the formal adoption of the Packaging Ordinance. The Packaging Ordinance and the Dual System were developed parallel to each other. At this stage, the political arena and the industrial arena were linked by two persons: the head of the waste division of the BMU, and the chairman of the AGVU. The latter had the support of the BDI and the DIHT.

Once the BDI, the DIHT, and the large retailers had agreed on the concept of a Dual System, thousands of companies had to be persuaded to join the industry scheme. The Minister's strategy to impose the burden on the retailers worked. *Large retailers* said that they would accept goods only from those packers and fillers who participated in the industry-led scheme. Some retailers tried to free-ride by offering other (cheaper) products to their customers, but they were persuaded to withdraw from their attempt to free-ride by other retailers and the DIHT. The threat of the retailers to accept only products with a green dot was part of the concept of the AGVU (see AGVU 1990: 111), but was never officially agreed on by the retailers, because the Federal Cartel Office (*Bundeskartellamt*) would not have accepted such a joint declaration that would effectively distort competition (Interview BDI 6.12.1996).

Confronted with the threat that their products would be sold no longer, the *filling and packing industry* had a strong incentive to join the collective system. The BDI and its member associations helped to persuade their members by subtle arm-twisting to join the club. The associations had a vested interest in self-regulatory measures. As the government saw packaging as the first waste stream to be regulated in the new manner, industry wanted to prove that imposed regulation or taxes would not be needed, that industry had the capacity to meet environmental objectives without take-back obligations or other more interventionist measures.

It was clear from the beginning that *the packaging industry and their material suppliers* also had to play their part. The packaging industry was - generally speaking - the weakest part of the packaging chain. It had to adapt to the demands of the retailers and the packers and fillers anyway. The retailers and the filling industry determined to a large extent the shape and composition of packaging by their demand. The packaging raw material producers had the knowledge and the technology to reprocess packaging waste. The material producers were also the ones most affected since the reduction of material and/or a switch to other materials had direct consequences for the size of their market. The interests of these groups varied, depending on the materials they provided. Whereas the glass, paper, aluminium, and tin plate industry took a more cooperative stance, the plastic material suppliers showed strong resistance against a Dual System and it took a lot of effort by the BDI to integrate them into a collective system.

The effectiveness of the strategy of the German government was striking. BDI and DIHT were very critical about the draft of the Packaging Ordinance. The BDI stated that it was '*absolut umweltschädlich*' (FAZ 7.8.1990) and the DIHT said that it is a '*dirigistischer staatlicher Eingriff*' that would result in '*unkalkulierbaren kostenmäßigen sowie struktur- und wettbewerbspolitischen Verzerrungen*' (FR 8.8.1990). But the peak associations were forced to cooperate with government to avoid the lesser evil.

'*Noch vor wenigen Wochen haben viele Unternehmen und ganze Branchen es weit von sich gewiesen, Verpflichtungen für den Verpackungsmüll ihrer Produkte zu übernehmen. Sollte es zu einer übergreifenden, geschlossenen Aktion der Wirtschaft kommen, geschähe das mit Investitionen von Milliarden unter dem doppelten Druck, den Umweltminister Töpfer und der Einzelhandel ausüben. Dieser Druck bewirkt, daß selbst Branchen, die sich noch vor unüberwindbaren Schwierigkeiten bei der Verwirklichung der dualen Abfallwirtschaft sehen, an ihrer Verwirklichung mitarbeiten*' (FAZ 24.8.90).

The draft Packaging Ordinance and the concept of a Dual System were already drawn up when the official public hearing took place on 7 August 1990. It comes therefore as no surprise that an official at the Environment Ministry said shortly before the hearing, that he did not expect substantial changes to the draft (TAZ 7.8.1990). The informal government industry negotiations were much more important than the formal institution 'public hearing'. Only minor changes occurred as the result of the hearing.

4.3.6 Bargaining in a federal setting: getting tough on quotas

The final outlook of the Packaging Ordinance was not clear at the time when the large retailers, the BDI and the DIHT developed the Dual System and tried to tie together all parts of trade and industry. Only the draft Ordinance had been adopted by the federal government. Article 14 of the 1986 Waste Act stipulated that the *Bundesrat*, had to give its assent, if the draft was to become law. This procedural provision which reflects the federal character of Germany, shaped not only the policy process but also the substance of the final Packaging Ordinance. Rather than keeping regulation to the minimum needed to ensure self-regulation by industry, the majority of the *Bundesrat* succeeded in increasing the intervention intensity of the regulation (see below).

The government's draft had been sent to the *Länder* in November 1990. The *Länder* had been already involved in earlier stages. In fact, there had always been regular and frequent consultations of the *Länder*, in particular in the *Länder* working group on waste (*Landesarbeitsgemeinschaft* Abfall, LAGA), where civil servants of the federal and *Länder* Ministries had met more or less monthly. The *Länder* were split themselves. The East German *Länder* had particularly serious problems with the local waste management systems. The East German recycling system, SERO, did not exist any more and East German citizens were unhappy about the local waste charges. They were not used to charges for waste disposal. The East German *Länder* hoped that the Dual System would help them to lower the charges (Henselder-Ludwig, 1992: 114). Other *Länder*, especially those led by a SPD government in coalition with the Green Party, wanted an increase in material recycling, more re-use of packaging and a greater say in the implementation of the Ordinance. The *Länder* proposed more than 80 amendments to the cabinet's draft of the Ordinance. The *Länder* succeeded in getting more power in implementation. Rather than the federal government, the individual *Länder* were responsible for providing the permit for the DSD on their territory. The *Länder* also initiated - at the request of the associations of local authorities - the requirement to co-ordinate the private Dual System with the local waste management systems (Interview Deutscher Städtetag, 6.12.1996). Most controversial in the bargaining were the technologys which were to be allowed under the label recycling and the quotas for refillables. These points will be discussed in more detail.

Material recycling instead of thermal recycling

The policy process prior to the *Bundesrat* stage was dominated by the question of whether take-back and mandatory deposits should be introduced and whether industry was able to establish a Dual System. Once the Dual System was announced, the question arose what requirements this collective system would have to meet. While the Ministry of the Environment emphasized the need for recycling quotas to ensure the environmental effectiveness of the regulation, the Ministry of Economic Affairs in concert with a large part of industry saw quotas as an unnecessary and rigid restriction on the functioning of markets (Interview BMWi). This question was not very controversial, however, since recycling was defined in very broad terms, including incineration with energy recovery, the so called thermal recycling. The majority of *Länder*-governments, however, feared that such a broad definition would result in massive burning of waste and would prevent industry from establishing a decent packaging waste collection, sorting and recycling infrastructure. The *Länder* majority succeeded in banning thermal recycling as a waste management option for used packaging collected by a Dual System. It is this ban on thermal recycling, which put powerful pressure on industry to establish the necessary infrastructure in particular for used plastic packaging. It also created unintended consequences beyond the German borders. Later in this book it will be shown, that due to a lack of domestic material recycling facilities, a lot of used packaging had to be exported to other EU Member States and beyond (see chapters 8, 9 and 11).

Increasing re-use

Besides the issue of the adequate method of recycling, the promotion of re-use was a second important topic in the negotiations among the *Länder* and between them and the federal government. The government had stipulated the increase of the share of refillables in its fixing of targets in 1989. The very first drafts of the Packaging Ordinance, however, did not include provisions in regard to standards for re-use. It was the final government draft that included the provision to maintain the level of re-useable drink packaging achieved in Germany. Seventy-two per cent of the aggregated market share of beer, soft drinks, mineral water and wine was to be sold in refillables.

Besides the interest of the Ministry of Environment in ensuring re-use by statutory instruments, the re-use quotas were also included in anticipation of the critique of many *Länder*-governments that the Ordinance favoured one-way packaging. The majority of

Länder-governments had always emphasized the importance of a high level of re-use (HB 14.2/22.3/11.4, 1991, see Spies 1994). The SPD-led *Länder* demanded a phased increase, *Dynamisierung*, of the re-use quota and Bavaria asked for a differentiation of the quota according to the types of drinks. Bavarian regional breweries usually sell their beer in refillables, and a high quota on refillables would protect their market share. Moreover, as the plebiscite indicated, waste policy was high on the political agenda. The SPD-led *Länder* which had a majority in the *Bundesrat* and Bavaria shared some interests with this majority. Therefore, it was likely that at the *Bundesrat* stage the provisions concerning re-use would be strengthened. The 're-use interests' (*Mehrweg-Lobby*) of German industry, i.e. the association of the German mineral water industry, the federal association of the German beer and drink producers and their retailers, and the Bavarian Brewer Association, used the *Bundesrat* as the arena to push their interests. They benefited from re-use quotas, since these quotas protected their markets against imports which are often sold in one-way packaging. The *Bundesrat* did not succeed in including a phased increase of the refillable quote and the differentiation according to types of drinks. The federal government argued that an increase in re-use would need the notification of the EC Commission since it could be seen as a restriction of the functioning of the European common market. A notification would delay the process and probably threaten the whole Ordinance. The federal government and the *Bundesrat* reached a compromise. First, the re-use quotas would not refer to the national aggregated market, but to those of the respective *Länder* levels. With this, different levels in the use of re-use containers could be taken into account, the high level in Bavaria as well as the low level in the East German *Länder*. Second, the *Bundesrat* asked the government to develop a new ordinance to increase the level of re-use of drink packaging. This ordinance was never adopted however. A first draft was made public in December 1991. It included the mandatory identification of re-use and one-way packaging, the standardization of re-use systems and the increase of re-use, differentiated according to types of drinks. The draft passed the stage of the public hearing (21.5.1992), but since then nothing has happened (Henselder-Ludwig, 1992: 91-95). The draft is 'sleeping' in the Ministry of the Environment. As will be elaborated on in chapter 9.4, the German government had already faced great adversity with defending the 72 per cent quota of the Packaging Ordinance against the criticism of the EU Commission. Under this circumstances it was impossible to adopt an ordinance that would not only protect the current level but would cause an increase in re-use.

4.4 Conclusion

The last section indicates that the *Bundesrat* was instrumental in increasing the strictness of the standards. The *Bundesrat* provided a supportive arena for the re-use lobby to voice its preferences. In addition, some of the green preferences, represented by the *Länder* Ministers of the Environment, coming from the Green Party, could be included in the final version of the Ordinance. The federal government was able, however, to prevent a mandatory increase in refillables. It used the EU Commission as the scapegoat and retained its autonomy towards the majority of the *Bundesrat*.

As described in this chapter, the German institution of federalism provided the *Bundesrat* with a veto in this case. This was, however, not the only national specific institution that shaped the process and the outcome of the packaging waste policy. Problem pressure was transmitted from the local to the federal level via the institutional channel of the Bavarian plebiscite. The proportional voting system facilitated the access of smaller parties, i.e. the Greens, to the parliamentary arena, though first the threshold of five per cent had to be overcome. A number of institutions facilitated the emergence and maintenance of environmental advocates in the federal Ministry of the Environment. Due to a low fluctuation within and between Ministries, the civil servants became socialized in their own divisions, and they identified with their tasks. Having a background in law, the civil servants were inclined to put the abstract principles as developed in the first environmental programme into practice. The application of the precautionary principle was particularly important: it triggered and justified high environmental standards.

Besides institutions, ideas mattered. In addition to the idea of precaution, the notion of ecological modernization was important. The belief that environmental protection and economic policy objectives are reconcilable, helped to sustain a pro-environment attitude in the federal Ministry of the Environment and its allies. In addition, the idea of the polluter-pays principle, advocated by the federal Ministry of the Environment, matched the 'privatization' frame of the Ministry of Economic Affairs. This provided a window of opportunity to place far-reaching producer responsibilities on trade and industry.

The reconstruction of the German waste and packaging waste policy helps us to understand why Germany adopted such a strict and formalized packaging waste policy. A fuller understanding of the German path can be provided by contrasting it with developments in the Netherlands and the United Kingdom. This will be done in the next chapters.

5. The Netherlands or the gentleman's approach

Dutch industry had to meet recycling targets that were lower than those in Germany but higher than those in Britain. Unique in Europe, however, was the fact that packaging waste growth was explicitly restricted by a ceiling. Industry had to undertake reducing - by the year 2000 at the latest - the quantity of new packaging introduced to the market below the quantity for the reference year 1986. Like the German approach, the Dutch one was quite ambitious, but the Dutch government was less impositional than its German counterpart. The Dutch approach can be seen as part of a long tradition of corporatist and consensus-oriented policymaking: the government and the Foundation of Packaging and the Environment (SVM) - representing large parts of industry - agreed on a series of measures in order to meet the ambitious prevention and fairly demanding recycling targets. The packaging chain was obliged to create recycling capacity, use recycled material in new packaging, and initiate pilot projects for the collection and sorting of packaging waste. In contrast to Germany, no separate dustbin for packaging was introduced in Dutch households. The Dutch policy was not laid down in a generally binding ordinance but in a covenant, an agreement between the government and the SVM, binding under private law, for those who signed. The following section will reconstruct the development of the Dutch packaging waste policy and will contrast it with the German development in order to answer - among others - the following questions: why did the Dutch government take a voluntary rather than an impositional route? And why were the standards, though different, almost as ambitious as in Germany?

5.1 The foundations of environmental policy

The environment as a major political issue reached the Dutch agenda at the end of the 1960s and the early 1970s. The time-scale corresponds with the German situation and that of other industrialized liberal democracies. In the Netherlands, however, ideas about the threatened environment diffused more deeply and more widely into society than in other countries. Environmental problems matched broader beliefs, particularly of young people, that the whole capitalist society had to be changed. Environmental problems were seen as consequence of the capitalist system with its inherent ideology of growth and material values (Goverde 1993: 54, Tellegen 1983). The study of Meadows and his group on '*The limits to growth*' (1972) served as a '*bible*' in this context. More than half of all copies of the book was sold in the Netherlands (Cramer 1989: 31). In the early 1970s, 20 per cent of the electorate saw the the environment as most important national problem (Aarts 1995, referred to in Bressers and Plettenburg 1997: 122).

The strength of environmental ideas and their connection to broader fundamental criticism of society can mainly be explained by three factors. First, the Netherlands was one of the most polluted countries in the world, due to a high population density and the reliance on polluting industries such as industrialized agriculture, bulk chemistry and transport. In no other OECD country was the production of pollutants and wastes per square meter higher than in the Netherlands (RIVM 1991: 41, RIVM 1993: 29). Second, the Dutch political system was comparatively successful in dealing with one of the major cleavages in modern societies, that between capital and labour. This created room for new political issues, among them the issue of environmental protection. The third factor is related to changes in Dutch political culture. Until the late 1960s the political culture was characterized by a passive political attitude of the population. This passive attitude served as one of the preconditions of the Dutch political system, which relied for most of the twentieth century on negotiations between the political elite representing the various pillars of Dutch society, namely a Catholic, a Protestant, and a socialist pillar. These pillars were in themselves homogeneous. Each pillar had its own political party, own newspaper, own schools and universities, own employer organizations and trade unions, etc. (see van der Heijden 1994: 228, Lijphart 1976, and Van Mierlo 1988, [who mentions the '*liberals*' as a fourth pillar]). This system of pillarization supported the corporatist political-administrative relationships in the 1950s. Since then, Dutch society started to depillarize, corporatistic practices were challenged by more open conflict and citizens demanded more transparency and democratic participation. The '68' movement in the Netherlands was more radical than in many other Western European countries (van der Heijden 1994: 230). The Dutch political system destabilized to some extent in this period. Between 1965 and 1973 three Cabinet crises resulted in new government coalitions (Putten 1982: 176).

During the second half of the 1960s and the beginning of the 1970s more than 600 environmental groups were founded. They were, as in Germany, mainly locally organized and dealt with local problems. Most groups existed for a short period, depending on the local policy issue at stake (van der Heijden 1994: 245, Van Noort 1988: 199).

The environmental concern of society was picked up by new progressive political parties founded in the 1960s. In particular the left-liberal *Democraten'* 66 (D66) and the green-leftist *Politieke Partij Radikalen* (PPR) put the environmental issue on their political agendas and on the parliamentary agenda (Leroy 1995: 38, Kriesi 1993: 170). Hence, the 'environment' was politicised earlier in the Netherlands than in Germany or other neighbouring countries (Leroy 1995: 37). This was related not only to the changing political culture and the decreasing relevance of the capital-labour cleavage, it had also to do with the Dutch voting system. The strictly proportional voting system facilitated access to parliament for new political parties and with them new political issues. A party only needed some 60,000 votes to enter parliament. Established parties had to be ready to respond to new socio-economic and cultural trends in order to maintain their share of seats (Gladdish 1991: 97). This institutional factor helps to explain why environmental protection reached the parliamentary political agenda in the Netherlands earlier than in the United Kingdom with its majority voting system or Germany with its 5 per cent threshold.

At the end of the 1960s, governmental action concerning the environment increased significantly. The first '*modern*' environmental law was the Surface Water Pollution Act (1969). One year later the Air Pollution Act was adopted (1970). As in Germany, the government changed its ministerial machinery to deal more effectively with environmental problems. With the cabinet formation in 1971 the Ministry of Social Affairs and Public Health was split in two; one of the new ministries was the Ministry of Public Health and Environmental Hygiene (*Ministerie voor Volksgezondheid en Milieuhygiëne*, VoMil). The old Ministry of Social Affairs and Public Health already had a division for environmental hygiene, but the new ministry established a general directorate for this task (*Directoraat-Generaal Milieuhygiëne*, DGMH, after 1982 DGM). The DGMH was a separate and specialized organizational unit for environmental protection issues. The allocation of environmental protection responsibilities to this Ministry indicated that - as in Germany - environmental problems were perceived as public health problems.

As in Germany, the Dutch government started to create the organizational infrastructure to deal with environmental problems. The National Institute of Public Health and Environmental Protection (RIVM) was founded to support the government with scientific information. In addition, a national advisory board was created (*Centrale Raad voor de Milieuhygiene* CRMH, later replaced by *Raad voor het Milieubeheer*), which comprised representatives of lower public authorities, employer organizations and trade unions, environmental and consumer interests and independent experts (Liefferink 1997: 224)

In 1972, VoMil published the Emergency Memorandum on Environmental Hygiene (*Urgentienota Milieuhygiëne*) which served as a basis for Dutch environmental policy during the seventies and the beginning of the eighties. The first part elaborated on the environmental problems in an '*apocalyptic framing*' (Hajer 1995: 177), pressing for immediate action. The problems were defined as pollution threatening public health. Problems of resource depletion were hardly considered. The report concluded that more legislation was needed to meet the current environmental problems such as the increased amount of waste. In the second part of the Memorandum, dealing with environmental measures, a compartmental perspective was taken. It concerned emissions into distinct environmental media: water, air, and soil. End-of-pipe technology was seen as a solution for environmental pollution problems, and command and control measures based on public law were preferred as governmental instruments (Goverde 1993: 52, Leroy 1994: 39-41). Based on this Memorandum many environmental regulations were adopted in the next decade which dealt solely with distinct environmental components.

The establishment of new *national* environmental protection organizations occurred, as in Germany, parallel to institutionalization of environmental policy. A number of environmental organizations were founded, which dealt with a broad range of environmental issues, including air pollution, the location of polluting industries, resource and energy problems and road construction (Van Noort 1988: 200). These organizations were more radical in their strategies than the traditional natural preservation organizations. Among them were the Dutch branch of

Friends of the Earth (*Vereniging Milieudefensie*, 1971) and the Foundation for Nature and Environment (*Stichting Natuur en Milieu,* SNM, 1972), an umbrella organization for a large number of environmental and nature protection groups. The SNM aimed at the co-ordination of national and provincial natural protection activities and also functioned as a think-tank. Four years later the national environmental consultation body (*Landelijk Milieu Overleg,* 1976) was established. This body served as a national body co-ordinating various large environmental organizations (Van der Heijden 1994: 247).

In order to grasp the driving forces of Dutch environmental policy one has to consider the intense personal and functional interrelationships of civil servants, in particular from DGMH, environmental scientists and activists from the environmental movement. In analogy to the concept of the '*iron triangle*' used by political scientists to characterize the structural interaction of governmental departments, parliamentary committees and interest associations in the U.S., Leroy has used this term to describe the interaction of civil servants, the academic environmental science and the environmental movement (Leroy 1995: 38). The environmental organizations were recognized by the government to a relatively large extent. They were morally supported by the General Directorate of the Environment. Moreover, as in Germany, a pattern of sponsored pluralism existed in the Netherlands. In 1974 fifteen per cent of the income of Milieudefensie was government funds. In 1990, more than fifty per cent was public money (Van der Heijden 1994: 261). The Stichting Natuur en Milieu is even financed up to roughly seventy per cent by governmental subsidies (Van der Heijden 1994: 258). VoMil also subsidized newly founded environmental research institutions. This was done mainly by the commission of contract research. According to Van Tatenhove (1993) the support of the environmental sciences also had the function of accelerating the institutionalization and professionalization process of the newly established DGMH (referred to in Leroy 1995: 38).

To a larger extent than in Germany, in the Netherlands an environmental advocacy coalition could be established which shared a common view on the character of the environmental problems, and possible solutions to it. In line with the '*Urgentienota Milieuhygiëne*' (1972), the environmental policy paradigm was made up of the following three elements: environmental problems are very serious, the sources of environmental pollution are locally identifiable and can be dealt with by technical devices, in particular end-of-pipe measures, and finally a strong government is needed, employing command and control measures. The rather optimistic view about the problem-solving capabilities of government was completed by expectations that environmental problems could be solved in five to ten years (Leroy 1994: 40)

The emergence of a pro-environment advocacy coalition was facilitated by the rule that each ministry was allowed to recruit its civil servants according to its own criteria. Therefore, a relatively large number of environmental experts and sympathizers of the environmental movement acquired a function in the Ministry of the Environment. This also enhanced the access to government by societal groups (Rosenthal 1989: 257, van der Heijden 1994: 224). Remember in this context, that ministries and even policy divisions are relatively independent from each other. Like other departments, VoMil also acted in a relatively closed policy community with few links to other departments.

5.2 The history of (packaging) waste policy

5.2.1 The Waste Substances Act (1977) and the Lansink Memorandum (1979)

Traditionally, the control of waste-disposal sites was to be based on the quite general Nuisance Act (*Hinderwet*), established in 1875 but with its origins in Napoleonic times. Through a licensing procedure, the act sought to control the danger, damage and nuisance caused by industry and increasingly also other installations (Liefferink 1997: 211). Most waste was disposed without control, however. A more specific national legal framework for the disposal of waste was created in the mid 1970s. The Waste Substances Act (*Afvalstoffenwet, Stb 455*) was drafted in 1975 and adopted in 1977, but its provisions did not come into force until between 1979 and 1985. The Act dealt basically with municipal waste. It specified tasks for central government, provinces and municipalities. Municipalities were obliged to collect household waste at least once a week. Provinces were to draw up waste plans, which were to be revised every five years. The plans had to include statistics of how much waste was generated and how it was to be processed. Provinces also had the task of granting the permits for landfills, waste incinerators, and other waste processing facilities. Central government had to supervise the implementation of the act (see Eberg 1997: 39, Houben and Leroy 1993:12). As far as waste prevention was concerned, the government was authorized by Article 27 to enact regulations to reduce the flow of waste.

The Waste Substances Act referred to packaging waste as an increasing problem. Accordingly, one of the provisions dealing with prevention concerned the reduction of packaging waste and the use of deposits to assure the return of packaging materials. Other issues addressed include the separate collection of certain waste types as well as the prohibition of the manufacture of products that create problems for environmentally sound disposal (Koppen 1994:153). The Waste Substance Act allowed for the introduction of obligatory deposit charges by means of Orders in Council (*Algemene Maatregel van Bestuur*, AMvB). Hence, as in the German case, the Dutch legal framework provided the opportunity for market intervention to prevent or recycle packaging waste. These legal means to take unilateral action were never implemented. Instead priority was given to voluntary agreements with the private sector.

As in Germany, the Dutch waste policy of the 1970s was primarily end-of-pipe-oriented and was also characterized by an implementation deficit. The sub-national policy implementers, such as the license issuers, lacked the necessary personal resources and they received too little support from the responsible administrators. Furthermore, the authoritarian approach of the Ministry of VoMil was not shared by the sub-national implementation agents. They displayed more sympathy with the interests of those to be controlled than with the civil servants in The Hague and they therefore often lacked motivation. As a consequence, implementation agents tended to respond to complaints instead of pursuing an active and systematic license-issuing policy. Moreover, '*(s)ystematic control was virtually non-existent*' (Bressers and Plettenburg 1997: 116).

A shift in the end-of-pipe policy paradigm was indicated by the Lansink Memorandum '*Motie Lansink*' (Tweede Kamer 1979-1980, 15800, XVII, nr 21). During the parliamentary debate on the 1979 budget, members of parliament put forward a memorandum which stated a hierarchy of waste management options. This hierarchy was as follows: 1) prevention; 2) product recycling (re-use); 3) material recycling; 4) incineration with energy recovery; 5) incineration and 6) landfill and controlled dumping as a last resort. This memorandum turned out to be very influential in the thinking about waste policy. It '*is still considered the political basis for Dutch waste policy today*' (Koppen 1994: 154). The "ladder of Lansink" informed the 1989 National Environmental Policy Plan (NMP), and the waste chapter of the 1994 Environmental Management Act (Wet Milieubeheer WM, 1994, Schuddeboom 1994: 100). As in Germany, packaging was the first product for which a specific prevention-oriented policy was formulated by government. Before that, however, industry avoided governmental interventions by self-commitments.

5.2.2 Self-commitments by industry

The issue of packaging and the environment has been on the Dutch political agenda for more than 25 years. The first phase of Dutch packaging waste policy lasted from the beginning of the 1970s until the mid 1980s. This period can be characterized by two elements. First, the government reacted in an *ad hoc* manner to the problem of packaging waste. Second, the packaging chain committed itself to action concerning packaging waste in response to pressure from environmental groups.

Already in the early 1970s there was an increasing public awareness about litter, which was used by environmentalists to start campaigns against one-way packaging. Locally organized groups boycotted supermarkets. Industry perceived the campaigns of the environmentalists against litter and against one-way packaging as a threat to its interests. Some parts of the packaging industry were afraid of losing their whole market. Producers of brands saw a threat to an important instrument of marketing. Retailers (distributors) were worried about their logistic system (Peterse 1992: 202/203).

Additional pressure for packaging waste reduction came from the Dutch parliament. In 1973 members of parliament put forward a proposal for legislation on packaging waste (Tweede Kamer 1973-74, 12304, no 1). Though the proposal was not adopted, it was clear that the issue of packaging and the environment had reached the national political agenda.

At quite an early stage a number of companies gave an organizational answer to the growing concern about one-way packaging. They established the Foundation of Packaging and the Environment (*Stichting Verpakking and Milieu*) in 1971. The initiative came from several large firms from all sectors of the packaging chain. Shell and Hoechst as raw material producers of plastic, Hoogovens as provider of tinplate, the packaging manufacturers Thomassen & Drijver-Verblifa, Elopak, and Tetrapak, fillers and packers such as Unilever and the large retailer Albert Heijn (Peterse 1992: 202). The SVM dealt solely with the topic of packaging and the environment. It included important members of all parts of the packaging chain. Hence it was a single-issue cross-sectional organization. The SVM is a foundation made up of affiliates rather than an association with members. This ensured that the organization could react flexibly in a dynamic environment. Their affiliates represented a large share of the turnover of the relevant Dutch industry. Estimates range from some 60 per cent up to roughly 80 per cent in the early 1990s (Interview SVM 9.11.1995). The general objective of the SVM was to serve as a counterweight to the public demand of setting limits to the use of one-way packaging (Peterse 1992: 202).

Given the important technical and economic functions of packaging, the industrial sector was surprised by the increasingly bad image of packaging. It did not understand the growing awareness of the environmental problems of packaging, and perceived the actions of local environmental groups as irrational. According to the packaging chain the problem of packaging was overestimated. Industry argued that as a basis of political action statistics were needed on both the share of packaging waste of the overall waste and its composition. The SVM was ready to provide the political actors in the conflict with more empirical evidence. They mobilized statutory trade associations to develop statistics (Peterse 1992: 203). These statistics would show, according to industry, that the problems of packaging waste was not as serious as claimed by environmental groups and some members of parliament (Mingelen 1995: 29).

Another strategy of the packaging chain was to pre-empt governmental action by committing itself to an active attitude towards the problems of packaging waste. The SVM stimulated initiatives to create recycling facilities. It was one of the initiators of a campaign to stimulate collecting used glass packaging by means of bottle banks and instigated the creation of a facility to separate tin plate from domestic waste (Peterse 1992:203). Another example was the commitment that industry would not advertise for one-way packaging. The SVM was able to enforce this commitment, as the following example shows. Once a producer of one-way milk carton packaging, who was not an affiliate of the foundation, planned a big campaign for its product. In order to prevent the campaign, the SVM organized a meeting with the firm and one of its biggest customers. This customer told the packaging producer that he agreed with the policy of the SVM, and that he knew another packaging producer who did not advertise for one-way packaging. As a result, no advertisements of the producer for one-way packaging saw the light of day (Peterse 1992: 203).

These commitments made by the SVM put the organization in a corporatist situation of interest intermediation. On the one hand, SVM represented the interests of the packaging chain concerning the issues of packaging and the environment *vis-à-vis* the government and the public. On the other hand, when commitments were made, the SVM was responsible for the implementation by its affiliates. It had to mobilize trade associations and individual firms for substantial assistance of implementation (Peterse 1992: 204).

5.2.3 Voluntary agreements on drink packaging

The top priorities of the Lansink Memorandum were the prevention of waste and the stimulation of re-use. The promotion of re-use was a threat to one-way packaging. The SVM, however, pleaded for material recycling as an alternative to the re-use of packaging (Peterse 1992: 203).

The SVM was able to prevent the government from taking unilateral action in the 1980s. The organization lobbied for material recycling instead of re-use in many contacts with members of parliament and civil servants. Countless sessions of consultation and negotiation took place between the SVM, other trade associations and high ranking staff of the Ministry of VROM. In this period the engagement of environmentalists and consumer organizations continued to be high. Milieudefensie frequently organized new campaigns against the use of one-way packaging. Target groups were: pupils, consumers, the general public, and the government (Peterse 1992: 204).

Joint Declaration and Code of Conduct on Drink Containers (1985)

An important result of the government-industry discussions was the Joint Declaration and Code of Conduct on Drink Containers (*Gemeenschappelijke verklaring en gedragscode van rijksoverheid en bedrijforganisaties in zake drankenverpakking,* 19. Dec. 1985). The Joint Declaration anticipated the European Directive on liquid containers which was adopted in the same year and had to be implemented by 1987. The European Directive was in line with the Dutch consensual approach in the sense that it explicitly allowed for implementation by voluntary agreements. Though the Dutch packaging waste policy development paralleled European developments, there are no indications that the Joint Declaration was determined by the European Directive. The Joint Declaration was signed by the Ministries of the Environment (VROM, the successor of VoMil, see below), Agriculture (LNV) and Economic Affairs (EZ) and by statutory trade associations (*[hoofd]produktschappen, bedrijfsschappen*), the SVM and the Foundation of Food Retailers (*Stichting Centraal Bureau voor de Levensmiddelenhandel*). They agreed to stabilize the amount of waste generated by drink containers (Klok 1989: 170).

The programme formation process of this Joint Declaration turned out to be more difficult than other agreements in the environmental policy field adopted during this period. This was due, according to Klok (1989), to the complexity of the problem, the large number of different packaging systems and the diversity of interests and situations in the industry affected. The VROM's alternative to the Joint Declaration, the establishment of a return system, was unacceptable for the private sector, given the burden of creating the logistical infrastructure and to standardizing packaging. This and other alternatives disappeared from the agenda after lengthy debates. Since industry and the Ministries for Agriculture and Economic Affairs did not perceive the packaging waste as very serious, there was not enough support to introduce more interventionist instruments (Klok 1989: 170-175).

The Declaration did not, however, achieve the paramount objective of stabilizing the amount of waste generated by drink containers. The amount of waste clearly increased, but the expected flood of packaging did not occur. The increase in waste was attributed by Klok to the free-rider attitude of some companies. Since a large number of enterprises were concerned, the decision of a single company had little effect on the overall amount of waste. It also became clear that economic aspects dominated the choice of packaging systems (Klok 1989: 180).

Code of Conduct on PET-bottles (1987)
Eighteen month after the Joint Declaration on Drink Containers, the Code of Conduct concerning PET-bottles (*Gedragscode inzake PETP-flessen, 15.7.1987*) was signed by VROM, LNV, EZ, the statutory trade association for soft drinks (*Bedrijfsschap Frisdranken),* and the foundation of food retailers (*Stichting Centraal Bureau voor de Levensmiddelenhandel*). This additional agreement was made because of the introduction of the one-way PET-bottle for soft drinks in 1986. The new packaging system caused fierce reactions from the environmental movement and other actors. Boycotts of the PET-bottle were organized by environmentalists and consumer groups. Pressure also came from parliament. Members of parliament made clear that they would not accept a worsening of the situation (Klok 1989: 175). The Code of Conduct was in line with the Joint Declaration on Drink Packaging. The overall goal was the reduction of waste from PET-bottles. No direct action was taken against the producer of these bottles. The Code of Conduct stipulated that empty PET-bottles - including imported ones - could be returned to the retailers against a deposit of 10 cents. The bottles were to be recycled for other purposes. The agreement preceded a pilot project in a Dutch municipality. The pilot project revealed (a) that more than 70 per cent of the consumers brought back the bottles and (b) that the bottles were appropriate for recycling. The committee which steered the pilot project concluded from these findings that the system could be introduced on a national scale (VROM press release, No 99, 17.7.1987).

In contrast to the general debate on drink containers, this case was better structured. The object of negotiation was clear and the structure of interests could be more easily identified (Klok 1990: 170). The crucial question was, however, whether the mandatory deposit (*retourpremie*) was high enough to ensure the rate of return of plastic bottles desired by government, who envisaged a return percentage of 95 per cent. Several participants had their doubts about this percentage, in particular those from VROM. The support of EZ for this collection system and the results of the pilot project, however, led to the broad acceptance of the rather low deposit of 10 cents (Klok 1989: 170).

The first results (1988) confirmed the doubts of VROM (Klok 1989: 170). In 1988, less then sixty per cent of PET-bottles were returned by the consumers. According to Aalders (1991: 89) this had to do with the bad functioning of the collection system. The major problem was the lack of commitment on the part of the retailers. In this case the internal control failed, owing to the fact that the retailers formed a diffuse group (Aalders 1993: 89). According to Klok, however, the 60 per cent was still a reasonable share for PET-bottles. The amount of waste increased less than without the policy measure. The most important reason for the small increase in plastic waste, however, was the continuing small market share of the PET-bottle. One of the reasons for this was the uncertainty surrounding the prospects of the collection system. There was a mismatch between the low return charge agreed to in the code of conduct and the high return percentage desired by government. This uncertainty affected the future prospects of the PET-bottle and contributed to its small market share. In this sense, the policy adopted was effective (Klok 1989: 180). In 1989 the Covenant was overtaken by developments because the one-way PET-bottle was replaced by a refillable PET bottle with a deposit charge of 1 guilder. One-way PET-bottles no longer exist in the Netherlands (Ingram 1997: 10, Schuddeboom 1994: 106).

5.3 Towards the Packaging Covenant (1991)

5.3.1 The second wave of environmentalism

In the second half of the 1980s, the issue of the threatened environment became a top priority on the political agenda of most OECD countries again. Events such as the Chernobyl disaster in 1986 and the study '*Our common future*' (1987) were the main reasons for this. Even Queen Beatrix's Christmas address of 1988 was devoted to the topic of the environment (Liefferink 1997: 218). As in the case of the first wave of environmentalism in the late 1960s and early 1970s, this second wave also turned out to be stronger in the Netherlands than in most other OECD countries. The share of the electorate that mentioned the environment as the most important national problem peaked to 45 per cent, a number unrivalled before or since (Aarts 1995, referred to in Bressers and Plettenburg 1997: 122). Between 1985 and 1990 the number of members and regular supporters of the four largest Dutch environmentalist groups increased from 459,000 to 1.532 million. Greenpeace alone increased the number of its supporters from 70,000 to 830,000 during this period (Van der Heijden 1994: 251). Almost 20 per cent of Dutch adults were members or registered supporters of an environmental organization, in contrast to 11 per cent in the United Kingdom, 2 per cent in Germany and a European average of 4,2 per cent (Hey and Brendle 1994: 642). In contrast to the 1960s and 1970s, the environmentalists were less radical and their criticism was less fundamental. The umbrella term '*sustainable development*' opened these groups towards new members, modified their objectives and moderated the political conflict to some extent (Goverde 1993: 54). These groups became increasingly professionalized and institutionalized. The SNM in particular was represented in a number of committees inside and outside of government (Bressers and Plettenburg 1997: 118).

The rise in environmentalism was partly caused by perceived shortcomings of the traditional environmental policy approach. In 1988 the RIVM published its report '*Concern for Tomorrow*' (*Zorgen voor Morgen*). Largely inspired by the Brundtland Report, the RIVM report was long-term-oriented, covering the period from 1985 to 2010. It described the state of the environment in the Netherlands in alarming terms and focused on the sources of pollution rather than effects. It reflected a shift from the traditional environmental approach, based on controlling emissions to air, water and soil separately, towards an integrated cross-media approach. As far as waste was concerned, industry and consumers were identified as key sources of the problem. High consumption of raw materials would cause environmental degradation and consumer behaviour was seen as being responsible for the increased generation of waste (VROM 1994: 6-7).

Waste prevention was high on the political agenda at this time. Household waste continued to grow by 5-10 per cent per year because, as the Ministry of the Environment admitted, sectoral source-oriented legal measures did not work. The disposal charges did not reflect the real long-term costs of waste disposal and were too low to provide an incentive to prevent waste (VROM 1988: 2). In 1987, the National Environmental Advisory Board pushed the need of waste prevention (CRMH no. A-87/605, 30.6.1987). The Advisory Board argued that in particular the societal, geographic and environmental hygienic limits to the dumping of waste demanded a substantial decrease in the amount of waste that was to be disposed of. In March 1988, a lengthy session of the environmental committee of the Dutch parliament took place. Again the topic was the prevention and recycling of waste. During the session, the Minister promised to formulate concrete plans for the prevention and recycling of waste (Tweede Kamer 1987-1988 20.200 chapter XI no. 69).

5.3.2 The government's approach

The need to deal with environmental problems in a more integrated way was acknowledged relatively early by the government. In 1982, the environmental division of the Ministry of Public Health and Environmental Hygiene was incorporated into the newly established Ministry of Housing, Physical Planning, and the Environment (VROM). This change reflected a shift from the anthropocentric public health-oriented problem definition to a broader physical environmental approach (Eberg 1997: 40, Goverde 1993: 52). The broader integration of waste and environmental policy ensued from the Environmental Policy Integration Plan (1983), which replaced the compartmental problem definition of the 1972 Memorandum *Urgentienota Milieuhygiëne* with a cross-media definition. The document aimed at the co-ordination of different governmental departments (external integration), and integration regarding administrative levels and target groups (Eberg 1997: 40-41, Leroy 1994: 43-44). The Environmental Policy Integration Plan was not the only plan that indicated increased government efforts to integrate environmental policymaking (see for more details Eberg 1997, 38-42). All these plans culminated in the National Environmental Policy Plan (NMP) '*To choose or to loose*' ('*Nationaal Milieubeleidsplan. Kiezen of verliezen*', NMP, Tweede Kamer 1988-1989a). Parallel to the NMP, the Memorandum on the Prevention and Recycling of Waste Materials was developed which can be read as the concretization of the NMP for the area of waste management. Before the Memorandum on waste is discussed in more detail, the main elements of the NMP will be sketched.

The National Environmental Policy Plan

The NMP established the key environmental quality objectives and set out a long-term programme of action. According to VROM, the plan was characterized by a management approach to environmental problems. In line with the analysis of the RIVM report, Concern for Tomorrow, eight environmental themes or areas of concern were identified, among them waste disposal in the broad sense, i.e. waste processing, waste prevention, re-use and recycling. General (quantified) objectives were set out for each theme. These overall objectives of the NMP were further broken down into numerous reduction targets for specified substances and waste streams. Responsibility for achieving these emission reduction targets was assigned to target groups, representing the key groups of polluters in Dutch society, such as 'industry' and 'consumers and retail trade'. This management approach came from Minister of the Environment, Winsemius, a former senior consultant of McKinsey and Company, who held office between 1982 and 1986 (see Hajer 1995: 186).

The NMP and its predecessors were partly the result of a learning process '*The NMP recognizes that a high quality environment cannot be achieved through conventional control measures alone. A mixture of new, clean technologies and structural changes of production and consumption patterns will also be required*' (VROM 1994: 5). From a cross-national perspective, the environmental strategy set out in the NMP was quite ambitious. It '*aims to achieve sustainable development in the Netherlands within one generation*', and '*offers the prospect of doubling Dutch GNP by 2010 while achieving emission and waste discharge reductions of 70-90% (except CO2)*' (VROM 1994: 5). The plan was widely hailed. Scientists, environmentalists and other national governments welcomed the integrated approach towards sustainable development as well as the adoption of concrete targets and the broader range of instruments proposed, including financial and social instruments, i.e. public-private agreements, (see Weale 1992b: 122-153). Moreover, the NMP served as a model for the Fifth EC Environmental Action Programme '*Toward Sustainable Development*' (Bressers and Plettenburg 1997: 109).

The Memorandum on the Prevention and Recycling of Waste Materials (1988)

The Memorandum on the Prevention and Recycling of Waste Materials was the government's reaction to the critical report of the Dutch Environmental Advisory Board (CMRH) and the discussions with the parliamentary committee. The Memorandum was written by officials of VROM together with experts from the RIVM. It echoed one of the central elements of the concept of sustainable development, that is to closure of material cycles as far as possible so as to curtail emissions and waste generation and to keep the loss of raw material to a minimum (see WCED 1987, VROM 1988: 3-4).

The Memorandum marked a new phase in Dutch waste policy, because it changed the '*piece-meal ad hoc approach to waste reduction*' (Koppen 1994: 154). It was oriented to the hierarchy of waste management objectives introduced by the Lansink Memorandum. Twenty-nine priority waste streams were identified. One of them was packaging waste. Targets were set for the reduction and recycling of waste for each of the waste streams. These targets were not discussed with target groups but based on research carried out by the RIVM. These quantified goals were not seen as obligatory, but as an indication of the preferences of government. The targets were intended to provide a context for discussion with the target groups. The definite targets and the measures to reach them had to be discussed with target groups who were listed in the Memorandum for each waste stream. The results of these '*strategic discussions*' were then to be laid down in implementation plans for each priority waste stream. The indicative targets set for each waste stream are listed in table 18.

Table 18 Indicative Waste Targets set by Dutch 1988 Memorandum

	all priority waste streams (28 mio t)		**one-way packaging waste (2 mio t)**	
	1986	**2000**	**1986**	**2000**
Prevention		5%		
Re-use	20%	35%		
Recycling	15%	25%	25%	60%
Incineration	10%	25%	25%	40%
Disposal	55%	10%	50%	0 %

Source VROM 1988: 19, 50

According to the Memorandum 50 per cent of packaging waste was dumped in 1986, while 25 per cent was recycled and another 25 per cent was incinerated. The overall target was to stop the dumping of packaging waste by the year 2000. Moreover, the rate of recycling was to increase from 25 to at least 60 per cent. The rest had to be incinerated (40 per cent). Note that all figures relate to the level of 1986. This means that in fact a ceiling was envisaged at the level of 1986. Hence, the absolute amount of one-way packaging generated was not to exceed 2 million tonnes in the year 2000, from which 1,2 million tonnes had to be recycled and not more than 800,000 tonnes were allowed to be incinerated. Implementation plans had to be adopted by 1 April 1991 (van Kempen 1992: 185).

The Memorandum not only set targets for preventive waste policy but also elaborated on the process of programme formulation. According to Koppen: *'The Memorandum 'institutionalized' to some degree the informal approach that had developed in practice. It established standard negotiation procedures, referred to as strategic discussions, as the official government procedure to regulate waste reduction. The agreements that are to result from the negotiations are from then on called covenants' (Koppen 1994: 154/155).*

The negotiation procedures established mirrored the general Dutch political culture of consultation, proportional representation, and consensus building. In contrast to the impositional approach favoured in Germany, the Dutch government emphasized the idea of partnership with the view of an internalization of environmental concerns into the daily activities of the target group (see Hajer 1995: 185-190, Le Blansch 1996: 25-31). The new approach also reflected the attempts towards deregulation which were prevalent in the thought of Dutch governments since a conservative-liberal coalition came to power in 1982 (Hajer 1995: 183, Hanf 1989). The consensual policy approach was elaborated on in a letter of Minister Nijpels that was attached to the Memorandum. The Minister wrote:

'I regard this Memorandum as the beginning of a process that will cover all aspects of waste policy to discuss with the concerned groups the targets, the measures and activities needed to achieve a more effective prevention and recycling of waste. It must be obvious that I expect more results from such an approach than from the one-sided use of legislation. Formulating an approach that is not carried by those concerned is most difficult to uphold and therefore hardly brooking of success' (Tweede Kamer 1988-1989c, 20877, no.1, p.1, quoted in Koppen 1994: 159). A civil servant of VROM explained the rationale behind this approach. '*The Ministry tried to get almost everybody around the table. A selection took place, but it was very broad. If a*

result could be achieved based on broad support, one could be sure that something would happen.' (VROM Interview. 4.12.1995).

5.3.3 The strategic discussions: talking, talking, talking

The Memorandum on the Prevention and the Recycling of Waste institutionalized the process by which the implementation plans had to be developed and gave an indication of the targets to be achieved. In 1989 the strategic discussion on packaging waste started. In the same year the government changed from a conservative-liberal government (CDA,VVD) to a conservative-social democratic coalition (CDA, PvdA). Minister Nijpels from the liberal party was replaced by Minister Alders from the social democratic party. It was known that Alders took a more critical stance on the voluntary approach than his predecessor Nijpels. But he was persuaded by his civil servants to follow the voluntary route (communication VROM 6.6.1998).

The Ministry of the Environment selected a broad range of actors for the strategic discussions including representatives of VROM, EZ, LNV, the National Associations of Provinces and Municipalities (VNG), the SVM, firms from each part of the packaging chain, as well as each type of packaging material, consumer and environmental groups. In addition, research institutes and consultants took part who provided assistance for both scientific and technical information as well as for the process of discussion (Koppen 1994: 164, Interview Milieudefensie 20.11.1995). The discussions took place in a general forum on packaging waste as well as in subgroups on specific packaging materials. These discussions had high priority for both VROM and the industry. In the general forum, which met on average once a month, some 60 persons participated while in the subgroups some 25 persons took part. All in all some 270 sessions took place (Mingelen 1995: 43). The representatives of private industry were high ranking managers. The scientific institutions and consultants were well paid. Everything was financed by the government (Milieudefensie Interview 20.11.1995).

The general objective was to get an overview of the interests of the participants, the possibilities and problems, and to enhance mutual understanding (Mingelen 1995). The strategic discussion concerned three topics: the targets set out in the Memorandum, the character and dimension of the environmental problem of packaging waste, and finally, the measures necessary to solve the problems and meet the targets (Mingelen 1995: 14). The discussion revealed that the perception of the present situation, in particular the environmental problems surrounding packaging waste, was very different. Environmental groups and industry often had absolutely opposing opinions about the '*facts*' (Milieudefensie Interview 20.11.1995). This was due to the material interests at stake, but also because of scientific uncertainty.

An important point in the discussion dealt with the prevention and recycling targets set out by the Memorandum. These targets were developed within the governmental machinery without consultation of the industry. During the strategic discussion, the industry parties tried to lower the targets. They argued, for instance, that the targets of the Memorandum for the year 2000 had to exclude the autonomous growth of waste, i.e. the increase of waste due to economic growth. The Ministry of the Environment maintained that the targets of the

Memorandum had to be treated as given, while environmental groups demanded a more ambitious reduction target. During the strategic discussions no agreement could be reached on this point (Interview Milieudefensie 20.11.1995).

There was also disagreement about the measures necessary to deal with the problem and to meet the targets. VROM planned either a legal provision aimed at the obligatory introduction of re-usable packaging for a great number of products or a mandatory deposit system. Quantitative prevention of waste was to be achieved by the reduction of material used for packaging (Mingelen 1995: 14). The industry parties strongly opposed in particular mandatory deposits on containers while environmental groups argued that a deposit system was the only assurance for a high return rate of containers. No agreements could be reached on this point (Koppen 1994: 165, Mingelen 1995:14).

After one and a half year of negotiations, several hundreds of sessions and dozens of environmental analyses, no full agreement could be reached about elementary aspects (Mingelen 1995: 14) The strategic discussion failed to reach consensus on two critical points: the introduction of mandatory deposit schemes and the reduction targets (Koppen 1994: 165). The deadline for the implementation plans which were promised to parliament, however, came closer (van Kempen 1992: 187, Mingelen 1995: 18) The Ministry of the Environment and the SVM decided to break off the plenary negotiations and continued bilaterally.

5.3.4 Bilateral negotiations: VROM and SVM

The SVM became the prime negotiation partner, because this single-issue cross-sectional trade association was able to represent a large part of the packaging chain and government believed in its capacity to commit its affiliates and other firms to an agreement made with the government. The disciplinary power of the SVM was due to the skilful aggregation of interests: differences in interests were tolerated, but shared interests were emphasized; no decisions were adopted by simply overruling a minority; no materials or types of packaging were discriminated against; proportional representation of interests and distribution of costs and benefits, at least in the long term and transparency of all process and decisions as far as possible (see also Mingelen 1995:33). The effectiveness of the SVM was also facilitated by the specific competitive structure of branches of the packaging chain in the Netherlands. There was either a small number of competitors with a high market share on the supply side, for instance, pre-manufactured products, milk packaging, or a few dominant firms on the demand side. The filling industry, for instance, had to do with large scale retailers/distributors such as Albert Heijn. Among others, this firm has shown social responsiveness and has taken a leading position in the aggregation of interests (Peterse 1992: 207, Koppen 1994: 165) The large firms used their market power to enforce the dominant strategy of the packaging chain. They could also be mobilized because of their strategic interest in co-ordinated industry action to prevent one-sided regulation. These factors resulted in positive experience from the past when the SVM successfully mobilized trade associations and individual firms to assist in implementation in a number of cases (Peterse 1992: 204).

The bilateral negotiations concerned basically three points: the targets for prevention and recycling, the question of whether more re-use systems had to be introduced and the form in which the implementation measures had to be adopted. The government was very determined to introduce strict packaging waste prevention and recycling targets. The strong emphasis of the Minister on prevention was due to the perceived urgency and seriousness of the problem concerning the disposal of waste materials (Letter of the Minister to the Parliament, Tweede Kamer 1990-1991, 21137, no. 49). In the updated National Environmental Policy Plan (NMP-plus, 1990) it had already been concluded that the overall amount of waste would increase by 2 per cent each year if no additional measures were adopted. As a result of this, the overall waste prevention objective stated in the Memorandum of the Prevention and Recycling of Waste was doubled to 10 per cent in the NMP-Plus. No consensus could be reached on this target however. As a consequence, the Minister of the Environment wrote two letters to the SVM (dated 17.8.1990 and 21.9.1990), in which he stated that in the year 2000 the quantity of packaging newly introduced into the market should be 10 per cent lower than in the reference year 1986, while 60 per cent of the remaining packaging waste must be recycled. By these letters, the Minister imposed targets on the industry. The introduction to the Packaging Covenant stated that these letters *'negated the procedure proposed in the "Memorandum on prevention and recycling of waste materials" namely of consulting the target groups about the achievability of the indicative objectives; as a consequence of this the objectives for prevention and recycling of packaging waste have been determined unilaterally by the Minister'*. It should be noted, however, that the real reduction on packaging waste by 10 per cent was not included as an obligatory target for industry, but 'only' as a target industry had to strive for.

Another issue on which no consensus could be reached in the course of the strategic discussions was the comprehensive introduction of re-use systems. The bilateral negotiations did not lead to a consensus. The decision on whether new re-use systems were to be introduced was made conditional on the outcome of life cycle analyses and market economic research. The respective provision in the Covenant was formulated at the '*last minute*' (Interview ten Heuvelhof 11.9.1997). Hence, the decision as to whether more re-use systems were to come was adjourned to the implementation phase of the agreement (see 5.3.3).

An important topic of the bilateral negotiations was the degree of formalization of the outputs of the process. In a similar situation, the German government chose for generally binding regulation based on public law. Why did the Dutch government not follow the German example? VROM had three options, (a) a rather vague implementation plan as proposed in the Memorandum on the Prevention and Recycling of Waste Materials, (b) a legal provision, i.e. an Order in Council which could be based on the Waste Substances Act (1977) and (c) a covenant. i.e. an agreement between the government and the industry to achieve certain policy goals. Examples of such covenants were the agreements on drink packaging and plastic bottles discussed above. It was known that Minister Alders was rather sceptical about voluntary agreements. Furthermore, *'contrary to recent policy practice, the NMP and its follow-up, the NMP-Plus, gave a limited role to covenants. Covenants are primarily viewed as temporary arrangements, prior to the enactment of formal regulations'* (Koppen 1995:

164). The NMP-Plus stated that a covenant could be adopted as an alternative for legal provisions only when the matter concerned could not be treated appropriately by legal measures. Where legal provisions are possible, they have priority owing to their advantages concerning transparency, democratic control and legitimacy.

Though Alders was initially critical about covenants he finally became a proponent of them for a number of reasons. An agreement with the packaging chain, signed by both, fitted into the consensus-oriented target-group approach of VROM. According to a civil servant a covenant created legitimacy, support and commitment among the packaging chain, since the industry had agreed to it (van Kempen 1992:188). Another civil servant argued that when the industrial parties signed an agreement, one could be sure that they would implement the measures agreed on. In the case of legal provisions the industry might argue that it was not possible to implement certain measures (VROM Interview 4.12.1995). Moreover, it was argued that a covenant provided the opportunity for a broader scope and more fine-grained regulation than legal provisions, since it was impossible to regulate each of the some 18,000 products offered in a typical Dutch supermarket. VROM also claimed that a covenant could be adopted faster than legal provisions. While legal provisions had to fulfil procedural requirements which made them time-consuming - in this case at least another year and a half was expected - covenants could be adopted within three months. The government also assumed that a covenant was more flexible than a legal provision. Since any change of the covenant depends only on the parties concerned, this could be done quite easily, assuming that it was easy to reach an agreement. Finally, a covenant would contribute to the overall trend of deregulation (van Kempen 1992: 188-189). The industry also opted for a covenant for mainly the same reasons as VROM. In particular, it gave the industry more influence on the content of the policy than in the case of a legal provision and more flexibility in its implementation (Interview SVM 9.11.1995).

The Packaging Covenant was different from the Code of Conduct and the Joint Declaration adopted in 1985 and 1987 in a number of respects. First, the covenant was the first agreement in the history of Dutch environmental policymaking which was explicitly binding under private law. In contrast to most preceding covenants the Packaging Covenant also included provisions for dealing with unforeseen developments and provisions in regard to the settlement of disputes. These provisions were a result of lessons drawn from the failure of earlier agreements which were criticized as being not enforceable.

An important point discussed in the negotiations was whether the SVM could represent affiliates and other firms and organizations and whether these companies were legally bound by the commitments of the SVM. Earlier agreements were perceived as rather vague at this point. In the case of the Packaging Covenant it was therefore stated that all organizations to which the agreement applied had to declare individually that they were a party in respect to the agreement (van Kempen 1992: 190).

5.3.5 The adoption of the Packaging Covenant

After an agreement had been reached on the content of the covenant on 16 May 1991, it was published and a press conference was held. The general reaction was positive, apart from environmentalists and consumer organizations. These groups summarized their position in a letter to parliament (30.5.91, Stichting Natuur en Milieu 1991) The letter was signed by the Stichting Natuur en Milieu, also on behalf of the Consumentenbond, Konsumenten Kontakt, and Vereniging Milieudefensie. These groups criticized that again no mandatory deposit schemes were introduced. It was also argued that the measures set out in the covenant did not guarantee that the targets would be met. A major point of critique was that the agreement had the form of a covenant, binding under private law. They criticized that there is an insufficient basis to bind individual companies legally and they also objected to the vagueness of the formulations in the covenant. The environmental movement and the consumer groups welcomed, however, the ambitious targets for material re-use. *'Business has made ambitious commitments on this point'* (Stichting Natuur en Milieu 1995).

The NMP-Plus required a public debate in parliament. The debate took place on 5 June 1991. It did not result in any changes of the covenant. The motion of the green party *'Groen Links'* to reject the Covenant, mainly for the same reasons as expressed by the environmentalists and consumer groups, was rejected by a vast majority (van Kempen 1992: 193). Therewith, the covenant acquired democratic legitimacy and it was signed by the Minister of the Environment on behalf of the government and by the chairman of the SVM on behalf of the packaging chain

5.3.6 Excursus: The Dutch politics of Life Cycle Analysis

The most difficult decision, which was postponed until the implementation phase, was the question of whether more re-usable packaging should be introduced into the Dutch market. The use of one-way packaging vs. re-usable packaging had been the crucial conflict since the end of the 1960s. The industry had always fought against the mandatory introduction of re-use systems. As an alternative they were prepared to engage in self commitments as well as voluntary agreements to increase recycling. This re-use issue was one of the main reasons why the strategic discussion failed, because no agreement could be reached on this topic. The bilateral negotiations, however, resulted in a vague compromise, which also reflected the scientific uncertainty about the environmental soundness of more re-use systems. Life cycle analyses and market economic research were to be paid for and done by the packaging chain, in order to show whether the replacement of one-way packaging by re-usable packaging would cause clearly less damage to the environment and whether there were any preponderant objections to such a change on market economic grounds. Only in this case would the packaging industry undertake to switch over from one-way packaging to re-usable packaging. In the supplement of the Covenant product groups were listed for which life cycle analyses (LCA's) had to be carried out, including: detergents and cleaning products, cosmetics, coffee-milk, milk and juices. Since the LCA's and the market economic research were a major episode in the Dutch packaging waste policy development and clearly reflect the Dutch consensus approach, it will be elaborated on below. It will be shown that although the whole process was carefully designed, the process and the outcomes were very contested. The main message of the LCA's was not controversial however: a comprehensive switch from one-way packaging to re-use packaging would not automatically result in less damage to the environment.

The implementation of the provision concerning re-use caused serious problems. As said in chapter two, life cycle analysis is a new and intricate method that aims to measure and compare the environmental effects of packaging systems. The results of such analyses depend to a large degree on the packaging systems compared, the environmental criteria chosen, and the assumptions made. The packaging chain agreed with VROM to use the method developed by the Centre for Environmental Science of Leiden University. This method was developed in cooperation with the Society of Environmental Toxicology and Chemistry, and belonged to the most advanced methods of life cycle analysis. Even though such an appreciated method was used, LCA was still associated with a tremendous degree of uncertainty, normative judgements, subjectivity, and dynamic developments in the field of technology, markets, law making, and environmental consciousness (SVM 1994a: 33, ten Heuvelhof *et al.* 1992: 14). Earlier LCA's of packaging systems showed the complicated character of the new method, and offered neither clear results nor a consensus on the most environmentally friendly alternatives.

In order to reach consensus on the results of the new LCA's, the SVM asked the Faculty of Administration Science and the Study Centre for Environmental Science, both at the Erasmus University of Rotterdam, to develop a standard for the process by which the LCA's were to be carried out. Six months after this request, in February 1992, the scientists came with their report. The report set out a process philosophy, including norms for proper behaviour, with the paramount goal of achieving consensus among all interested parties, including consumer groups and environmental organizations, on the outcomes of the LCA's. The scientists emphasized (a) that the assumptions on which the analyses would be based had to be clear and accepted by all parties; (b) that the analyses had to be done on a high qualitative level, controlled by independent 'top' scientists, and (c) that the judgement of environmental analyses and market economic research were to be made by all interested parties as well as by external experts (ten Heuvelhof *et al.* 1992: 17). In order to enhance the possibility of consensus the research group proposed creating a broad steering committee rather than having the process controlled by the SVM (ten Heuvelhof *et al.* 1992: 16,17). Next, a scientific advisory board should be created. The individual LCA had to be managed and assessed by project groups, one group for each product, comprising representatives of industries and societal interests and headed by relatively more independent persons such as former politicians and other heavyweights. The report also mentioned provisions of checks and balances in order to facilitate consensus within this institutional arrangement. Finally, the report described elements of the LCA's and the market analyses, circumscribed the structure of the report, and set out rules for the choice of research institutes.

The recommendations of the group were accepted by VROM and the SVM. The steering group, the scientific advisory body and the project groups were established. They comprised high ranking politicians, such as a former Minister of Defence, and representatives of industry, as well as scientists and consultants, including B&G, a group that had strong ties with consumer organizations, and was well known for its strong criticism of earlier LCA's carried out by the industry. Hence, industry paid for one of its major critics. Environmental organizations were also asked to participate but they refused to do so. The SVM, however, asked a scientist who had some ties with the environmental movement to join in. By this, at least an indirect representation of environmental interests could be guaranteed (SVM 1994a: 33-34). The chairman and the secretary of the steering group were members of the Rotterdam group who had developed the process standard.

Thus a complex institutional arrangement was established based on project groups, supervised by a steering committee, and advised by a scientific committee. The process soon showed a considerable delay. The steering group met for the first time in September 1992, three months before the LCA's and the market economic research were to be finished. The whole process took more than eighteen months longer than agreed to in the Covenant. Despite this rather careful design, the process and the outcomes of the LCA's were very much contested. Officially, the delay was traced back to the difficulties of life cycle analysis. But conflicts about the right questions to ask, the adequate assumptions and the right conclusions also contributed to the delay (Trouw 1.12.1994). There was, for instance, controversy about which kind of alternatives were to be included in the assessment. While the representative of the B&G group wanted the relatively new polycarbonate bottle for milk to be included in the assessment, the representative of a large filler of dairy products was very much against this (Interview Chairman of the Steering Committee 11.9.1997), probably because he expected that this new type of packaging would be more environmentally friendly than the packaging he used for his products. Another conflict concerned the size of the project groups. The SVM wanted every part of the packaging chain to be represented in each product group (Interview Chairman of the Steering Committee 11.9.1997), but the steering group argued that this would involve too many participants than would be manageable.

Consensus was also very difficult to reach, though, according to the SVM, a lot of time was spent in trying to achieve it. On a number of occasions the representative of the consumer groups formulated a dissenting opinion, sometimes supported by the expert with affinity for the environmentalists and the independent expert for LCA's (Stuurgroep Milieu-analyses Verpakkingen 1994). The lack of quality was also acknowledged by the Steering Committee, which stated that the quality of the reports suffered from time pressure, multiple interests, and a lack of data (Stuurgroep Milieu-Analyses Verpakkingen 1994: 45).

The Packaging Committee, which was established to monitor the implementation of the Covenant, was particularly critical on the process and the outcomes of the LCA's and the market economic research. The LCA's suffered, according to the Committee, from several serious shortcomings. The Rotterdam group had interpreted the conditions under which producers did not have to switch to re-use packaging even though this would cause clearly less damage to the environment, in a rather broad way, privileging the packaging chain interests. In order to examine whether there were '*preponderant objections ... on market economic grounds*' the Rotterdam group proposed comprehensive cost benefit analyses, as well as an assessment of the impacts of a change towards re-use systems on competitive position and on employment. These aspects were not included in the original meaning of preponderant objections on market economic grounds, which was confined to something like "*not saleable to the target group*" (Epema-Brugman 1996: 1). The Rotterdam group also included the cost effectiveness of waste avoidance measures in its standard for the market economic research. The inclusion of this criterion, in particular, was according to the Packaging Committee, inadmissible ('*niet toelaatbaar'*, Commissie Verpakkingen 1994a: 4). Also, the LCA's deviated from standards set out by the Rotterdam group. The LCA's for the various packaging systems did not proceed in a similar way. Some LCA's strongly emphasized the energy consumption and waste generation, others also took a broader range of aspects into account, such as water pollution, environmental and human toxicology, global warming etc. What is more, the same aspects were not interpreted in the same way. Whereas some institutes focused only on energy consumption, other institutes traced them back to the sources of energy (oil, natural gas etc.). Also the assumptions made differed between the LCA's. The Packaging Committee criticized the overall accessibility of the reports. Some reports showed a lack of clarity and were difficult to read. Unlike the recommendation of the process standard, the assumptions made were sometimes not clear or even not presented at all.

As far as market analyses were concerned the Packaging Committee criticized that different proceedings were followed, that the analyses went beyond the criteria intended by the Covenant (e.g. including cost-efficiency) and neglected some other aspects like weight and user friendliness. Again underlying assumptions were not stated clearly (Commissie Verpakkingen 1994a: 5).

The Packaging Committee also criticized the SVM for drawing the general conclusion that the replacement of one-way packaging by re-usable packaging was not necessary according to the Covenant criteria. The Commission argued that conclusions had to be made for each product group separately.

The difficulties of the implementation process can be explained by several factors. The terms as set out in the covenant were very vague. No criteria were developed concerning the decisive terms '*cause clearly less damage to the environment*' and '*no preponderant objections to such a change-over on market economic grounds*'. Also, the environmental analyses were based on life cycle analysis which was still a young and poorly developed method. The factor of uncertainty was at least 25 per cent, for some aspects even more than 100 per cent. Furthermore, the packaging industry had a significant interest in specific outcomes of the LCA's, and tried to manipulate the process as much as possible (Interview Teuninga, member of the LCA advisory board). Finally, a lot of organizations were involved in the process. For each product a specific bureau carried out the LCA, which was then checked by peer review by another bureau. This resulted in different concepts and processes, even though a steering group was established as well as a scientific body.

According to the Chairman of the Steering Group, the criticism of the Packaging Committee, had also to do with different aspects. While the Packaging Committee emphasized the standardization and the uniform proceedings and criteria of LCA's, the main interest of the Steering Group was to ensure consensus. '*The Steering Group did not look into the details of the working of the project group as long as there was a consensus within the project group about the way to follow'* (Interview Chairman Steering Group 11.9.1997).

Even though the Packaging Committee was very critical of the process and the outcomes of the LCA's and the market economic research, it argued that it was more important to proceed with the implementation of the Covenant than to make the LCA's more perfect. The Committee also acknowledged that, though the LCA's formed a ponderous task, the familiarity with life cycle analysis had increased. Next, and in agreement with the SVM and many other participants, the Committee stated that the LCA procedure increased the knowledge about environmental aspects of their products and possibilities for their improvement. In contrast to the SVM, however, the Packaging Committee concluded that for certain products, such as jam, vinegar and distilled drinks, the replacement of one-way packaging by re-usable packaging would cause clearly less damage to the environment. Since the respective market analyses were qualitative rather than quantitative, the Commission recommended pilot projects in order to estimate the economic effects of this step. The Committee also argued that for certain other products substantially less damage for the environment might be expected, therefore a closer judgement should be carried out (Commissie Verpakkingen 1994a: 23-25).

The Ministry of the Environment and the SVM did not follow the advice of the Packaging Committee, however. Moreover, the issue of the comprehensive introduction of re-usable systems disappeared from the political agenda. The industry had no interest in expanding re-use systems, VROM emphasized the packaging optimization of one-way packaging achieved by new knowledge generated by the LCA's. Moreover, after the election of 1994 a new Minister of the Environment was appointed who showed less interest in this topic as compared to other issues. Furthermore, the environmental organizations could not mobilize consumer support any more, since the public was uncertain about the environmental advantage of re-usable packaging, due to contradictory results of LCA's (Interview Milieudefensie 20.11.1995).

5.4 Conclusion

The Covenant is a compromise. Once again the industry was able to prevent interventionist measures such as mandatory deposits or a ban of certain materials or types of packaging. In exchange, the companies which signed the covenant accepted the unilaterally imposed prevention and recycling targets and committed themselves to do everything they could to achieve these targets, among other things, by using less packaging and by creating more recycling facilities. The programme formulation shows that, even though the government made great efforts to reach a consensus, this approach had its shortcomings. *'The strategic discussion about packaging waste has failed to resolve some of the major conflicts between industry and the consumer and environmental groups. Difficult decisions have been postponed until the implementation phase....' (Koppen 1995: 169)*

As far as the allocation of responsibilities was concerned, the dividing line between industry and local authorities was not drawn very clearly. During the programme formulation process of the covenant, there was no discussion about the distribution of responsibilities concerning the organization and financing of the collection, separation, transportation and treatment of waste. The *Vereniging van Nederlandse Gemeenten* (VNG), representing the local authorities, which were basically responsible for waste collection, neither played an active role during the strategic discussion nor took an active part in the bilateral negotiations between VROM and the SVM. In short: in order to reduce the complexity of the negotiations, the number of participants was dramatically reduced, problems were divided into smaller ones, difficult decisions were postponed to the implementation phase and compromises were sought by vague formulations.

As in Germany, nationally specific institutions played an important role in the process and the outcome of the Dutch packaging waste policy. The strictly proportional voting system helped to 'green' the parliamentary agenda at quite an early stage. Due to a lack of narrowly defined and standardized recruitment criteria, environmentally-minded persons, many of them natural scientists, became civil servants at the Ministry of VoMil (later VROM). The relatively high autonomy of Dutch ministries facilitated the creation of a broad pro-environment advocacy coalition including environmental groups and scientists. The general openness of the Dutch political system is also reflected in the recruitment of the former business consultant, Winsemius, as Minister of the Environment. His period in office (1982-1986) coincided with the recognition of shortcomings of the command and control approach to environmental problems. He introduced a new management approach into the Ministery, which included among other things quantified targets, and most notably the concept of internalization, which was, in the Dutch case, a moral concept rather than an economic concept as in Germany.

Chapter seven will provide a more elaborate and synoptic account of the ideas and institutions shaping the Dutch path towards the covenant. The next chapter will deal with the third country of this cross-national comparison: the United Kingdom.

6. The United Kingdom or where is the problem?

The German and the Dutch governments made great efforts to bring about an effective packaging waste policy. Both governments arrived at binding measures which required a change in target group behaviour. The British government chose a hands-off approach. Only a vague non-statutory target was announced and industry was asked to avoid unnecessary packaging and to increase recycling. Why did the UK government undertake a comparatively modest approach to the problem of packaging waste? What were the crucial determinants which made the British path so different from the German and the Dutch one?

6.1 The foundations of environmental policy

As in the Netherlands and Germany, the "environment" became an important issue in the United Kingdom at the late 1960s and early 1970s. This environmental wave was, however, less significant than in Germany and the Netherlands. There are a number of reasons for that.

First, at that time, the United Kingdom already had comparatively sophisticated institutions and instruments to cope with pollution. It was the first industrialized country of the world. Accordingly it was also the country where the side effects of mass-production and distribution became obvious at a very early stage albeit mitigated by geographical factors. In particular, smoke produced by burning of coal in factories and households constituted a clear threat to public health (Vogel 1986: 31-37). There was a long tradition of using common law for protecting individuals from specific nuisances due to pollution. Ashby and Anderson quote a case from 1691, *'when Thomas Legg of Coleman Street in London complained of the smoke from his neighbour's bakehouse, the baker was ordered to put up a chimney 'soe high as to convey to smoake clear of the topps of the houses''* (Ashby and Anderson, 1981, p.1). The national Alkali Inspectorate dates from 1869 (Weale 1992a: 161).

Secondly, due to characteristics of the physical environment the actual nature of public perception of environmental problems differed from that in Germany and the Netherlands. The United Kingdom has considerable advantages in being an island. Its physical environment is more capable of absorbing and carrying pollution than is the case in Germany and even the windy Netherlands. Air pollution is quickly dispersed in Britain, and water pollution is soon carried out to the sea. Therefore the public is less exposed to pollution than in Germany and the

Netherlands (Richardson and Watts 1985: 32). Moreover, the United Kingdom produces fewer pollutants and wastes per square meter than Germany or the Netherlands (RIVM 1991: 41, RIVM 1993: 29). Accordingly, the protection of nature through land use planning rather than pollution issues as such prevailed in the public debate. Nature conservation issues were also carried by established environmental organizations, such as the Royal Society for the Protection of Birds (1889) and the Council for the Protection of Rural England (CPRE, 1926). These national groups '*are run by middle and upper class professionals well attuned to the norms of political behaviour and sympathetic to the prevailing policy style. The leadership of groups like the CPRE and the National Trust contains a significant aristocratic element - an important factor in terms of access to the policy process and in terms of the lobbying style of the groups*' (Richardson and Watts 1985: 13). These long established and traditionally conservative organizations continued to operate in the 1970s but without challenging established conventions and institutions and prevailing production and consumption patterns. However, new groups were founded in the course of the late 1960s and early 1970s such as Greenpeace Great Britain (1976) and Friends of the Earth (1969) which attracted many people. Unlike the traditional environmental groups, the new ones believed that environmental problems could not be solved unless fundamental problems of economic and population growth would be attacked. They belonged to the new social movements and strongly criticized the prevailing values as well as consumption and production patterns. However, compared with Germany and the Netherlands, the relative importance of the new groups was lower in the UK.

Not only were the agents of new environmental ideas outnumbered by traditional groups, but what is more, it was difficult to get access to Parliament. Due to the majority voting system, an environmentally-oriented party was not a serious competitor and therefore the traditional parties were not forced to adapt to new ideas. It is worth noting that the House of Lords has traditionally been more open to environmental ideas, albeit mostly nature protection ones, not least of all due to the aristocratic characteristics of the traditional environmental protection groups (Richardson and Watts 1985: 13).

In short, societal forces pressing for a move in environmental protection policy were weaker than in the case of the Netherlands. Moreover, there were also no strong pro-environment advocates within the government as was the case in Germany.

Apart from the absence of public pressure, the following reasons were responsible for this. Britain was the first state to set up a Department of the Environment (DoE, 1970), with a wide range of responsibilities dealing with the people's environment in a broad sense. The Department emerged from Ministry of Housing and Local Government, and the Ministry of Public Buildings and Works. Until 1974 the Department of Transport was also included. One might argue that the broad scope of activities of the DoE reflects the idea that all environmentally relevant functions should be brought together. However, it always contained functions like local government finance, as central features of the work, which were unrelated to problems of environmental policy (see Weale *et al.* 1996: 264). In 1993/1994 for instance, only 10 per cent of the employees dealt with environmental protection issues (OECD 1994: 23). Moreover, many key environmental areas were left in the realm of other ministries. A study conducted in the early 1970s revealed that over 42 per cent of parliamentary questions which could be reasonably classified as 'environmental' were answered by departments other than the Department of the Environment (Kimber *et al.* 1974, quoted in Richardson and Watts 1985: 9). It is important for this study that the Department of Trade and Industry (DTI) rather than the DoE was responsible for (packaging) recycling. Hence, in contrast to the Netherlands and Germany, a ministry of economic affairs was the leading department in regard to a crucial aspect of packaging waste policies.

In contrast to the Netherlands and Germany, no policy entrepreneurs arose from this Ministry who advocated a strict environmental protection policy. This can be related partly to institutional factors. In the large Ministry of the Environment only a small proportion of people and money was devoted to the task of environmental protection. Also, the civil servants were generalists and came from the former ministries whose tasks were mostly unrelated to environmental protection, whereas in the case of Germany and the Netherlands many civil servants were specialists on environmental issues, and many came, in particular in the Netherlands, from environmentally-oriented scientific institutions. In contrast to their Dutch and German colleagues it was also unlikely that the British civil servants became socialised into environmental issues, because civil servants rotated regularly within the DoE, and took on tasks in other divisions that were unrelated to environmental protection. For instance, the civil servant who was head of the packaging unit in the early 1990s moved to the local government finance division, where she had nothing to do with environmental issues any more (Interview DoE, July 1996).

In the absence of policy entrepreneurs, it is understandable that no efforts were made to create capacity and support to develop national environmental policies to the extent of the German and the Dutch approach. Environmental protection policy remained to a large extent the task of specialized agencies and local authorities. There was, for instance, no equivalent to the specialist national institutions such as the German Federal Environment Agency (UBA) or the Dutch National Institute for Public Health and Environmental Hygiene (RIVM). British civil servants were dependent on the scientific expertise of the Royal Commission of Environmental Pollution (RCEP), founded in 1969, and other advisory committees. These advisory bodies, however, were comparatively important in the UK. Britain developed a vast network of official advisory committees. In the context of a highly centralized state those bodies channelled and institutionalized groups' contact with departments. They had at least two major functions. First of all, advisory bodies co-opted specialized knowledge into the policy process, which is very important given the dominance of generalists in the British civil service. In addition, the consultation machinery grants societal groups the feeling that they are somehow part of the existing processes (Richardson and Watts 1985: 11)

There were also no substantial attempts to generate more environmental consciousness among the general public and to translate this into public policies. The government did not support the new environmental movement or environmentally minded scientific institutions by providing organizational or financial support as was done in the Netherlands and, to a lesser extent, in Germany. Those groups which were challenging the prevailing consumption and production patterns,were at the periphery of the decision-making process, either without access to the relevant committees or with access only to less important ones (Richardson and

Watts 1985: 18, Vogel 1986: 52-53). What is important in the context of this study, is that although Friends of the Earth was recognized by the DoE, it did not belong to the policy community of the DTI (Richardson and Watts 1985: 18) which was the leading department for recycling.

In contrast to the Netherlands and Germany, there was also neither a general evaluation of the state of the Environment comparable to the Dutch Emergency Memorandum from 1972, nor a principle-based environmental programme such as the German Environmental Programme of 1972. The Dutch and German documents reflected a pro-active view on environmental protection policy. In contrast to Germany, the British approach to environmental policy was not oriented to *ex ante* formulated abstract principles. The reluctance in Britain to a principle guided perspective shows parallels with the English system of case law. The system of common law is informed by an inductive logic in which principles are derived *ex post* from individual cases. By contrast, in a codified roman law system, cases have to be judged according to abstract legal principles developed beforehand in statutory law (Bridge 1981: 357, Weale 1992b: 82, Weber 1994: 178-179) Weale has argued that in regard to environmental protection '*policy is conceived as a series of problems, constituting cases that have to be judged on their merits. General norms are to be avoided if the decision can be left to the exercise of continuous administrative discretion. There is a desire to avoid programmatic statements or expositions of general principles governing particular areas of policy*' (Weale 1992b: 81) . Richardson and Watts added in regard to the environment that '*there is no particular priority accorded to anticipatory solutions - and the emphasis on negotiations itself inhibits radical changes the British style is reactive*' (Richardson and Watts 1985: 11). The most extreme example of the reactive approach is probably the Deposit of Poisonous Waste Act which was 'rushed' onto the statute book. It was written within a weekend after a well publicized scandal involving children who were playing with cyanide at a disposal site (Haigh 1990: 127, Vogel 1986: 57-60).

6.2 The history of (packaging) waste policy

6.2.1 The Control of Pollution Act 1974

The United Kingdom had not only favourable physical and geographical conditions that mitigate water and air pollution, but also favourable conditions for waste disposal, i.e. a large amount of suitable landfill sites. The country is rich in clay pits which make impermeable rubbish dumps. As in the Netherlands and Germany, there was no comprehensive legislation for waste before the 1970s. Local authorities, however, had had the power to control waste as an aspect of public health for a long time. The Public Health Act (1936), which consolidated much older legislation, empowered them to remove house and trade refuse and to require removal of accumulation of noxious matter. The Town and Country Planning Act (1947) required planning permission for the development of any new waste disposal site or plant (Haigh 1990:127, RCEP 1985: 14).

Since the 1960s there was growing concern about the environmental effects of waste. This concern led to the formation of a working group on waste disposal in 1967. There were no substantial outcomes of this working group in the first years of its existence. In 1972, however, the Deposit of Poisonous Waste Act (1972) was 'rushed' (Haigh 1990: 127) onto the statute book, forced by a well publicized scare about the dumping of toxic waste (see above; see also Vogel 1986: 59). Two years later the Control of Pollution Act was adopted. It provided a more sophisticated piece of legislation. The 1972 Act was repealed in 1981, when the more elaborate system embodied in the 1974 Act was fully in operation.

The Control of Pollution Act (COPA) addressed a wide range of environmental problems including water pollution, noise nuisance, air pollution and waste disposal. The act contained the obligation for waste disposal authorities (.i.e. county councils in England), to produce waste disposal plans for all household, commercial and industrial waste. Such plans had to include information on a) the type and quantities of waste; b) of the methods of disposal, for example reclamation, incineration, or landfill; c) the sites and equipment being provided, and d) the costs. It is worth noting that the Act required disposal authorities to consider what arrangements could reasonably be made for reclaiming waste materials. The plans did not require the approval of central government, but a copy had to be sent to the DoE. The Act introduced a comprehensive licensing system for the disposal of waste over and above existing planning controls. A license by the Waste Disposal Authority was needed for any land to be used as a deposit for waste or as a location of a waste disposal plant. The authority had to maintain a public register with particulars of all disposal licences, sometime called site licenses. The responsibility for waste collection was assigned to the Waste Collection Authorities, i.e. the district councils in England (for more details Haigh 1990: 128-130, RCEP 1985: 125-138).

The administrative practice of British municipal waste management in the late 1970s and the 1980s was largely determined by the action of sub-national actors, since waste disposal was to a large extent a local government function. In this it reflected a general characteristic of British environmental policy, namely the decentralized structure of policy implementation (see Gray 1995). DoE activities were mainly confined to developing the overall policy, promoting research, and issuing advice (Haigh 1990: 130). Another general aspect of environmental policy that affected the administrative practise of waste management, was the fragmented structure of the implementation machinery. An application for a license, for instance, had to be referred to the regional water authority. Disputes between this authority and the waste disposal authority, i.e. the County Councils, were frequent. Such conflicts had to be referred to the Minister of the Environment as the final arbiter.

The administrative implementation of a number of provisions of the Control of Pollution Act took considerable time, in particular section two, which was concerned with waste disposal plans. It was probably due to the EC Waste Framework Directive of 1975, that the implementation accelerated later. Whereas the Control of Pollution Act said nothing about the period within which waste disposal plans had to be adopted, the Directive required those plans as soon as possible. *'The Directive must have provided pressure for not deferring implementation of Section 2 of the Act (concerned with waste disposal plans) much longer. There is no knowing how long Section 2 would have been delayed but for the Directive' (Haigh 1990: 136).* Section two was brought into force on 1 July 1978. Four years later, however, only three plans had been completed. Ten years later one-third of the waste disposal plans were still incomplete (Haigh 1990: 135).

Besides the development of waste disposal plans, licenses for landfill sites had to be filed and enforced. However, the legal requirements for waste disposal had always been less strict than in most other West European countries. Many farmers earned extra money by allowing local authorities to dump waste on their grounds. The 'relaxed' attitude towards waste disposal was also reflected in the usage of the North Sea as a dustbin. In December 1985 the Royal Commission on Environmental Pollution published a comprehensive and critical report on waste with the title '*Managing Waste: the Duty of Care*'. The report criticized the delays in the development of waste disposal plans, but also the lack of control of the operation of landfill sites and the aftercare of these sites (RCEP 1985).

The practise of waste policy implementation was affected by the general policy of the Thatcher government. The aim to cut public expenditure resulted in the substantial reduction of funding for pollution control agencies (Vogel 1993: 262). Moreover, the implementation of the above mentioned. Section two of the 1974 Control of Pollution Act was postponed as a cost cutting measure. Furthermore, the Waste Management Advisory Council was abolished as consequence of the overall goal to streamline the public sector. The increasingly stringent controls on local government expenditure also limited the potential for new locally funded waste management initiatives (Gandy 1994: 46).

The government did not accept all of the proposals of the RCEP but it announced new legislation including a duty of care to be imposed on waste producers who would be obliged to take all reasonable steps to ensure satisfactory disposal of their wastes, a registration scheme for waste transporters, extra power for waste disposal authorities over site licensing, and the enforcement of conditions and the aftercare of disposal sites.

6.2.2 Recycling in the absence of national packaging policies

There was no distinct British packaging waste policy in the 1970s and 1980s. The DTI, responsible for recycling, saw recycling as an industrial process that responded to market mechanisms. There were, however, no strong market incentives to engage in recycling. The beneficial geological factors, the modest environmental standards, the lack of resources of the implementation agencies and the relaxed attitude of the implementers resulted in low landfill costs. In the early 1990s for instance, landfill costs ranged between five and twenty pounds per tonne (OECD 1994: 80). This was some 10 per cent of those in Germany (Key Note Market Review 1993b: 71).

In Germany and the Netherlands material recycling was done in the absence of national policies not only for altruistic reasons but also for economic ones. The value of the secondary material and the avoided costs of waste disposal were weighed against the costs of collection and recycling of waste. Since the waste disposal costs had always been low in the UK, this calculation resulted most of the time in the cheap disposal option, preventing the UK from establishing a waste collection and waste recycling infrastructure as in Germany and the Netherlands. Accordingly, comparative figures on recycling show Britain at the lower half of the European Community recycling league. In 1983, for instance, there was 1 bottle bank per 22,250 people, while the corresponding figure for Germany was 1 per 2,000, and for the Netherlands 1 per 2,300 (RCEP 1985: 55). In 1989, the recycling performance of West Germany and the Netherlands was on a per capita basis twice as good as Britain's for paper and aluminium and three to five times as good for glass (ENDS Report, October 1990: 27).

Gandy (1994: 146) described in a detailed study the waste management under the Greater London Council. He shows that the Labour administration, following the 1981 elections, sought a strategic policymaking role in recycling and waste management. It promoted recycling workshops by the provision of financial assistance, and granted access to civic amenity sites. The Council's waste management plan '*No time to waste*' (1983) announced the expansion of bring-back systems of onstreet collection facilities, along with the recovery of landfill gas. The Council also emphasized the need to tackle the production of waste at the source with a nationally-oriented waste management strategy. It lobbied both Whitehall and the EC Commission for the reduction and control of packaging and the increased use or returnable and re-usable beverage containers. Recycling activities of the local authority of London and other local authorities increasingly suffered, however, from more stringent controls on local government expenditure. These controls limited the potential for new locally funded waste management initiatives (Gandy 1994: 46).

The *laissez-faire* approach of the government did not change with the 1985 EC Directive on Containers of Liquids for Human Consumption. The plans for the directive met the resistance from the British government. The British government did not think that a directive was needed. The government believed in line with British industry that market forces were already promoting savings in energy and raw materials and reduction in waste generated. It also opposed any barriers to trade in containers and objected to any discrimination between different types of containers or between different types of packaging. Unlike the government, the House of Lords reported in favour of the Directive after it was clear that no legally binding targets were included. The RCEP, Local Authority Recycling Advisory Committee (LARAC), Friends of the Earth, and other environmental groups also supported the planned directive. LARAC, for instance, argued that a 50 per cent recycling target for each packaging material had to be achieved by the end of the century by means of voluntary agreements between the government and container manufacturers (Haigh 1990: 167, RCEP 1985: 48-50).

In order to implement the Directive, the DTI, after consultation with some sixty-five organizations, submitted a programme to the Commission. This programme covered the six programme items under four separate types of material (glass, metals, plastic, paper/composites) outlining the measures to be taken and by whom. Guidelines stressed that the programme was based on what was feasible under present economic, industrial and market conditions and that the programme was to be implemented by voluntary agreement. Hence in contrast to Germany, where an ordinance was adopted to control the waste generated by plastic bottles, and to the Netherlands where government and industry agreed on a rule of conduct on drink containers, the British approach was more modest and did not require serious industry action. Local authorities and environmentalists had no effective institutional channels to challenge the *laissez-faire* approach of the British government (Haigh 1990: 167).

6.3 The minimal approach to packaging waste

6.3.1 The second wave of environmentalism

The second international wave of environmentalism affected not only Germany and the Netherlands but also the United Kingdom. Chernobyl, the Brundtland Report '*Our Common Future*', and other events resulted in increased media attention to environmental issues and caused an upsurge in political salience in 1988 and 1989 (Rüdig, Franklin, Bennie 1993: 14). Membership of environmental groups rose significantly. In particular the environmental groups founded in the late 1960s and 1970s were able to stengthen their base. Friends of the Earth more than quadrupled its membership between the mid 1980s and 1991 to some 200,000. And the number of supporters of Greenpeace rose by almost 500 per cent, from 80,000 in 1987 to 380,000 in 1991 (see Grove-White 1994: 192, 198). By the early 1990s more than 10 per cent of the electorate were members or registered supporters of an environmental organization; this was five times as many as in Germany, but half as many as in the Netherlands (Hey and Brendle 1994: 642).

The increase of environmental consciousness of the late 1980s and the early 1990s also affected consumer behaviour in the UK. Surveys indicated that an increasing number of people was prepared to pay more for goods if they were more environmentally friendly. Companies with a green image such as the Body Shop increased their turnover significantly. A poll conducted by MORI showed that 51 per cent of the population claimed that they bought products made of recycled material and 52 per cent said that they bought products in recycled packaging (Key Note Market Review 1993b: 19). Studies commissioned by Friends of the Earth concluded that consumers were willing to use collection facilities for recycling activities if an adequate infrastructure were provided.

Eventually the new environmentalism also affected party politics and elections. To many observers it came as a surprise that the British Green Party achieved 14.5 per cent of the votes at the elections for the European Parliament in 1989. It is worth noting that this result was the best performance of any green party in any EC Member State at this election (ENDS Report, June 1989: 11; for details on the Green Party: Rüdig, Franklin, Bennie 1993). The success of the Green Party had no direct effect on the established party system, since the British first-past-the-post system prevented the British Green Party from sending any members to Brussels. In addition, the party success was perceived not only as the expression of green preferences but also as a protest against the general policy of the Thatcher government.

The increase in politically relevant environmentalism in Britain was less strong than in Germany and the Netherlands, however. Research on the importance of environmental issues for the elections of the European Parliament in 1989 indicated that the environment was not as high on the political agenda as in Germany and the Netherlands. While 72,8 per cent of the Dutch respondents and 53,8 per cent of the German respondents ranked environmental protection among the three most important issues, in Britain only 29,5 per cent did so (Kuechler 1991, quoted in Hey and Brendle 1994: 643). The increase in membership and support of the new environmental groups - though substantial - was lower than in Germany and the Netherlands. Moreover, most of the registered members of an environmental groups were still part of the traditional nature conservation organizations such as the National Trust. Even green consumer preferences were weaker as compared to Germany and the Netherlands (ENDS Report, June 1989: 13).

One reason for this comparatively low degree of environmentalism was the prevailing importance of economic issues in the 1980s. There was a rise in unemployment and the manufacturing base had been eroded, which, incidentally removed some of the more obvious sources of environmental pollution. The average rate of unemployment between 1974 and 1985 was significantly higher in the UK (8,3 per cent) than in Germany (4,8 per cent), and slightly higher than in the Netherlands (7,9 per cent) (Schmidt 1989b: 60 based on OECD data). The political debate was heavily concentrated on the difficult unemployment situation (Richardson and Watts 1985: 19). The idea of ecological modernization, which gained prominence in Germany and the Netherlands in the 1980s, was virtually absent in the UK. Accordingly, the potential of environmental policy for the creation of employment was not recognized. '*British debates never escaped the belief that there was an inevitable tension between environmental protection and economic development*' (Weale 1992b: 88).

Another reason for the comparatively lower degree of environmentalism was that while acid rain was perceived as a particularly dramatic environmental problem in the 1980s in Germany and the Netherlands, it only gradually emerged on the public agenda in the United Kingdom as more scientific evidence was generated and diffused (Boehmer-Christiansen and Skea 1991, Hajer 1995).

Besides the comparatively lower degree of importance assigned to environmental protection issues, waste disposal in particular was lower on the agenda than in Germany or the Netherlands. A survey done by Gallup International Institute revealed that waste disposal was not seen as one of the most important problems facing the nation. Respondents of all three countries said that air pollution was the most important environmental problem. But while in Germany and the Netherlands waste disposal was ranked second, mentioned by 21 per cent in Germany and 18 per cent in the Netherlands, it was not among the three most mentioned environmental issues in the UK (Dunlap, Gallup and Gallup 1992: 9).

Still, the increasing environmental consciousness in general, and the Green Party success in particular, resulted in a greening of British party politics, at least for a short while. It also affected the policy of the government.

6.3.2 The government's approach

The first hint that the government reacted to the international environmental policy debate and the national increase of environmentalism were Thatcher's '*green*' speeches in 1988. The Thatcher government was throughout most of its period in office hostile to, or at best uninterested in, what may be regarded as the modern constellation of environmental problems: acid precipitation, ozone depletion, pollution of the seas, river quality and waste management (Weale *et al.* 1996: 265). Thatcher's speeches indicated at least rhetorically a change in the thinking of the government. The pro-environment stance was particularly represented by the Minister of the Environment Patten[13]. While his predecessors, like Heseltine, Baker and Ridley, either displayed more interest in other aspects of the department's work, for example urban renewal in the case of Heseltine, or demonstrated hostility to the growing consciousness of environmental problems, as in the case of Ridley, Patten showed a genuine interest in issues of environmental protection and pollution. He was, however, a victim of the broad task of the DoE. During his brief term in office his attention was absorbed by the controversy surrounding the community charge (the local finance instrument replacing the poll tax), leaving him little opportunity to focus consistently on environmental questions (Weale *et al.* 1996: 265).

Still, it was very much due to the effort of Patten that the British government produced the first comprehensive overview of the state and objectives of British environmental policy in the White Paper '*This Common Inheritance*' (HM Government 1990). The White Paper was based on two preliminary reports prepared by the DoE for the Rio Earth Summit in 1992. The White Paper, like the Dutch National Environmental Policy Plan (NMP), frequently referred to the Brundtland report. Both documents display on their covers the earth: the blue planet as it appears from the outer space. The introductory remarks to the Summary of the White Paper (HM Government 1990a) emphasize a strong moral commitment to 'save the earth', but also reveal that the British interpretation of sustainable development emphasized the need for economic growth as a precondition for environmental protection.

'We have a moral duty to look after our planet and hand it on in good order to future generations. That does not mean trying to halt economic growth. We need growth to give us the means to live better and healthier lives. But growth has to respect the environment. And it must be soundly based so that it can last. That is what is meant by sustainable development'(HM Government 1990a: 1). The White Paper also stated that '*safeguarding the environment can be very costly*' and *'We must do it in the most effective way*' (HM Government 1990a: 3). The cost aspect was very much emphasized in the British document. This emphasis indicated again that the prevailing idea in the United Kingdom was that environmental protection was difficult to reconcile with economic growth, while the idea of ecological modernization as presented in Germany emphasized that reconciliation.

As far as the precautionary principle was concerned, the British government took a more reluctant view than the German one. While in Germany environmental measures were also

[13] The functional equivalent to the German and the Dutch 'Mnister' s the British 'Scretary of State'. In this study, however, the term Minister will be used for the British Secretary of State in order to prevent misunderstandings.

legitimized in the absence of clear scientific evidence, the White Paper emphasized the need for scientific information and cost benefit analyses: '*We must act on facts, and on the best analysis of likely costs and benefits*' (HM Government 1990a: 2). Under the heading "*Facts not Fantasy*" the White Paper stated that: '*Precipate action on the basis of inadequate evidence is the wrong response*' (HM Government 1990: 11).

The most important differences to the Dutch NMP was that the White Paper did not state long-term-oriented and quantified reduction targets for pollutants and that the many measures were couched in rather general terms and were mostly procedural, i.e. co-ordination of policy within government, rather than substantial. Though the White Paper was subtitled '*Britain's Environmental Strategy*' it was not as strategic as the Dutch NMP. It reflected the general and traditional reluctance for anticipatory action but also for the setting of medium and long-term goals. The latter confirms the view of Hayward that there '*are no explicit, overriding medium or long-term objectives*' in British politics (Hayward, 1974, 398-399).

Packaging waste management according to the White Paper

The White Paper stated that the government aimed at encouraging the best use of valuable raw materials, and safe and efficient disposal of waste. This meant minimizing waste, recycling materials and recovering energy, tight control over waste disposal, and action against litter. This cannot be read as a hierarchy of waste management options, however. In contrast to Germany and the Netherlands, the British government doubted the scientific justification for any rankings of waste management options.

The UK government favoured a '*voluntary approach*' to waste management. There was no scheme aimed specifically at packaging waste but the government set the target to recycle 50 per cent of the recyclable household waste by 2000. The 50 per cent household recycling goal '*was an initiative of Chris Patten, who saw the need of clear and ambitious targets. In the 1980s, the environmental awareness was much higher than ever before or after*' (Interview DoE 12.7.1996).

The government stated in its White Paper that probably 50 per cent of all household waste was recyclable. This implies a recycling target of 25 per cent. From a cross-national perspective this target was not very ambitious. In the British context, however, it was quite ambitious. Based on official statistics Friends of the Earth estimated that only 2.6 per cent of household waste was recycled in 1990 (Stupples 1993: 64). The government itself stated in its White Paper a current level of 5 per cent (HM Government 1990: 190). The target, however, was not a statutory requirement. There were no legal obligations or sanctions provided for the case that the target was not be achieved (see Fischer 1993: 30, Stupples 1993: 59). The recycling target was not as clear as the official at the DoE suggested. The government said that composting was included as a recycling method, but it was initially not clear whether incineration with energy recovery was treated as a recycling method. Later it was agreed that incineration with energy recovery should not be treated as 'recycling', at least for the purpose of the target (ENDS Report 1992: 14). If one compares this approach with the Dutch and the German objectives, it is significant that neither explicit plastic waste prevention targets were formulated as in the Dutch case, nor mandatory recycling goals as in Germany, nor specific packaging recycling goals as in both countries (see chapter 3 for the comparison).

The relatively pro-environment strategy of the Minister of the Environment was constrained by the distribution of responsibilities within the government. He could not initiate measures directly aimed at steering industry's recycling activities, because this task belonged to the domain of the DTI. The Ministry, therefore, had to stimulate industry indirectly through its competencies in regard to local authorities. Accordingly the White Paper stated that the government would work together with local authorities to assess the effectiveness of experimental recycling projects and to encourage more recycling banks and further action to stimulate recycling, including recycling plans and recycling credits. The legal framework for this policy was to be provided by the Environmental Protection Act. This Act was adopted in 1990 and replaced the Control of Pollution Act (1974). The regulatory change and the new administrative practice will be discussed below, since it is a relevant part of the British packaging waste policy.

Given the general pro-business stance of government and the lack of competencies in this area for the DoE, the challenge to the industry was rather weak. The White Paper stated that the government would work out a scheme for labelling recycled and recyclable products and would look for other ways of encouraging companies to extract recyclable products from domestic waste and recycle them (HM Government 1990: 191). With regard to packaging, the government announced that it would encourage industry to reduce unnecessary packaging and to discuss with industry targets for reduction of packaging and similar measures (HM Government 1990: 189). Government and industry interaction and the resulting industries' codes of practise will be discussed in more detail, after the national recycling policy in regard to the local authorities has been sketched.

6.3.3 Local authorities' recycling plans and recycling credits

The government's waste policy in regard to the local authorities was based on the Environmental Protection Act (EPA, 1990). The EPA replaced and extended the provisions of the Control of Pollution Act (1974), on waste collection and disposal. Since the local authorities were legally in charge of these tasks, the EPA provided a new legal framework for their activities. The Act established a new waste management infrastructure for household waste. It contained provisions that either directly or indirectly affected the treatment of used packaging. Among them were the duty of local authorities to submit recycling plans, a system of recycling credits, and the tightening of waste disposal standards. One reason for the tightening of waste disposal standards was, according to the White Paper, '*to make waste disposal much more expensive*" in order to "*provide a strong incentive for industry to cut down the volume of waste it produces*'(HM Government 1990: 187).

Recycling Plans

The Environmental Protection Act obliged the Waste Collection Authorities to prepare draft recycling plans, to be submitted to the Minister by 1992. The plans had to specify systems for the collection and processing of recyclable material, identify markets for the material collected and set local targets for collection. Local authorities were, however, not actually required to meet any commitments made in the plan. The implementation of the plan was not obligatory (see Stupples 1993: 60). A DoE Waste Management Paper (No. 28) set out guidelines for WCA's on the preparation and the contents of the plans. The plans had to be set up to achieve the government's target for household waste recycling of 25 per cent. This was rather ambitious. The Economist commented: *'When the Government published its white paper on the environment last autumn, it set a bold target: to recycle a quarter of all household waste by the end of the century. That goal is far off: hardly any English local authorities, which empty the nation's rubbish bins, recycle more than 2 per cent of the contents'(*The Economist 6.4.1991*).*

Accordingly, a first national survey of council recycling plans by Friends of the Earth have shown that only a third, i.e. 35.9 per cent, of councils said that they were planning to meet or exceed the government's target by the end of the century. Most local authorities have been unwilling to commit themselves to the target because they lack the money needed to invest in recycling schemes and because of the lack of markets for recycled materials. Even those authorities which are '*going for it*' said they could not achieve 25 per cent without government action to tip the balance in favour of recycling (Stupples 1993: 60).

Recycling credits and supplementary credit approvals

The Environmental Protection Act also established a recycling credit scheme which placed a duty on waste disposal authorities to pay credits to waste collection authorities or other bodies for material which was diverted from the household waste stream for recycling. These credits represent the savings made in disposal costs by the WDA if the local authorities send waste for recycling instead of disposal. The value of the credit is dependent on the avoided cost of landfilling the waste. Besides this credit scheme for recycling, a waste collection credit scheme was established, which paid anybody who recycled waste before it reached the rubbish bin. The payments between different kinds of local authority were compulsory; payments by local authorities to other bodies were left to the discretion of the local authorities (The Economist 6.4.1991). In addition to the recycling and collection credits, a scheme of supplementary credit approvals was introduced in England that was to support the establishment of recycling infrastructure. One of the aims of the government was to encourage public-private-partnerships, whereby supplementary credit approvals were combined with the authorities' own capital receipts and private sector contributions. An often cited example for such a partnership took place in the local community Adur. It was a pilot project scheme for household waste recycling (Haigh and Mullard 1993: D3, Institute of Grocery Distribution 1992). The main factor that constrained the effectiveness of the recycling credits were the low disposal costs (Stupples 1993: 61). The recycling credits were too small to outweigh the financial advantages of disposal.

Tightening waste disposal standards
The government realized some of their promises made as a reaction to the critical RCEP report on waste management (see above). The EPA increased the waste disposal standards through a duty of care on all those who handled wastes. The EPA delegated local authorities' waste disposal operations to "arms' length" companies, controlled by the newly created local Waste Regulation Authorities (WRA, i.e. County Councils in England). As a result waste disposal operators were no longer responsible for checking their own standards. Moreover, applicants for a license would have to prove that they were fit and proper persons to qualify as an operator. The EPA also required restoration and regular inspection of those landfill sites where tipping was stopped. The provision had to be implemented by 1 April 1993. As already said, this policy was seen by the government as an indirect economic instrument to stimulate recycling. The Minister of the Environment Patten explained this at a conference shortly before the White Paper was issued. He said there was a tendency to regard waste disposal as cost free, partly because *'diffuse and inconsistent systems have served to obscure actual costs'*. Separating their disposal operations into free-standing companies would *'lead to significant increases in the cost of disposal to landfill, providing an increased incentive to recycle'* (ENDS Report, March 1990: 11).

6.3.4 The government's challenge to industry

The government had not only created a framework for local authorities to stimulate their recycling activities, it had also asked manufacturers and retailers to take on their share of the responsibility for what ultimately happens to the products which they place on the market. In its White Paper the government stated that it is encouraging industry to reduce unnecessary packaging and that the government would therefore discuss with industry and retailers proposals for *'targets for packaging reduction; measures to reduce packaging through, for example, environmental auditing; and arrangements enabling consumers to discard or return packaging at the point of sale'* (1990: 189). The government also said that it would encourage improvements in packaging design which would minimize waste and promote recycling or re-use. The request to trade and industry was accompanied with a threat *'While the Government intends to proceed by voluntary means, it will if necessary consider the introduction of regulatory measures, such as deposit schemes, and will review measures applied in other countries'* (1990: 189). The rationale behind this challenge was *'the public pressure, the general feeling that there was too much overpackaging. There were pro-environmental companies like the Body Shop who were making those points very clear'* (Interview DoE 12.7.1996).

The tone of the White Paper was rather modest. Shortly before its publication, the Environment Minister Patten had chosen a more aggressive tone. He had called for a *'an all-out attack on excessive packaging with effort put into genuine reductions rather then defending the status quo'*. He pointed out that other countries had responded to the problem of packaging and the environment by banning one-trip containers and imposing mandatory deposits and packaging taxes, and these would be assessed with an eye to the environment White Paper (ENDS Report March 1990: 11).

The industry's reaction to the government's challenge was rather modest. The British packaging chain organization INCPEN was a strong opponent of Government interventions. In line with similar organizations in Germany, in the Netherlands and on the European level, it argued that the packaging waste problem was generally overestimated (INCPEN factsheet 11/1995: 2, see also chapter 2). The arguments were, however, more radical than those of its German and Dutch counterparts. INCPEN claimed, for instance, that '*packaging saves far more resources than it uses by preventing products from being damaged or spoiled.*'(INCPEN 1995: 20). INCPEN even claimed that the packaging systems of unrestricted markets '*use less material than those in countries where certain types of packaging have been discriminated against*' (INCPEN factsheet 11/1995: 4).

Based on the above mentioned arguments, trade and industry were reluctant to respond to the government's initiative. This caused the Government to repeat its requests several times. In 1992, the Government showed that it was losing patience with the packaging industry. In an address to INCPEN on 22 January, the junior Environment Minister Tony Baldry recalled the White Paper's promise to discuss a series of measures with packaging manufacturers and retailers and he added: '*We expected that the packaging industry, its suppliers and customers would prefer to take the lead in solving the problem of reducing the amount of packaging waste which goes to final disposal, rather than being told what to do by the Government. The Government had gone to great lengths to give industry its chance but its response had proved disappointing...We still have no clear indication that responsibility for the impact of packaging waste has been accepted by industry.*' (quoted in ENDS Report Jan. 1992: 11). In contrast to Germany and the Netherlands, however, there was no close and frequent interaction between government and representatives of industry. Government officials only met industrialists occasionally in this matter. Nothing comparable with the Dutch strategic discussion, or the Dutch and German bargaining between industry and government, occurred in the United Kingdom.

Parallel to the challenge to industry, the government also commissioned studies on several economic instruments. This reflected the general British approach emphasizing the need for scientific proof and sophisticated cost-benefit analysis before policies are initiated. However, the scientific reports had not only improved the knowledge of government about the probable effects and efficiency of economic instruments, but served also as an additional threat to industry, showing that government was ready to take action. It is somehow ironic that a major impetus for the research came from an industry organization, namely the Advisory Committee on Business and the Environment. In November 1991, this organization recommended that economic instruments should be introduced to promote recycling of waste and to establish deposit refund systems (Haigh and Mullard 1993: D2). Earlier the Advisory Committee had already argued that the British waste disposal costs were too low (Stupples 1993: 60).

One study commissioned by the DTI and prepared by Environmental Resources Ltd dealt with deposit and refund systems. It concluded that a deposit of 5p per container should ensure a rate of return to retailers of 95 per cent compared with 30 per cent for a collection system without refunds. More than one million tonnes of plastic, metal and glass would be diverted from the waste stream. This would be 6 per cent of total household waste and would contribute a quarter to the government's recycling target. The costs were estimated at 2 pence per container (ENDS Report Jan. 1992: 12).

The repeated attempts of the government to encourage industry to come up with its own policy initiatives finally resulted in a number of initiatives, which were, however, rather modest and rather vague. The three most important will be discussed below.

6.3.5 COPAC's Business Plan

The most substantial reaction of industry to the challenge from the government was the business plan of the Consortium of the Packaging Chain (COPAC). COPAC was established for this precise purpose by five packaging manufacturer associations, as well as the Food and Drink Federation and the British Retail Consortium. In its business plan the Consortium proposed a specific target for packaging. COPAC aimed at diverting 50 per cent of all used packaging from landfill, hence not only used packaging from household sites, but also from industrial and commercial sites. The Consortium said that dealing with used packaging in household waste alone would not be enough for achieving the target. The Consortium believed that higher targets could not *'at this time, be shown to be achievable, economically supportable, or environmentally desirable'* (COPAC Action Plan quoted in Haigh and Mullard 1993: D9). COPAC claimed that the lack of end markets limited the targets which could be set for the recycling of packaging materials. The 50 per cent diversion target was set only under the condition, however, that appropriate collection and sorting facilities were in place. COPAC prepared a forecast of the likely diversion rates, provided that adequate collection and sorting facilities were in place. The plan envisaged the recycling of 3.6 million tonnes out of a total of 8.5 million tonnes of household and commercial packaging to be generated in 1999. In support of the plan, the packaging manufacturers and suppliers forecasted making an average capital investment of 50 million pounds per annum up to 1999 on recycling and recovery operations.

Table 19 COPAC's forecast of UK diversion rates for packaging

Year	Household	Industrial and Commercial	Total
1987	1	35	20
1991	11	43	27
1995	19	45	32
1999	33	51	42

Source: COPAC Action Plan to Address UK integrated Solid Waste Management, 26 October 1992 (percentage of overall packaging waste in the UK).

The COPAC plan had a number of shortcomings. First, the targets and the business plan were conditional on action of other actors, in particular the government. The plan depended on an increase in the collection of source separated materials from households and a number of high density bank systems to cover half of the UK households by 1999. The Consortium also believed that further commitment was needed by government to overcome such problems as countering the impact from other countries of subsidized materials, in addressing current planning and licensing procedures, and encouraging the development of energy-from-waste systems.

Second, the business plan failed to make an equitable contribution to the *household* waste *recycling* target set by government (ENDS Report Oct. 1992:1). If one argues that more or less all packaging materials are recyclable, the figures reveal that only glass and metals (51 per cent resp. 45 per cent) came close to meeting the government's targets of 50 per cent of recyclable household waste, while the total of 33 per cent, clearly fell below government's target. Strangely enough, the COPAC plan was based on diversion and not on recycling. This implied the assumption that 25 per cent per cent target could be achieved not only by material recycling and composting but also by energy recovery. This was a misinterpretation of the government's target, however. It was clear that incineration with energy recovery would not be considered as recycling.

Third, the plan lacked incentives, underestimated the costs, and said nothing about the distribution of costs. An industrialist from a major packer and filler and in a prominent position at INCPEN questioned in an interview the seriousness of the cost assessments. Further, he continued, *'a voluntary organization, COPAC, took a voluntary look at how it thought recovery and recycling would develop in the UK. There were no stimuli in there, other than were identified as natural market sources. Nothing dramatic was done to stimulate the market. It produced a report with some numbers in it. And this was very much driven by the packaging manufacturers. It was not driven by the retailers or by the packers and fillers'* (Interview Industrialist 24.6.1996).

Moreover, it was argued, the plan did not provide an adequate answer to developments in other Member States and on the European Community level. To quote the industrialist again: *'This report was produced around the same time the German ordinance came in. When the German ordinance came into being, it was obvious that some sort of voluntary natural solution would not be sufficient. Something more stimulating was needed'* (Interview Industrialist 24.6.1996).

The DTI objected in particular to the voluntary character of the plan (Interview DTI 15.5. 1996). The government was under pressure from two quarters. It needed to ensure progress to recycle half of recyclable household waste by the year 2000. And it needed to have a defensible position for the forthcoming negotiations on the EC Directive on packaging waste. This will be discussed in chapter seven and eight.

6.3.6 Codes of Practice

The Packaging Standards Council

Besides the COPAC business plan, industry came up with two other main initiatives, one being the Packaging Standards Council (formerly the Packaging Council, PSC), which was set up in June 1992. This was an initiative of INCPEN and was mainly financed by it. The Council was to handle complaints about industry's behaviour in terms of living up to a one-page Code of Practice. The code, which was produced by the Council, set out general guidelines for good packaging practice in the industry and covered legal, environmental, social and economic factors. The code required all packaging to be designed with *'due regard to its reuse and/or recycling, and to its ultimate safe disposal in accord with relevant waste management targets'*(quoted in Haigh and Mullard 1993: D7). It also stipulated that the packaging system should use *'the minimum amount of energy and raw material consistent with the requirements of protection, distribution, preservation and presentation of the product, and should minimize emissions'.* The Council was primarily concerned with consumer packaging rather than secondary and transport packaging. It was chaired by Lord Clinton-Davis, a former EC Environment Commissioner. It was a small, not well resourced and rather unknown organization. It received not more than 30 to 50 complaints per year. INCPEN said that the PSC *'provided a unique channel for consumers to encourage industry to improve packaging and to get impartial information about packaging. In the last three years it has demonstrated that it can work to the benefit of both consumers and industry'* (INCPEN factsheet 10/95: 2). Friends of the Earth responded to the setting up of the PSC by criticizing the code as '*unenforceable*'. The ENDS Report said that '*the one-page code certainly cannot be said to be draconian'* (ENDS Report 203, June 1992: 13).

The Packaging Standard Council was forced to stop functioning in 1996. Although it only needed a funding of 75,000 pounds a year, it ran out of money (PSC press release 3.6.1996). Some weeks earlier, INCPEN, the main source of money for the Council, wrote in a letter that it '*believe(s) that it has a vital role to play both for consumers and industry'* and that INCPEN *'is extremely keen to see the PSC continue'* but INCPEN regretted that *'we cannot commit any more funds this year beyond what we budgeted for'* and therefore *'we think it regrettably prudent that the PSC and INCPEN prepare to close down the PSC'* (INCPEN 1996 Letter to PSC). In one of the last meetings of the PSC, which industry members had bothered to attend, members complained about those sectors of industry who had benefited from the good PR without putting anything in it. The consumer representative was disappointed about the *'generally bad job industry had done in explaining the role of packaging to consumers'* (PSC Notes on the Meeting 2.5.1996).

The Guidance Notes of the British Retail Consortium

In May 1992 the British Retail Consortium (BRC) issued guidelines to help retailers and their suppliers develop policies on packaging. One year later an updated version was published (BRC 1993). The notes had been prepared by the BRC Environment Committee *'after full consultation with the Food and Drink Federation to offer guidance to retailers and their* *suppliers in determining their own company policies on packaging with the aim of achieving the optimum balance between best environmental practice and these vital functions (*safety of consumers, M.H.)'(BRC 1993: 1). The guidance notes stressed the importance of life cycle analysis and urged industry to develop *'realistic but rapid'* targets to achieve environmental objectives which include reduction, re-use and recycling of packaging materials (BRC 1993: 2). The document emphasized that landfill was not seen as a primary route for packaging disposal. Incineration with energy recovery was the preferred alternative when more environmentally appropriate than reduction, re-use or recycling. Under the heading '*non-discrimination'* the document stipulated that *'these Notes do not discriminate between alternative packaging materials or systems unless impact assessment carried out by accepted methodology shows that environmental benefits exist'*(BRC 1993: 2).

The guidelines recommended the use of marking and labelling systems to differentiate between logos for products containing recycled materials and products capable of being recycled. Guidance was given concerning product safety, marking and labelling, volume reduction, outer packaging, outer wrapping disposed of at store level, customer pack design, plastic degradability and general considerations. The general idea of guidance notes was to suggest measures avoiding certain packaging, for example, packaging made from multi-layer laminate materials; to use recycled materials; or to re-use packaging whenever it had environmental benefits and was not hindered by a number of other aspects like practicability, hygiene, safety, and product performance.

The Guidance Notes, in particular in the updated version of 1993, were more concrete than the code of practice initiated by INCPEN. The Guidance Notes could help to reduce the adverse impacts of packaging. They strongly discouraged the use of multi-material packs and laminates, for example (ENDS Report June 1992: 13). The rather concrete character of the Guidance Note can be explained by the fact that INCPEN, the driving force after the PSC, had to take into account the interests of the packaging manufacturers more seriously than the British Retail Consortium. In addition, the context of the British national packaging waste policy increasingly changed in the period 1992 to 1993. The EU Commission had come up with its proposal for a European Directive, and the unintended consequences of the policies of other countries became more visible and painful to Britain (see chapters 8 and 9). Because of these external developments, the need for government intervention grew and *'through issuance of the guidance industry hoped to avoid regulatory action by the Government'* (Haigh and Mullard 1993: D9). The effectiveness of the guidance depended, however, on the efforts taken by the retailers, since the BRC itself was rather weak. The retailers were the strongest actors in the field. '*Five or six of them do control 75 per cent of the market. They do more or less dictate the behaviour of the suppliers. They are not used to do things themselves'* (Interview Specialized Journalist 23.7.1996).

The general reaction of industry to the government's challenge was rather modest. This can partly be explained by the very character and function of British trade associations. British trade associations are generally less prepared to become involved in corporatist agreements than are their Dutch and German counterparts. To quote an industrialist who then held a prominent position at INCPEN: *'Trade associations are servants of their sector, usually held together to defend specific interests and specific issues. It is very difficult for trade associations to turn around and tell their constituting companies that they have something to do that these companies do not particularly want to do. Trade associations themselves tend to be providing a resource on behalf of the companies to look after their interests. Not to promote political activities in a voluntary environment' (Interview June 1996).* And an official at the DoE explained: '*Essentially, trade associations are very weak bodies. They do not have resources. If you want to get things done, you need intellectual heavyweights. People with a political sense. Trade associations are bureaucratic, more like civil servants. You need people who operate in the real world. People who are prepared to lead'* (Interview DoE 1996).

The modest market-led approach of the United Kingdom which resulted in rather low recycling activities came under increasing pressure from initiatives at the European level and unintended consequences of the policies of other Member States. How the UK reacted to these challenges will be discussed after the analysis of the policy process and the content of the European Directive on Packaging and Packaging Waste.

6.4 Conclusion

As in the German and the Dutch cases nationally specific institutions to some extent shaped the development of packaging waste policy. The majority voting system blocked the access of a green party to parliament. The administrative arena was also relatively closed to critical environmentalists. An important competence related to packaging policy, namely recycling, was allocated to the Department of Trade and Industry, which had the function of promoting British trade and industry (Hennessy 1990: 432) rather than advocating environmental protection. This ministry had almost no ties with environmental groups. Moreover, it was not likely that the generally- trained civil servants moving between different tasks within the broad Ministry of the Environment would develop and sustain a pro-environment attitude. More generally, the British institutional system insulated the government from societal pressure as long as it had a majority in Parliament. There were no institutional channels by which critics of the government's *laissez-faire* approach, such as environmental groups, but also local authorities, could effectively challenge Whitehall. There was neither a *Bundesrat* nor a developed system of judicial review as in Germany. Moreover, the financial dependence on the central government frustrated the development of local recycling policies.

Within the institutional setting of a central state, the ideas of central government mattered most. The perceptions of the government and the British political debate in general were framed - more than the German and Dutch ones - framed by economic growth *versus* environmental protection. This made business-related environmental arguments weak. This was the case in particular during the economic recession in the 1980s. The comparatively low priority of the precautionary principle as compared to the principle of sound empirical evidence and the need of cost-benefit analysis, was one of the reasons why the British government did not adopt a clear hierarchy of waste management options and quantified minimum and/or maximum targets for these options.

The government's pro-business, anti-regulation and small public sector ideas affected the packaging waste policy in different ways. The government reduced public expenditure for implementation tasks. This made landfill a cheap waste management option and reduced the incentives for recycling. Moreover, the government did not put enough money into the recycling credit scheme to outweigh the costs of this option as compared to landfill. The pro-business and anti-regulation stance made the government's threat to impose packaging regulations on business not very credible. It was this lack of a credible threat which resulted, in combination with the weakness of trade associations, in the minimal response of the business community to the government's challenge. The following chapter will provide a more systematic and deeper understanding of the factors and mechanisms that shaped the variations in national packaging waste policies.

7. Explaining variations: ideas, interests, institutions

In the late 1980s and early 1990s, Member States differed in their policies towards the problem of packaging waste. The differences between Germany, the Netherlands and the United Kingdom are striking. The German Packaging Ordinance, based on public law, imposed quite ambitious recycling targets and a strict timetable on industry. Moreover, it introduced take-back requirements for used packaging on the retailers. The private sector was forced to set up its own collection and recycling system (DSD), in order to avoid this requirement. Rigid sanctions could be adopted if the DSD were to fail. In the Netherlands, the government and large parts of the private sector agreed on a bundle of measures in order to meet a waste prevention target and fairly high recycling targets. The negotiated agreement was binding under private law. The United Kingdom favoured a strictly voluntary and market-led approach. Central government required local authorities to develop recycling plans and set up small recycling credit schemes. The government set only a general recycling target for municipal waste which was not legally binding. Industry developed a business plan and codes of practice which were not binding and therefore not deemed very effective.

In chapter three the differences and similarities between the national approaches have been summarized. The national histories which led to the different approaches were outlined in chapters four to six. Important institutions and ideas that mediated the perception of external conditions and events, and which shaped the interests of actors have been identified. The chapters became increasingly comparative. While the German case was discussed more or less from a single-country perspective, the Dutch developments were compared at crucial points with the German ones. The British chapter entailed cross-references to both other countries, in order to highlight the distinct developments there. In this chapter the three cases will be compared systematically. The variations in national policy approaches on the five policy dimensions will be examined and an attempt will be made to explain them.

It is striking how differently the problem of packaging waste was perceived by policymakers and the public in the three countries under study. This difference is crucial in explaining the different policy approaches, but it affected a range of policy elements. The factors and mechanisms which shaped the national problem perception and accounted for variation will therefore be analysed in advance. The main part of the chapter is organized according to the different policy elements. The differences and similarities in national policies will be explained in terms of ideas, interests, institutions and external conditions. In addition, a number of characteristics of the packaging waste issue will be included, because these specific features of the problem shaped the policy approach to some extent. The chapter will also allude to more general historical and cultural aspects of the respective countries. This is not done in order to suggest a direct causal link between these general features and the specific characteristics of the packaging waste policies developed. However, ideas, interests and institutions are embedded in this context. Looking at these aspects sheds light on the normative and historical underpinnings of ideas, interests and institutions, and therefore helps to provide a deeper understanding of the various paths on which packaging waste policies have developed.

7.1 Problem perception and agenda setting

Policy problems do not exist as such. The nature and the extent of problems depend on the perception of actors and the way people communicate about the problem (Hajer 1995 and Luhmann 1988 respectively). In the early 1990's the German public was afraid of a waste avalanche. Pictures of mountains of waste covering cities were on the front pages of newspapers and journals. In the Netherlands, the waste issue also reached a prominent place on the political agenda, while in the UK it was not a salient issue at all. The reason for this lies neither in the amount of waste produced, nor the level of economic and technological development, since these are roughly similar in these countries.

One explanation possible could be differences in the geophysical conditions which make landfill more or less costly. A landfill site is typically an old excavation site where gravel, and/or cement have been extracted from the earth. In order to prevent the leaching of waste into the ground water, the landfill site has to be lined with an impermeable layer of, for instance, clay (Key Note Market Review 1993b: 3). The United Kingdom has slightly better conditions than Germany and the Netherlands in this respect. It has many old mining sites. '*The bricks from which the Midlands and the south-east of England are built come from clay pits that make impermeable rubbish dumps. The Netherlands, most of it lying at or below sea-level, find it hard to dig holes for waste*' (The Economist 29.5.93). But the differences in geophysical conditions are not very large. In the Netherlands, landfill sites can benefit from salt layers which are very good protectors against leaching. Former mining sites exist not only in the UK but also in the East and South of the Netherlands and in many places in Germany. Price differences of six hundred per cent in disposal costs (Key Note Market Review 1993b: 71), as they existed between Germany and the United Kingdom in the early 1990s, cannot therefore be explained solely by reference to geophysical factors.

The most important reason for the differences in price of landfill lies in the strictness of standards for the development, operation and aftercare of waste disposal sites, the transport of waste and the qualification of waste operators. Generally speaking the strictness of standards and the rigor of enforcement are higher in Germany than in the UK. Hence, both geophysical and policy factors shaped the capacity and the price for landfill. In addition, though implementation deficits were also recognized in Germany, the general adherence of the

German civil servant to an abstract concept of the public interest, prevents him from being as lax with control and enforcement as his Dutch and British counterparts with their greater understanding for the interests and constraints of the target groups involved. Moreover, the implementation of the German 1972 Waste Act preceded the implementation of the British 1974 Control of Pollution Act and the Dutch 1977 Waste Substances Act. Finally, due to local protests it was increasingly difficult to develop new capacities. And the (legal) processes involved in planning a new site also increased the costs. Variations in these largely political constructions of scarcity account for differences in problem perception between the United Kingdom on the one side, and the Netherlands and Germany on the other side, and made alternatives other than landfill more necessary and attractive in the latter countries.

Waste incineration as one alternative, however, was seen, in particular in Germany, as a bad solution, due to the relatively high costs, again partly produced by strict standards and cumbersome planning procedures, and a high sensitivity to the problem of hazardous substances such as dioxins. Hence the high level of environmental regulations made not only landfill but also waste incineration expensive alternatives. This created an urgent waste problem in Germany.

Hence, geophysical conditions, the level of environmental norms and standards, the strictness of their enforcement, and the lack of public support contributed to the costs and the available capacities, and pushed waste reduction higher on the agenda in Germany and the Netherlands than in the UK. How can this high level of environmental regulation in the Netherlands and Germany be explained? Surveys quoted in the nationally-oriented chapters indicate that the politically relevant environmentalism was stronger in the Netherlands and Germany than in the United Kingdom. This environmentalism was nurtured and supported by environmental advocates in the legislative and the executive arenas in the Netherlands and Germany. In contrast to the UK, in Germany and the Netherlands determined advocates for strict environmental policies had emerged. Civil servants as policy entrepreneurs developed the programmatic and organizational basis for strict environmental policies in the early 1970s. The institutional factors which either facilitated or hindered the emergence of the advocates have been discussed in the respective chapters. In the Netherlands and Germany, these pro-environment interests of the executive were supported by environmental groups, stemming from the middle class, which criticized the prevailing consumption and production patterns. The relative political power of these environmental ideas was shaped by political institutions. Facilitated by the proportional voting system, new environmental ideas entered the Dutch parliament already in the late 1960s and early 1970s. Germany followed a decade later. The British majority voting system blocked access of a green party to parliament.

The German and Dutch pro-environment interests remained strong in the 1980s despite economic problems. One of the reasons for this was the policy paradigm of ecological modernization, which was the idea that economic growth and environmental protection could be reconciled. Pro-environment advocates argued relatively successfully that environmental protection could create economic growth and employment. They were not only supported by environmentalists, but also by parts of industry. "Green" industries, such as waste management firms, benefited from high environmental norms and standards. They became an economically important sector in Germany, due to the strictness of standards based on technology pushing principles such as '*best available technology*'.

The factors mentioned above help to understand why landfill and incineration became costly and unpopular and the need for more prevention, re-use and recycling emerged on the political agenda. It does not explain, however, why so much attention was devoted to packaging waste, which made up less than 5 per cent of the overall waste stream. The negative symbolic function of packaging and the visibility of used packaging to the consumers made it politically attractive to the environmental ministries concerned with the development of packaging waste policies. For the Dutch Minister Alders, the German Minister Töpfer, and, to a lesser extent, the British Minister Patten, packaging provided a window of opportunity to implement the idea of producer responsibility, and to burden trade and industry with responsibility for products in the phases after consumption. Packaging waste was seen by the pro-environment advocates as a pioneer case for policies in other areas such as cars, electronic goods, white goods (e.g. refrigerators) or batteries.

7.2 Formalization

The three countries differ in the degree of legal codification of their policy programmes. While Germany adopted an ordinance, the Dutch government came to agreement with large parts of the packaging chain on a covenant which was binding under private law. The British government did not introduce any measures that were binding for any part of the trade and/or industry. The **German** approach very much reflected the traditional legalistic approach that guides most of German public policymaking. The specific characteristics of German historical development have provided Germany with a distinctive normative underpinning to for its regulatory policies and processes. Dyson argues that nowhere does the German state formation manifest its significance more obviously than in the strength and distinctiveness of the German legal culture. Central to this culture is the idea of the *Rechtsstaat,* by which public action should take place through lawful procedures and with explicit justifications in terms of legal principles (Dyson 1992: 9, Weale 1992b: 74). The pivotal role of law dates back to the early nineteenth century and the reforms that followed the catastrophic defeats of the Napoleonic wars. Essentially, the concept of the *Rechtsstaat* served as a principal legitimization of the autonomy of Prussia's public bureaucracy (Dyson 1992: 12).

Later in the nineteenth century, Germany sought to achieve its aim of controlling state power by creating special administrative courts to develop and apply a body of administrative law (Blankenburg 1997:44). After the abuses of power in the Third Reich, the Rechtsstaat was seen in the post-war period as a political and normative concept, spelling out the fundamental principles of collective life. The post-war state was adapted to living with the legacy of the Third Reich '*by embracing a markedly activist conception of the central role of law in social, economic and political life*' (Dyson 1992: 9).

In other words: '*German attitudes towards regulation appear notably supportive. Regulatory action enjoys a high degree of legitimacy*' (Dyson 1992: 1). '*Regulations reflect the importance attached to such values as predictability, reliability and 'public regarding' behaviour*' (Dyson 1992: 10). Procedures matter and tend to be elaborated by, for instance, ready recourse to administrative law. Law also involves constitutional rights and the possibility for judicial review. Law is internalized within discrete policy sectors. It is built into the working world of policy. '*This formal, legalistic approach to policy is central to the regulatory culture of Germany*' (Dyson 1992: 9-10).

The general legalistic tradition is echoed in the case of environmental policy. Environmental protection has its origin in the regulation of public health aspects of production, which was traditionally based on licensing. The '*juridification of politics*' (Weale 1992a: 160) was reinforced by the fact that environmental protection was, for a long time, located at the Ministry of Internal Affairs, where even more civil servants were lawyers than the already high average in German ministries. Moreover, all Minister Töpfer's predecessors had a legal background. It is typical for the German approach that the 1977 voluntary agreement on drink containers was not perceived by the civil servants as superior to statutory measures, but a second-best solution imposed by powerful countervailing political and economic forces of the time.

The choice for a generally binding ordinance was shaped not only by the German legalistic culture, but also by negative lessons drawn from the failure of the 1977 voluntary agreement on drink containers and the positive lessons from the 1988 Ordinance on plastic bottles. The government learned that trade and industry could not be committed by voluntary agreements to control the generation of waste from used drinking bottles. Note, however, that the Dutch voluntary agreements also did not reduce the share of drink containers in household waste bin but it was acknowledged that no flood of one-way packaging occurred. Accordingly, the Netherlands mantained a more positive approach to voluntary agreements. This difference indicates that one can draw different lessons from similar experience. In Germany, the glass was half empty. In the Netherlands it was half full.

The issue of packaging and the environment was as prominent on the political agenda in the **Netherlands** as it was in Germany. But in contrast to Germany, the Netherlands did not choose a statutory generally binding agreement but rather a covenant, i.e. an agreement binding under private law for those who signed. Why did the Dutch government conclude a voluntary agreement? And, related to that and equally important, why was the covenant signed by a large part of the important companies of the packaging chain? The first reason is that the Netherlands has a less legalistic tradition than Germany. This is not to say that the Netherlands avoid making strict and detailed laws. In fact, as already noted, each ministry has its own regulatory culture. And a rather legalistic culture was typical for the Ministry of Public Health and Environmental Hygiene. However, the implementation deficits indicated that direct regulation did not fit with the more salient elements of Dutch legal culture which emphasizes pragmatism. As in Germany, the Netherlands has codified law, but its usage is more flexible and informal as typified by the Dutch opportunity principle. This allows for toleration of a lesser evil, e.g. soft drugs, in order to fight a greater evil, e.g. hard drugs, more effectively (Blankenburg and Bruinsma 1994: 64-66, van Waarden 1995: 342). Moreover, the Dutch system of administrative law and judicial review developed rather late and is not as elaborate as in Germany (Blankenburg and Bruinsma 1994: 17-20).

Against the background of a pragmatic legal tradition, it is not surprising that the Dutch government employed the instrument of covenants more often and more widely than Germany used functional equivalents. The Dutch government reasoned that covenants are an ideal instrument to implant the idea of internalization of environmental concerns into the daily practice of target groups. The target group would feel more committed to achieving the targets, if it had signed the document. This idea was introduced by the liberal Minister of the Environment, Winsemius, who came from a business consultancy firm. It was elaborated by his liberal successor Nijpels and - though hesitantly - accepted by the social democrat, Alders. The idea of internalization was a reaction to the perceived lack of effectiveness of the command and control approach. It is typical for the less legalistic Dutch culture, that the Dutch government weighed effectiveness criteria more heavily than the legal weaknesses of the instrument, such as the limited legal protection of third parties.

The belief in the effectiveness of a covenant was based on a trust in the commitment of the packaging chain. This trust was justified for a number of reasons. The same reasons explain why the SVM, its affiliates, as well as a number of other companies, signed the covenant. First, the government's threat to impose unilateral action was widely feared in the business community. Environmental and consumer interest groups as well as members of parliament put pressure on the political process demanding effective reduction of waste in particular by increasing re-use of packaging. The instrument of consumer boycott was generally recognized and feared by industry. Interventionist measures were on the political agenda such as mandatory deposit charges and take-back responsibility. This discussion was very much informed by the German debate which was followed very closely by all actors involved. In addition, the Minister of the Environment, Alders made packaging waste a top priority and it was known that he initially had a predilection for legal instruments rather than agreements with target groups. Industry was not sure whether he would continue the approach developed by his liberal predecessors (see Aalders 1993: 75). Besides, the government's willingness to follow a more impositional route, the skills of the SVM made the effectiveness of the Covenant more likely. The rules employed by the SVM to accommodate interests, are embedded in the rules of the political game which date back at least to the period of strong pillarization and which are associated with the politics of accommodation. This includes a) the toleration of differences in interests, b) no adoption of decisions by simply overruling a minority, c) the more important the problem, the higher the level of decision-making, d) proportional representation of interests and distribution of costs and benefits, at least in the long term and e) depoliticization of sensitive problems by 'objective' economic, technological or other principles (Lijphart 1976: 128-146, van Praag 1993, quoted in van Deth and Vis 1995: 44). The SVM tried to make all processes and decisions as transparent as possible. This was in contrast to the practice of policymaking in the period of pillarization when secrecy was a rule of the game. The government's belief in the commitment of industry was also informed by the general Dutch industrial culture which is characterized by a cooperative and pro-active attitude of industry. This point will be developed further below.

The adoption of non-binding local authority recycling plans and voluntary business plans and codes of practice in **Britain** is not surprising given the strong deregulation stance of the British conservative government and the comparatively low political silence of the packaging issue in the UK. The government's attitude is imbedded in legal culture in which written statutory law plays a less important role in political life than in Germany or the Netherlands. Britain is less legalistic. Vogel (1986), for example, characterizes the regulatory culture in Britain as being generally informal and voluntary. When laws are enacted they are usually very broad, providing a framework for the discretionary authority of the implementers. And civil servants are rather reluctant to bring firms before the court in case of non-compliance. The reliance on informal measures is related to the general constitutional framework. *'Surely it is more than coincidence that self-regulation is widespread in a state whose non-codified constitution is heavily reliant on unwritten rules and conventions. It seems reasonable to suggest that the general reliance upon convention can be seen as manifesting itself in a more specific manner in the policymaker's preference for voluntary agreements and informal self-regulatory systems'* (Baggott 1989: 443). Accordingly, higher civil servants, most of them coming from Cambridge or Oxford, usually have no legal background.

Feick (1992) explains the non-legalistic legal tradition as follows. *'A high degree of legal formalization almost necessarily leads to debate in parliament, lifting even technical issues of detail to a highly politicized level. But in Britain the parliamentary level ... is reserved for the more general and fundamental discussions. Additionally, the executive is supposed to govern and manage, giving it quite some leeway in the details of policymaking'* (Feick 1992: 78).

This comparatively low significance of statutory law was reinforced by the ideology of the Thatcher government. One should not confuse rhetoric and reality, however. British conservative governments were, as all other governments, prepared to enact legislation to further their political objectives, such as the reregulation associated with the privatization of industry (Dunleavy 1989). The absence of legally codified measures pointing at business had, therefore, very much to do with the comparatively low profile of environmental issues in the UK. The reasons for this low profile have been elaborated on in chapter six and summarized in section 7.1.

7.3 Strictness of standards

The comparison between the three countries revealed variations in four dimensions of packaging waste standards: a) waste management objectives, for which targets were set, i.e. reduction, re-use, recycling, incineration; b) the kind of targets, i.e. global or specific, statutory or voluntary; c) the level of targets; e.g. high percentages or low percentages; and d) the time scales to achieve the targets. Taking the values along these four dimensions together, Germany set the strictest standards for packaging waste. The Dutch standards were more modest than the German but clearly more ambitious than the British (see table 12 for an overview). How can this variation be explained? The first factor is of course the variation in the perceived problem pressure as elaborated on above. This does not explain, however, why the Dutch standards were more modest than the standards set in Germany, since the degree of problem pressure was rather similar in these two countries. Moreover, the general reference to problem pressure, does not explain the variation in the different dimensions of the standards set. The following part will provide a more differentiated explanation.

German policy was very much driven by rather abstract legal principles. These principles embodied general ideas and legitimized public action. The principle of precaution legitimized the government to go 'beyond science'. Accordingly, the government set recycling standards which were both strict and rather arbitrary. Germany set uniform targets for the recycling of each material (64 per cent), with the exception of tinplate and aluminium for which a target of 72 per cent was set. These targets were not based on sound scientific evidence, i.e. environmental or economic assessments. The German standards were particularly strict in regard to sales packaging made from plastic. Industry had to ensure a recycling quota of 64 per cent within four years, up from a level of only one per cent in 1989. The fact that this had to be done by material recycling rather than incineration with energy recovery, made this target so strict. This ban can be traced back to the environmentalist interests voiced by the Green Party and channelled via the *Bundesrat* into the national decision-making process.

The idea of precautions that legitimized high standards without sound scientific evidence to support them in the environmental policy field is embedded into the general idea of the state as sole defender of the public interest. The mission of the bureaucracy as the autonomous guardian of public interest goes back to the nineteenth century Prussian *Ordnungspolitik* (Dyson 1992: 10). It has been reinforced by the idea of the welfare state, which legitimized the right of the bureaucracy to intervene widely in social and economic affairs on behalf of the public interest (Dyson 1992: 15).

This fundamentalist German approach contrasts with the pragmatic **Dutch** approach. The Netherlands only set a global mandatory recycling target, and left it to business to divide the targets across the different packaging materials. In order to create a context for this, the covenant included 'obligatory effort' targets for each material, ranging from 50 per cent for high grade plastic bottles to 80 per cent for glass. This greater differentiation reflects the more pragmatic Dutch approach. The greater degree of pragmatism is also indicated by differences in the time horizon between the Dutch and the German approach. German trade and industry had to achieve the recycling targets within four years after the adoption of the Ordinance. The Dutch counterparts had nine years to reach the global recycling target. The flexible and pragmatic Dutch approach can also be traced back to the less legalistic culture in the Netherlands (see 7.2.) and the consensus-oriented approach towards the target group (see 7.5).

The compromise concerning re-use reflected an informal rule of the political game, namely that problems have to be solved by depoliticizing them. This rule dates back to the period of pillarization. In order to neutralize sensitive problems and to defend compromises *vis-à-vis* the members of one's own pillar, the political elite used complicated arguments and difficult calculations which were not easily understood by the public at large. Political problems were reduced to questions to be solved by 'objective' economic, mathematical or other principles (Lijphart 1976: 135-137). While Germany solved the problem in regard to re-use with the imposition of a quantified quota for refillable packaging, the Covenant parties agreed on elaborated life cycle analyses and market economic research.

The **United Kingdom** pursued only a minimalist approach to the problem of packaging waste. There was only a rather vague household target and no specific targets for packaging waste. The first reason was the low profile of the problem of packaging waste problems (see above). But there is more to it. The traditional British emphasis on sound scientific evidence prevented the British government from adopting a clear hierarchy of waste management options and also from adopting clear quantified targets for these waste management options. This in contrast to what Germany did in its 1986 Waste Act, the 1989/1990 fixing of targets and the 1991 Packaging Ordinance, and to what the Netherlands did in its 1988 Memorandum on the Prevention and Recycling of Waste Materials, the 1989 National Environmental Policy Plan, and its various covenants. To the lack of sound scientific evidence contributed the poor data available on the amount, composition and treatment of municipal waste in general, and packaging waste in particular. Since, in contrast to Germany and the Netherlands, packaging was not defined as a specific waste stream, no data had been collected on this fraction in the past. Generating these data would be difficult for two reasons. First, government would be reluctant to ask industry for the data, because it did not want to put extra bureaucracy on business, and business might not be willing to provide the data for the same reason. Second, even if business would have been willing to provide the data, it would have been difficult to generate them, due to the weakness of trade associations, and the fear that such data might be abused by competitors. Underlying the lack of will and capacity of industry to provide the data is an individualistic industrial culture. This argument will be discussed in section 7.6.

7.4 *Allocation of responsibilities*

The German, Dutch and the British governments emphasized that trade and industry had to take responsibility for the environmental aspects of the products they produced and what happened to them after consumption. The idea of producer responsibility reflects a conceptual learning process in the three countries or in fact most OECD countries, namely that end-of-pipe solutions are not sufficient to cope with environmental problems. But while the end-of-pipe measures such as waste collection and landfill can be organized by public actors, packaging waste reduction, either by packaging prevention, packaging re-use or packaging recycling can be done more efficiently by business itself.

The idea of producer responsibility was particularly strong in **Germany**. Producer responsibility reflected the polluter-pays principle. This is one of the main principles in German environmental policy, though the German government did not always act accordingly. That this principle could be applied in such a comprehensive way had to do with the perceived problem pressure, and the lessons drawn from the public organization of end-of-pipe approaches. Moreover, the German Minister of the Environment Töpfer, who was trained as an economist, was a strong advocate of this principle which derived from environmental economics. Töpfer was quite a powerful Environment Minister because of his expertise. It was also important that the idea of producer responsibility matched the idea of the privatization of waste management, which was advocated by the Ministry of Economic Affairs. These favourable conditions provided a window of opportunity to implement the idea of producer responsibility in the most radical way, i.e. trade and industry were not only responsible for the recycling of packaging waste but also for the collection, transport and sorting.

That the monitoring, assessment, licensing and enforcement remained a public task is a reflection of the fact that the policy is codified in the form of an Ordinance (see section 7.2). That these responsibilities lay with the *Länder* rather than the federal government reflects a standard operating procedure in Germany which is embedded in the constitutional guaranteed allocation of implementation functions to the *Länder* in the German system of federalism. In this specific case, however, the allocation of implementation authority was a compromise between the federal government and the *Länder*.

The polluter-pays principle was not as radically introduced in the **Netherlands** as in Germany. The allocation of responsibilities in the Netherlands was more pragmatic. Pilot projects revealed that a dual system would not be efficient and not necessary to achieve the global recycling target set in the Covenant. The collection responsibilities for industry were therefore confined to paper/cardboard and glass. These materials were collected by bring systems. No separate dustbin for used packaging has been introduced in Dutch households. The comparatively modest form of producer responsibility is also related to the general consensus-oriented approach (see chapter 7.6 for details). The government did not want to put too much burden on industry, in order to ensure the adoption of an agreement.

That a system for monitoring, assessment, and dispute settlement was explicitly mentioned in the Covenant was the result of a learning process. The government drew lessons from negative experience with earlier covenants which were rather vague in this respect and therefore raised difficult questions about the control and evaluation. The function of monitoring and assessment of the agreement was allocated to the Packaging Committee. This Committee was made up of two delegates from the Ministry of Environmental Affairs and two delegates from the Packaging Chain. It was chaired by a more independent person who was appointed by both parties together. This allocation of responsibilities reflects the idea of partnership as formalized in the covenant (see section 7.2). The settlement of disputes was given to the system of arbitration as institutionalized in Dutch civil law. This choice reflects the fact that the Covenant was based on private law.

The allocation of responsibilities in the **United Kingdom** has to be seen against the background of the minimalist approach to packaging waste. The government employed a *laissez-faire* and market-led approach *vis-à-vis* industry. Local authorities remained primarily responsible for the collection and recycling of used household packaging. This reflects the decentralized character of British environmental policy more generally. In addition, the government provided financial resources to stimulate public-private partnerships. Since the government did not employ statutory instruments *vis-à-vis* trade and industry, no provisions for licensing, enforcement or dispute settlement.

7.5 Modes of interest integration

All three governments integrated in one way or another economic and societal interests in the decision-making process. This similarity has to do with the character of the issue. The problem of packaging and the environment is characterized both by scientific and technical complexity and by a lack of societal consensus on important matters. The government, therefore, integrated business interests in particular in order to gain relevant knowledge and to build a basis for support. Though all three countries integrated interests in the policy process, the intensity of integration, the type of interests integrated, and the specific purpose of integration varied.

The consultation of interests was not as elaborate in **Germany** as in the Netherlands. In Germany there was *ad hoc* and informal interaction with a broad range of interests, including environmental interests. The provision of the Waste Act, on which the Packaging Ordinance was based, required a formal hearing. This hearing, in which some 130 groups participated, took place at a rather advanced stage of the decision-making process and did not result in fundamental changes. The comparatively less intensive contact with non-governmental interests in the consultation phase can partly be explained by the fact that the German government was less dependent on information. Quite some data were already available, provided by the Federal Environment Agency among others. Moreover, there was less need for data as compared to the other countries, because the precautionary principle legitimized the adoption of standards in the absence of accurate evidence. The contact with non-governmental interests intensified at the stage of negotiation. The German government interacted with representatives of peak trade associations. This procedure reflects more general patterns of corporatist policymaking. Since the German tradition bears resemblance to the Dutch tradition they will be discussed together (see below).

The **Dutch** government intensively consulted a broad range of actors at an early stage in the policy process. This was done in the rather formalized way of a strategic discussion. Not only representatives of business and societal interests participated, but also consensus-searching process mediators. The intensive contact with a wide variety of interests can be explained by the need for information and legitimization. The broad willingness of interests to participate in the process was facilitated by the general tradition of pragmatic tolerance. Ideological differences are respected or at least tolerated. This principle dates back at least to the period of pillarization and still serves as an informal rule in Dutch policymaking. Effective government was possible only when 'confessional' differences were accommodated (Lijphart 1976: 130-132, van Praag 1993, quoted in van Deth and Vis 1995: 44).

The Dutch and the German government engaged in close and intensive negotiation with trade and industry. In order to do so, the government needed representatives from industry who represented all parts of the packaging chain and who had the authority to commit their affiliates to promises made to the government. In all three countries the government dealt with trade associations, thus on an aggregated level of interest representation. In Germany, the peak associations of trade and industry were instrumental in the development and enforcement of the DSD. The self-regulatory capacity of industry was reinforced by the legally authorized and credible threat to impose more painful measures unilaterally by the government. In the Netherlands the focal point was the SVM, which was quite effective in aggregating business interests (see chapter 5 and section 7.2 of this chapter).

The negotiation with selected interests of highly aggregated interests in the **Dutch** and the **German** case reflects general patterns of interest intermediation in these countries. They have frequently used organizations of civil society to assist in policy formulation and implementation. Government relies on segments of society that perform a dual function of representing interests of the packaging chain concerning the issues of packaging and the environment, and of supporting policy implementation. Accordingly, the state actively supports the emergence and existence of interest associations and their mutual agreements. Interest associations have important public functions, for instance, in the area of the labour market or vocational training (van Waarden 1995: 339).

Hence, it is a standard operating procedure in Germany and the Netherlands to negotiate with trade associations rather than individual companies. What Shonfield observes about modern capitalism in Germany can be easily extended to the Netherlands, namely close organizational and institutional links between government and peak associations in which bargains can be struck about performance targets (Shonfield 1969). Both countries can still be described as associative states because, '*business associations play a key role as intermediaries between business and the state*' (Grant 1993: 15).

These general features date back to the seventeenth and eighteenth century. In Germany, '*the licensing and regulation of guilds, trade associations, cartels and later trade unions led the bureaucracy into a highly structured relationship with industry*' (Abelshauser 1984, quoted in Dyson 1992: 15). Peak associations in Germany represent the summit of a hierarchically integrated system of associations rather than being mere umbrella organizations as in Britain. The sectoral and subsectoral associations are well resourced, have high membership densities, and have clearly demarcated areas of responsibilities. In addition, there is obligatory membership of the regional boards of trade and industry (chambers of commerce) and of artisan organizations (Grant 1993: 15).

The central elements of the idea of partnership and horizontal relations were also already present in the Netherlands in the seventeenth century. A strong civil society existed then and government dealt sparingly with public affairs. Important tasks were delegated to semi-private establishments, such as the water boards. Well-organized groups such as trade guilds had public functions. At present, organizational channels for self-regulation and participation of social actors in the preparation and execution of public policy are, among others, the Social Economic Council, a tri-partist formal advisory council, and a number of statutory trade associations, organising about one-third of the Dutch economy (van Waarden 1985: 204). These organizations were created by the Act of Statutory Trade Associations (*Wet op de bedrijfsorganisatie,* 1950), a *'milestone along the road to cooperation between different organized interests in society'* (Putten 1982: 169). Statutory trade associations were established for a broad range of branches and products. They have a statutory monopoly in their domain, statutory affiliation of all firms in the sector, tax authority, statutory authority to enact binding legislation within their domain, and their decisions are public law. The organizations cannot autonomously change their structures; rules and decisions are subject to formal state approval; and the state is represented on the governing board (van Waarden 1985: 204).

The **British** mode of interest integration has to be seen against the minimalist approach to packaging waste in the early 1990s. Government had *ad hoc* contacts with a broad range of interests. But in contrast to Germany and the Netherlands, this was an arm's length relationship. The generally weak response of trade associations to the challenge of the Ministry of the Environment matches a more general pattern of the British system of interest integration. Generally speaking, in the United Kingdom peak associations have been quite weakly developed and they are constrained by internal divisions. Neither the CBI - the traditional voice of the manufacturer- nor the Institute of Directors, is particularly strong. '*The balance of powers between individual interest groups and peak associations has always been heavily biased in favour of the former'* (Dunleavy 1989: 264). The British state has been described as a company state. '*In a company state the most important form of business-state contact is the direct one between company and government. Government prioritises such forms of contact over associative intermediation'* (Grant 1993: 14). The development of a company state dates from the mid-1970s. Earlier attempts to invigorate business associations in Britain '*largely ended in failure'* (Grant 1993: 14; see section below).

The further development of British packaging waste policies confirms the thesis of the company-state. Government focused on individual companies rather than on trade associations in order to arrive at a more effective policy (see chapter 9.2).

7.6 Orientation towards target group

The three countries differ in their orientation towards trade and industry. The German government was the most impositional. The Dutch one emphasized the accommodation of different interests more, while the British state took a *laissez-faire* approach.

The **German** government was very determined in its efforts to force trade and industry to internalize the environmental costs they caused. Tremendous public pressure to reduce the mountains of waste and the failure of voluntary agreements made the government's threat to impose both a take-back obligation for used packaging on industry at the point of retailers and mandatory deposits very credible and forced industry to come with its own packaging waste collection, sorting and recycling system. This impositional orientation towards the target group contrasted with the consensus-oriented approach in the Netherlands and the *laissez-faire* attitude in the United Kingdom. The active role of government represents a tradition that dates back to the roots of Germany's industrial development. As in many other countries which industrialized rather late, the state played an active role in shaping the industrialization process. In Germany, state interventionist practices pervaded all major aspects of economic life. Schmidt even argues that '*relative to the comparatively low level of economic development, there was hardly another country in the second half of the nineteenth century in which the government role in economic development was more visible and more powerful than in Germany*' (Schmidt 1989b: 65). A high degree of political control of the economy has remained a characteristic of the German experience, regardless of the nature of the political regime (Schmidt 1989b: 65). State intervention is embedded in an industrial culture where there is '*less hostility to the concept of firms or other social and economic organisations as enmeshed within a web of institutional interests*' than in the Anglo-American tradition (Dyson 1992: 260). German industrial culture gives priority to a 'public' conception of the firm as interlocked with a network of institutional interests over a 'private' conception of the self sufficient firm (Dyson 1983).

The interventionist attitude was legitimized by leading economists such as Adolph Wagner. Basing their thinking on the notion of the political leadership's responsibility for the well-being of their subjects, they opted for an economic order in which the state would play an active role in the economic process. '*That option, with varying nuances, has been the choice of the vast majority of German political leaders - irrespective of the political regime - for the last century*' (Schmidt 1989b: 67).

In the **Netherlands**, the Ministry of the Environment employed an *'authoritarian style'* during the 1970s and early 1980s (Bressers and Plettenburg 1997: 116). This approach was gradually replaced by a consensus-oriented approach, which fitted much better the general political culture of the Netherlands. The consensus-oriented approach was pushed to an extreme in the case of the strategic discussions, when more than 270 sessions were held in order to achieve an agreement on the packaging waste policy to be adopted. The consensus-oriented approach is also reflected in the participation of process mediators. Not only government but also business followed a consensus-oriented approach. This is indicated by the fact that the SVM asked the University of Rotterdam to develop a standard for the process of life cycle analysis with the paramount goal of achieving a consensus on procedures and outcomes among all interested parties. Moreover, more than in Germany or the United Kingdom, Dutch advisory committees have the function of creating commitment rather than of increasing knowledge. Support or '*draagvlak*' is certainly one of the most frequently used words in the Dutch policy discourse.

The consensus-orientation is deeply embedded in the history of the Dutch state formation. The need for commitment and consensus goes back to the fight against the water, which was seen as a common enemy. Commitment and consensus were necessary in order to carry out large infrastructure projects, such as dikes (Zahn 1989: 41). Unanimity of all delegates was needed in the Estates-General of the Dutch Republic (1579-1795). This implied a time-consuming search for compromises between the members of the Estates General and between the members and their home base (van Deth and Vis 1995 22-23). The difficulties involved can hardly be imagined, given the absence of modern transport and telecommunication. Moreover, this pluralistic society with its well-organized private interests contained considerable differences in perceptions and values. In such a situation, without a strong central state, policymaking in the Netherlands developed itself into a certain policy style. Daalder (1966) has written of the former Dutch Republic: '*The need to adjust conflicting interests fostered a tradition of compromise and an acceptance of disagreement... Effective political power at the centre depended on the ability to balance carefully widely varied particular interests... A climate of constant reciprocal opposition fostered a habit of seeking accommodation through slow negotiations and mutual concessions*' (189). The Napoleon intermezzo did not result in a substantial change of the emphasis on compromise. The Dutch constitution of 1848 was '*governed by the idea that government policies should be formulated by deliberation and consultation... It is reasonable to argue that the constitution and the laws based on it reflect the Dutch ideal of decision-making by deliberation, consultation and mutual agreement*' (Putten 1982: 168, 169). The search for consensus prevailed also in the period of strong pillarization. According to the informal rule of pragmatic tolerance, minorities may not be simply overruled (Lijphart 1976: 132-133, van Praag 1993, quoted in van Deth and Vis 1995: 44). The search for consensus is embedded in a conception of government that, in contrast to the German conception, emphasizes that government is not seen as being superior to other actors, because it defends the public interest, but merely as one regulatory ordering institution (*ordenende instantie*) besides others. Accordingly, interest groups are not seen as actors who undermine state power but as legitimate guardians of the public interest (Zahn 1989: 262).

The **British** government was determined not to intervene into market processes. This was due to the low problem pressure and the *laissez-faire* ideology of government. Moreover, while Dutch industry by and large sought to cooperate with its government, British industry was much more passive and reluctant to take responsibilities. The general unwillingness of large parts of British industry can be explained by the lack of a credible threat by the government and the weak authoritative power of trade associations. The passive attitude of industry has its roots in British industrial culture. Dyson emphasized that British industrial culture stresses a 'private' concept of the autonomy of action and self-sufficiency of the firm (Dyson 1983) and Judge argues that industrial development in Britain occurred within a context of ideas and social and political relations which all stressed the separation of enterprise from government. '*British success was ... believed to emanate from individual initiative, the self-sufficiency of the firm and resolute leadership of individual entrepreneurs*' (1990, pp.3-4).

This industrial culture was reflected on and reinforced by the prevailing attitude of the Conservative party throughout the twentieth century, to police the public/private divide as restrictively as possible - with the notable exceptions of farming, Empire, defence and law and order (Dunleavy 1989: 260). One should not push cultural arguments too far, however. From the 1940s until the late 1970s state intervention was considerable, stable, and permanent in the United Kingdom. And the pattern of intervention in both the welfare state and the mixed economy was consistently statist, committed to direct government control of enterprises or service provisions (Dunleavy 1989: 243). What is rather distinctive about the British system in general, is '*the lack of any broad political consensus underlying virtually all the interventions undertaken*' (Dunleavy 1989: 285).

7.7 Ideas, Interests, Institutions

Ideas have been very important in the development of national packaging waste policies. Variations in the general ideas about the role of the state and the firm in society have been stated above. In this section, the issue-specific policy paradigms will be revisited. All three countries initially framed environmental problems as *public health* problems. Accordingly, command and control measures were adopted. In those days environmental protection and waste policy was perceived as a burden on economic growth. The oil crisis of 1973 helped people to perceive waste not only as something that had to be discarded, but something with an economic value. The idea of *ecological modernisation* helped in keeping environmental protection on the political agenda in Germany and the Netherlands during the economic crisis of the 1980s. As a result of a conceptual learning process, end-of-pipe solutions were replaced by more *preventive approaches*, such as measures aiming at less waste generation. The implementation of ideas such as the *polluter-pays principle* and the generation and implementation of the idea of *internalization of environmental behaviour* in the Netherlands have to be seen as the result of this learning process. The *precautionary principle* helped to legitimize strict targets in regard to packaging recycling in Germany, while the British emphasis on *sound empirical evidence* was one of the factors that prevented Britain from doing the same.

Generally speaking, the range of *business interests* was quite similar in the three countries under investigation. They did not substantially deviate from the interests as described in chapter two. As discussed above, variations occur in the degree of aggregation of interests and the business attitude towards government. *Environmental interests* varied in one important respect. Environmentalist criticism of the prevailing consumption and production patterns was stronger in the Netherlands and Germany than in the UK. In all countries studied, ideas created new interests. Environmental values of consumers made retailers change their product list. Brand producers were eager to gain and maintain a good environmental image. 'Green' companies such as the Body Shop performed well economically. But there were also differences between the countries. The idea of ecological modernization helped to redefine the interests of some industry sectors in Germany and the Netherlands. Moreover, the idea of precaution led to strict standards in Germany which in turn stimulated emergence of a considerable *environment industry* in Germany, including waste management and recycling industries, which soon became market leaders in the world. These interests helped to counterbalance the demands of some sectors for lower standards in the wake of the economic crisis in the 1980s in Germany.

Ideas and interests were mediated into the political process by institutions. The proportional *voting system* in the Netherlands allowed green ideas to enter parliament in the late 1960s and early 1970s. Germany, having a 5 per cent threshold, followed a decade later. The British majority voting system prevented the UK from having a green MP. In Germany, the institution of *federalism* strengthened green ideas. Though the Green Party was not part of the federal government, it could use its power as part of a number of *Länder*-governments via the *Bundesrat.* Moreover, the constitutionally guaranteed possibility of a *plebiscite* in Bavaria, in conjunction with the required joint decision-making in the German system of federalism, allowed for a bottom up transmission of green ideas. The Christian-democratic Bavarian government, indirectly forced by a powerful citizen's group, joined the social democrat/green *Länder*-governments against the Christian-democratic/liberal federal government and used the veto point to increase standards for the re-use and recycling of packaging waste beyond the levels preferred by the federal government.

Within the government machinery *institutions governing the recruitment and career path of civil servants* shaped packaging waste policies to some extent. The Dutch system was very open, allowing pro-environment advocates to become civil servants. The British and German systems were more closed, favouring either generally trained "talented amateurs" from Oxford and Cambridge, or lawyers in the case of Germany. In Germany, however, the low degree of job rotation in the civil service allowed for a certain socialization and identification with environmental tasks and created environmental advocates. The *distribution of responsibilities within government* was also important in this respect. While German and Dutch civil servants could identify with environmental protection issues within their defined units in the ministries, British civil servants often switched to places where they had nothing to do with environmental protection. Note that in contrast to Germany and the Netherlands, in Britain the Department of Trade and Industry was primarily responsible for one important waste policy aspect, namely recycling. The DTI, however, perceived its role as a sponsor of industry rather than an advocate for environmental protection. Recruitment criteria were not only important for civil servants but also for Ministers. The Dutch practice of recruiting experts as ministers, often from universities, rather than politicians, opened the system to new ideas. In this case the recruitment of Winsemius, a charismatic business consultant, helped to change the VROM impositional approach towards more partnership and also resulted in the establishment of new ideas, i.e. the internalization of environmental behaviour. In Germany, the recruitment of Töpfer, an expert on environmental and regional economics, was rather the exception to the rule of having lawyers as Minister of the Environment. The recruitment of Töpfer was instrumental, however, in implementing the polluter-pays principle by imposing take-back responsibilities on trade and industry.

More generally speaking, differences in *legal systems* were important. The German and the Dutch systems informed by codified abstract law, are more principle-oriented than the British case law system. Principles such as the German polluter-pays principle or the precautionary principle typify this approach. In the German legalistic tradition, the transformation of abstract legal principles, i.e. precaution and polluters pay, through general legal provisions, i.e. Article 14 Waste Act, into Ordinance such as the Packaging Ordinance, is perceived as a logical rather than a political process. Government action is legitimized by the argument that it merely applies abstract principles to concrete policy problems. Industry usually perceives a threat only when the principles are embodied in ordinances, hence at a rather late stage. This restricts their opportunities for opposing the policy. The legal institutionalization of principles make policies based on this principles less vulnerable to external changes. The other side of the coin is, however, a certain rigidity, this in contrast to the flexibility and pragmatism prevailing in the Netherlands and the United Kingdom.

7.8 Conclusion

National governments differ in their national policies for coping with similar problems. The reasons for this are complex and can only be identified by a reconstruction of the policy process. The findings for the case of packaging waste reflect more general characterizations of the distinctive national policy styles. Germany applied a re-active and impositional approach to the problem of packaging waste, while Britain employed a re-active and *laissez-faire*-oriented strategy. The Dutch style is comparatively more active than the German and the British style, but more consensus-oriented than the German one. The Dutch approach is well characterized in the definition of the Dutch word for policy '*beleid*'. In Van Dale's Dictionary of Modern Dutch (Sterkenburg, 1994) '*beleid*' is defined as a) a target-directed, systematic way of handling a task and b) consultation, deliberation.

Ideas and institutions are embedded in a larger national history which to some extent determines their meanings and points towards a certain stickiness of these arrangements. Since the national packaging waste policies are shaped to some extent by these ideas and institutions, they will not easily change. Different policy approaches in the field of business regulation bear the danger of trade distortions, however. A lack of harmonization may threaten the functioning of the European common market. It is precisely this tension between national diversity and European integration which will come to the fore in the second part of the empirical study. Will European integration result in a convergence of national packaging waste policies, or will variation persist?

8. The European Directive on Packaging and Packaging Waste

This study focuses on the interrelationship between national and European policymaking[14]. In the first empirical part, I have analysed the development of the German, Dutch and British packaging waste policies until the early 1990s from a more national perspective. This second part traces the developments following these national policies. This chapter will concentrate on the European level and will focus on the development of the European Packaging Directive. The next chapters are then devoted to national developments in the context of this major piece of legislation.

European institutions have a direct impact on national policies. As elaborated on in the theoretical part, there are two mechanisms by which this direct influence is exercised: negative integration and positive integration. Member States are sometimes forced by the European Commission and the European Court of Justice to modify or abolish legislation that is not in compliance with the European Treaties. This is called negative integration. Also, the European Union adopts regulations and directives to create a common market or to pursue other objectives such as environmental protection. This process of positive integration has an impact on national policies when these pieces of legislation have to be implemented. Both mechanisms of integration may bring about a convergence of national policies.

It has also been said earlier, that Member States have various possibilities for protecting their nationally distinct policies and regulatory styles against influences of the EU. When Member States are threatened by negative integration, they may try to make the issue concerned a subject of positive integration. In other words, Member States may try to translate their own policies into European regulation or directives. Furthermore, Member States may try to prevent positive integration when this would threaten their existing policies and regulatory styles. A Member State can prevent the adoption of European regulations and directives by using its veto in cases where a unanimous vote is required, or by organising a qualified minority in the case of majority voting. When a blockage is not possible, or from the point of view of the Member State not preferable, Member States can try to shape the draft according to their own national preferences and practices (see Héritier *et al.* 1994, 1996). Another opportunity for Member States to weaken the adaptation pressure concerns the implementation of European regulations and directives. European legislation, in particular

[14] The study does not include a detailed description of the European institutions and procedures. The relevant aspects for the purpose of this study will be mentioned as they become important for the argument. See for an overview on the inter-institutional dynamics Wallace (1996).

directives, provide discretionary space for Member States in the implementation phase. This national leeway can be filled according to national preferences. Finally, and less elegantly, Member States can weaken adaptation pressure by lax and delayed implementation. It is not self-evident, that Member States try to Europeanize their *existing* policies and regulatory styles. Member states may use the European arena to change domestic programmes either as the result of a learning process (Majone 1996b: 228) or to escape from constraints imposed by domestic actors. Golub (1996: 333), refers in this context to the Belgian privatization and the Italian monetary discipline case.

Hence, a number of steps of Europeanization can be distinguished. These steps provide the rationale for the organization of this chapter. How and why did the issue appear on the European agenda? (8.1.1) Why has the Commission chosen to deal with it by positive integration rather than negative integration? (8.1.2). How did the policy process proceed ? What was the role of the Commission and the European Parliament? Did Member States try to shape the substance of the Directive according to their own national practices? (8.2., 8.3), and what is the remaining discretionary space for national governments? (8.4).

How the remaining discretionary space has been filled by the Member States, i.e. the legal implementation of the European Packaging Directive by Germany, the United Kingdom and the Netherlands, will be discussed for each of these countries separately. In those chapters developments independent from the EU will also be discussed, since the EU Directive may not be the only driving force of national policy changes.

8.1 The Europeanization of packaging waste policies

8.1.1 Pressure for European action

The analysis of the Dutch, German and British policies revealed a diversity of packaging waste policies and regulatory styles. Given increasing trade between advanced industrial economies, diversity in a field of business regulation which is closely related to the free movement of goods can hardly be without any effects beyond national territories. Germany is the largest European economy and has adopted the most ambitious and interventionist packaging waste policy. It comes as no surprise that it was in particular the German approach that caused external effects.

A number of Member States and their business communities felt negatively influenced by the German Packaging Ordinance. First, foreign manufacturers feared that their exports into Germany would be hindered. This problem relates to the German system of funding the collection and sorting of used sales packaging. It is financed by a license fee, symbolized by a green dot. It was feared that most of the German retailers gave privilege to 'green dot' products. In this case foreign manufacturers would have to comply with an additional system before being able to sell their products in Germany (The Economist 22.8.1992, ENDS Report June 1991: 13, Simonsson 1995: 14). Foreign manufacturers also claimed that they did not have sufficient information about fees, procedures and acceptance requirements. The DSD therefore established a technical barrier to trade. Moreover, foreign drink producers feared that they would be forced to sell their products in refillables, due to the German refillable quota. The German Ordinance stipulated that mandatory deposits and take-back obligations would be introduced when the aggregate market share of drinks sold in refillable containers fell below 72% (see for details chapter 4). German retailers could be forced to privilege refillable containers. It was argued that this is discriminatory because it is difficult to export refillable containers given the logistics needed (ENDS Report June 1991: 13). As a matter of fact in 1995 only 12 per cent of drinks imports came in refillable packaging; in contrast 75 per cent of domestically produced drinks were sold in refillable packaging (Long and Bailey 1997: 218).

Member States also argued that subsidized packaging waste from Germany threatened their emerging waste collection and sorting infrastructure. Used plastic and paper/cardboard packaging waste had been increasingly exported to other countries, such as the United Kingdom, Belgium and France, because the German DSD was very successful in the sense that much more packaging was collected than initially expected. The mountain of collected and separated packaging waste grew faster, however, than the domestic recycling capacities. As a consequence, and because, the collection, transport, and sorting of this waste was subsidized by the 'green dot', it could be offered on the recycling markets of other Member States for a lower price than their own collected waste, even when transport costs were taken into account (The Economist 22.8.1992: 54, ENDS Report June 1991: 13, Oct. 1993: 214; Nov. 1993: 13, see chapter 9.4 for details). Given this advantage, recycling companies increasingly chose German secondary raw material rather than their own used packaging. As a consequence, the emerging waste collection and sorting infrastructures of these Member States was threatened.

National governments began considering the ban of waste imports. This would, however, touch the free movement of goods throughout the Community, and it was more than uncertain whether the Commission and the European Court of Justice would accept such a ban. In the case of the Walloon waste import ban (1992), the ECJ allowed for national restrictions on certain types of waste (C- 2/90, ECR 1992, p. 4431). The judgement, however, can be seen as an exception rather than the rule. There are several reasons to believe that the Court would not allow a ban of trans-national waste shipments destined for recycling. Moreover, the restrictive provision was seen as 'an exceptional, temporary measure' preventing Wallonia from becoming the dustbin of Europe (see also Gerardin 1993: 185, Golub, 1996: 316). Besides the consideration of import bans, the Member States, such as the UK, also tried to adopt national measures to cope with this problem (see chapter). In addition, a number of Member States went to Brussels to protest against this unwanted movement of goods. They were supported by companies and trade and industry associations. INCPEN, for example, submitted a formal complaint to the Commission of the European Community regarding the three problems mentioned above (ENDS Report June 1991).

8.1.2 The Commission's choice for positive integration

The Commission came under strong pressure from Member States and businesses who pointed at the trade effects of the German packaging ordinance. In a situation in which national policies threatened the functioning of the internal market, the Commission has basically two alternatives. As the guardian of the Treaty it can file an infringement procedure (Article 169) forcing a Member State to abolish policy or modify it so as to bring it in compliance with the European Treaties. The other possibility for the Commission is to use its right to initiate European legislation in order to harmonize divergent national policies. The Commission chose the latter, the way of positive integration. In order to understand the Commission's choice to propose a Directive on packaging and packaging waste, the history of European activities regarding packaging and packaging waste will be elaborated on. This history reflects two divergent frames of the packaging and packaging waste issue, namely 'environmental protection' and 'distortion of the single market'. This 'janus-headed' character is reflected in two developments that go back to the 1970s.

The first development, reflecting the environmental dimension, was the initiative of the Commission to adopt a European directive to reduce the amount of packaging from drinks in households and to promote refillables. This initiative started in 1975 and finally resulted in the 1985 Directive on Containers of Liquids for Human Consumption (85/339/EEC, OJ No L (85) p. 176). The second issue, reflecting the 'single market' aspect, is the Commission's and Court of Justice's involvement concerning the interventionist Danish packaging waste policy that caused a threat to the functioning of the internal market.

The 1985 Directive on Containers of Liquids for Human Consumption

Though the European Economic Community was essentially an institutional arrangement devoted to market integration, the Commission recognized, as early as the beginning of the 1970s, that the establishment of a common market and the transnational character of pollution required the enactment of common environmental regulation. Environmental policy, therefore, had been on the agenda long before it was formally included in the Treaty by the Single European Act (1986). The first environmental legislation was adopted in the 1970s, and in 1973 the Council of Ministers adopted the first official Environmental Action Programme. Further programmes followed in 1977, 1982, 1987 and 1992. The European Court of Justice (ECJ) has stimulated these developments. In its case law, the ECJ has declared environmental protection essential to the Treaty of Rome, even though the term 'environment' was not mentioned in the Treaty (see Koppen 1993: 133-136).

The Commission began to develop a proposal for a directive on containers of liquids for human consumption as early as 1975, aiming at the promotion of refillable packaging. This initiative can be seen in the context of the Framework Directive on Waste, that the Council adopted in the same year (75/442/EEC, OJ No L 194/41). This Directive dealt, among other things, with the re-use of waste. Article 3 stipulates that *'Member States shall take appropriate steps to encourage the prevention, recycling and ...any other process for the re-use of waste'*. The preparation of the directive on drink containers took no less than ten years, not in the least because of vigorous protest by industry (Gehring 1997: 346). Nine drafts were developed before the Commission presented its proposal in 1981. The Directive was based on Article 235. This article states: '*If action by the Community should prove necessary to attain, in the course of the operation of the common market, one of the objectives of the Community and this Treaty has not provided the necessary powers, the Council shall...take the appropriate measures'*. It is striking that this Directive was based solely on this general enabling clause. Most of the 120 environmental regulations and directives which were adopted before environmental protection was formally included in the Treaty, were justified by Article 100 of the Treaty (harmonization of national regulations) in combination with Article 235. The sole reliance on Article 235 reflects the high political salience of the *environmental* problems posed by one-way drink packaging. Accordingly, the proposal was entirely environmentally motivated and did not refer to possible single market implications. '*It failed to establish anything like watertight rules to prevent Member States from erecting barriers to trade...*'(ENDS Report, August 1990: 21). The Directive, finally adopted in 1985 (85/339/EEC, OJ L (85) 176) could, however, be conceived *'at best as the first step on the long way toward a Community policy on packaging waste'* (Gehring 1997: 346). Rather than developing a coherent and uniform European packaging waste policy, as initially intended by the Commission, the Directive merely established a European framework for the elaboration of national packaging waste policies. Member States had to develop annual programmes for reducing packaging in household waste and for increasing the share of refillable and/or recyclable packaging. Measures focused on consumer education, the facilitation of refilling and recycling and promoting the selective collection of non-refillable containers and recovery of materials used. The Directive was couched in fairly general terms, lacked specific targets, and contained escape clauses in a number of key provisions (ENDS Report, August 1990: 21). Interventionist measures such as a ban on ring-pull cans or mandatory recycling measures which had initially been proposed by the Commission were dropped after discussions with industry and governments, even before the official Commission proposal saw the light of day (Haigh 1990: 166, Klages 1990: 47).

The modest character of this Directive is indicated by the comparatively low degree of legal codification required for national implementation. The Commission initially wanted the objectives set by the Member States to be legally binding. The Commission had in mind the legally enforceable *contrats de branche* in France (Haigh 1990: 166). It dropped this requirement in the course of the process. The Directive explicitly allows for the implementation by voluntary agreements between government and business. At least until the early 1990s, this Directive has been the only one which explicitly allows for the voluntary route (Sevenster 1992: 76). There were, however, certain requirements for such voluntary agreements, including a certain degree of detail regarding the required objectives and measures. The need for these minimum requirements has been confirmed by the European Court of Justice in a case against France (C-255/93, ECR 1994, p. I-4949).

The Danish bottle case

While the 1985 Directive on drink containers pointed to the environmental dimension of the issue of packaging and packaging waste, the Danish bottle case dealt with the consequences of packaging policies for the internal market, in particular the free movements of goods. The free movement of goods as one of the essential elements of market integration was guaranteed by Article 30 of the Treaty of Rome. The article stipulates that '*quantitative restrictions on imports and all measures having equivalent effect shall, without prejudice to the following provisions, be prohibited between Member States*'. In the Dassonville case the European Court of Justice has interpreted this clause rather extensively and considers '*all trading rules enacted by Member States which are capable of hindering, directly or indirectly, actually or potentially, intra-Community trade ... as measures having an effect equivalent to quantitative restrictions*' (8/74 [Dassonville], ECR 1974: 837)[15]. In Article 36 the Treaty provides for a number of legitimate exemptions regarding, for instance, measures '*justified on grounds of ... the protection of health and life of humans, animals and plants*'. The Court has made clear that Article 36 must be strictly interpreted and that it does not extend to justifications not mentioned in the Article 36 (Geradin 1993: 179). Environmental protection as such does not appear in Article 36. In the *Cassis de Dijon* case (120/78, ECR 1979:649-75), the ECJ recognized, however, that Member States may, when applying measures equally to domestic and imported products, restrict imports for motives other than those specifically recognized by Article 36, the so called Rule of Reason. Environmental protection was recognized as one of these motives, also called mandatory requirements, by ECJ case law (see [Waste Oils], ECR 1984: 531).

The task to balance Article 30 against Article 36 and mandatory requirements respectively is finally assigned to the European Court of Justice on a case-by-case basis. The ECJ has declared in a series of decisions numerous national legislative acts incompatible with European law, and has therefore contributed to national deregulation (see for examples Gehring 1997: 341, Koppen 1993: 138-139). In other cases, however, the Court decided in favour of national regulations. An influential decision dealt with packaging: the Danish bottle case.

While Germany and the Netherlands engaged in voluntary agreements with industry to control the waste generated by one-way drink packaging in the 1970s, the Danish government went a step further. In 1977 it prohibited the marketing of soft drinks in one-way bottles and cans. Moreover, in 1981 the government declared that the marketing of beer and soft drinks was henceforth allowed only in licensed refillable containers (Klages 1990: 49). These provisions clearly constituted an obstacle to trade, since refilling systems implicitly privilege local producers. The Commission received protests from beverage and packaging producers and trade groups located outside Denmark. Their protest was supported by the United Kingdom and other Member States (Gehring 1997: 347). In 1984 the Danish government introduced a derogation from its rules as a reaction to the Commission's demand to provide

---

[15] The ECJ has decided in the Keck Case that marketing modalities ('*modalités de vente'*) do not fall under Article 30 any more (Case C-261/91 [Keck], ECR 1993, p. I-6097, see also Mortelmans (1994)

for exemptions from licensing for foreign producers and importers. Non-approved containers were allowed except for any form of metal container, either within well-defined limits (3 000 hl per producer per annum) or in order to test the market, provided that a deposit-and-return system is established (ECR 1988: 4609). The Commission was not satisfied with this provision and instituted an infringement procedure that reached the Court in 1986.

The Commission was in a difficult position. On the one hand, the recently adopted Directive on liquid containers for human consumption explicitly encouraged Member States to stimulate refilling systems. On the other hand, the interventionist measures adopted in Denmark clearly presented a case of market distortion. Despite the 1985 Directive the Commission challenged the Danish system. The Commission said that Denmark *'has failed to fulfil its obligations under Article 30 of the EEC Treaty'*, (ECR 1988: 4609), that the Danish collection system discriminates against imported products and that - even assuming that the system is not discriminatory - it is not proportionate to the intended objective of protecting the environment. The Commission referred in this context to the 1985 Directive, saying that this Directive shows that this objective is achievable by less restrictive means (ECR 1988: 4609-4611).

The Court did not follow the Commission's position. It accepted the basic provisions of the Danish systems, including the mandatory deposits and the ban on metal cans, saying that these provisions must be regarded as necessary to achieve the objectives pursued in relation to the protection of the environment so that the resulting restrictions on the free movement of goods cannot be regarded as disproportionate (ECR 1988: 4628-4633; see for a discussion of the case Geradin 1993: 183-185, Koppen 1993: 140-141, Krämer 1997: 186-192).

The Danish bottle case indicates that *'(d)espite Court driven negative integration under the Cassis de Dijon jurisdictions, the Member States enjoy an almost unlimited freedom to choose their own level of environmental protection as long as specific Community measures are absent'* (Gehring 1997: 342) This view has been confirmed by the - already mentioned - Court decision in 1992, in which the Wallonian prohibition of waste imports was justified for environmental reasons despite its adverse effects on the internal market, as far as non-hazardous waste was concerned (see 8.1.1, C 2/90, ECR 1992: 4431-81).

The Danish bottle case explains why the Commission did not bring Germany before the Court to force it to modify or abolish its Packaging Ordinance. The Case has shown that the internal market as established by the EC Treaty and enforced by the Commission and the Court does not seriously hinder the Member States from pursuing their own environmental policy. The Court judgement '*effectively shelters national environmental measures in areas that are not subject to European harmonization legislation against the threat of Court-driven deregulation*' (Gehring 1996: 16). It was this restrictive effect of the Court decision on the control of Member State action which became apparent when Germany announced its ambitious and comprehensive packaging waste policy in 1990. Given the limited possibilities of negative integration, the European Commission choose the way of positive integration and by doing this, DG XI picked up the chance to revitalize its own efforts to develop a comprehensive European packaging waste policy.

8.2 Towards the European Directive on Packaging and Packaging Waste

8.2.1 The Commission's draft: getting tough in packaging waste issues

The Commission is seen as quite an influential actor in the European policy process, though its real power is subject to an intensive debate (see Marks, Hooghe, and Blank 1996, Moravcsik 1993, Sandholtz 1993). The Commission has the sole right to initiate legislative proposals. It uses this agenda setting power as a policy entrepreneur, engaging in a dialogue with interest groups, Member States and using its extensive network of experts to shape a draft and control the process so as to serve its own interests (Marks, Hooghe, and Blank 356-361, Peters 1992: 86-87). The General Directorates have remarkable autonomy in developing their proposals. Moreover, they also have their own organizational cultures (Cini 1995)

In comparison with the 1985 Directive, the legal base for European environmental protection policy was strengthened. The Single European Act (1986), which revised the Treaty of Rome, declared in Article 130r the preservation, protection and improvement of environmental quality to be a Community objective in its own right, rather than a subordinate aim of economic development. Article 130s provided the base for EC environmental action. Environmental action could also be based on 100a. This Article, designed to accelerate the creation of the Single Market, assumed an environmental protection component. Article 100a (3) states that the Commission must *'take as a base a high level of environmental protection'* in its proposals concerning the Single Market. The Single European Act did not provide any distinguishing criterion between Art 100a and Article 130s (Gerardin 1993: 167).

Both changes strengthened the Commission. Article 130r probably gave the Commission more scope and confidence in advancing environmental protection proposals. While Directives based on this Article still need the unanimous assent of the Council, decisions made under the new Article 100a need only a qualified majority. This means that single Member States cannot veto Commission proposals any more. Majority voting often allows the Commission to frame its proposal in a way that lies as close as possible to their interests so long as it secures a coalition with a majority of Member States (Keleman 1995: 311).

As the 1985 Directive indicates, the Commission was aware at a very early stage of the negative symbolic value of packaging waste for many consumers and environmentalists. Prior to the announcement of the German and the Dutch regulations, the Commission was already preparing proposals on plastic waste and metal packaging (SEC (89) 934 final). This selected approach to packaging was replaced immediately after the plans of Germany and the Netherlands to adopt a more comprehensive packaging waste policy became public in 1990. The Directorate General of the Environment (DG XI) saw a window of opportunity and jumped on the bandwagon. As the Dutch Environment Minister Aalders and his German counterpart Töpfer before, the Commission addressed the complex but also politically attractive issue of packaging waste in a comprehensive manner, including all types of packaging and packaging materials. DG XI could take advantage of the Council resolution on waste policy in which the Council '*welcome(d) and support(ed)*' a Commission's communication on waste (90/C122/02, OJ No C 122, 18.5.1990, p.2). The Council declared - among other things - that it would be desirable, from the point of view of prevention, recycling and re-use, to establish action programmes for particular types of waste and therefore invited the Commission *'to establish proposals for action at Community level'*.

The legal base for the proposed Directive on packaging and packaging waste was Article 100a. The reasons for this choice will be discussed later. Directives based on Article 100a were to be adopted under the cooperation procedure under the SEA (Article 149). The important formal aspects of this procedure will be mentioned when the decision-making process towards the Directive is discussed. It is worth noting, however, that after the Commission had issued its proposal and the European Parliament had adopted a great number of amendments in its first reading, the Treaty of the European Union (TEU) came into force (November 1 1993). According to the TEU, directives based on Article 100a have to be adopted by the - newly introduced - co-decision procedure (Article 189b). Under this procedure the directive needed the assent of the European Parliament. When the amendments of the Parliament are not accepted by a qualified majority in the Council the article provides for a conciliation procedure. A conciliation committee made up of an equal number of members of the Council and the Parliament tries to develop a joint text. If the Committee succeeds, the text becomes law after it is approved by an absolute majority of the European Parliament and a qualified majority of the Council. This is only one of the many possible ways a proposed directive may take under the co-decision procedure, but it is the way the proposed Directive on packaging and packaging waste took.

The Directive has been very controversial. Several newspaper reports even cited it as '*the most lobbied directive in the EU's history*' (Porter 1995b: 15, Golub 1996: 314). This perception might be slightly exaggerated, given other controversial areas of regulatory activities such those concerning the liberalization of energy supply (Eising, forthcoming) or car emissions (Arp 1993). Notwithstanding this, there has certainly been a high degree of conflict and strong pressure politics involved in the more than four years of the formulation of the Directive. Moreover, in the course of the policymaking process, the substance of the Directive changed dramatically. The initiative moved from an environmental protection issue to an internal market issue, from a high level of environmental protection targets, to a medium level and from a rigid approach to a more flexible approach, implying less convergence of national policies than originally intended by the Commission and also on a lower level. Table 20 provides an overview of the metamorphosis of policy objectives and targets of the Directive from the first official outline proposal (April 1991) to the final adopted version (December 1994). The remainder of this chapter will detail the process and the outcome.

Table 20 Policy objectives from pre-drafts to the the adopted version of the Directive

	DG XI April 91	Com 1 15.7.92	EP EC 1 8.6.93	EP 1 23.6.93	Com 2 9.9.93	Council 4.3.94	EP EC 2 7.4.94	EP 2 4.5.94	Final 20.12.94
hierarchy of waste management options	yes	no	yes	yes	no	no	yes	no	no
per capita limit of packaging waste generation	yes (150 kg)	no	no	no	no	no	no	no	no
five-year target									
recycling (global)	60	no	no	no	no	25-45	25	25-45	25-45
recycling (each type of material)	no	no	40	40	40	15	25	15	15
recovery (1)	90	no	60	60	60	50-65	50	50-65	50-65
ten-year target									
recycling (each material)	no	60	60	60	60	no	no	no	no
recovery	no	90	90	90	90	no	no	no	no
maximum landfill	10	10	10	10	10	no	no	no	no
exemptions for front runners	no	no	yes, broad	yes	yes	yes	yes	yes	yes
exemptions for laggards	no	no	yes, limited	very limited	very limited	yes	yes	yes	yes

(Based on official documents, Golub 1996, and ENDS Report several issues) DG XI draft proposal, Com1=Commission official proposal, EPEC1= First report by the Env. Com. of the EP, EP1=First reading EP, Com2=Amended Commission Proposal, Council=Common Position of the Council, EPEC2=Second report by the Env. Com. of the EP, EP2=Second reading of the EP(1) Recovery includes recycling, incineration with energy recovery and composting.

Modes of interest integration

The European Commission has only restricted resources to pursue its right to initiate proposals. The Commission is therefore even more than national administrations dependent on actors who control crucial resources. DG XI, who developed the first draft of the Directive, was aware of the intricate character of the packaging issue, the broad range of interests involved, the lack of data and the high degree of scientific uncertainty. The lack of data was a tremendous problem. DGXI used figures about the quantities of packaging waste and the quantity of recycled packaging waste which were calculated almost entirely by extrapolation from a single Dutch study, which had to deal with different and inconsistent national methods of measurements and of estimations (ENDS Report August 1990: 20). It comes as no surprise that these figures were criticized by other DG's, interest groups and Member States (see Porter 1994).

DG XI commissioned scientific reports and consulted a broad range of actors in order to increase its knowledge and to build up legitimacy by those actors potentially affected. In this case the need for expertise and legitimization was so high that '*Independent experts appear to have the lion's share of responsibility for drafting the legislation, though*' (ENDS Report August 1990: 19). There was a conscious attempt *'to gain information from as many of these as possible through a sort of informal issue network'* (Porter 1995b: 15). Public and private interest groups were all eager to be part of the game. A network analysis carried out by Porter (1994) revealed that between 1990 and 1993 the number of entities lobbying DG XI alone totalled about 280, the majority from trade and industry. (Porter 1995b: 16).

Confronted with the broad range of interests and overwhelmed by this enormous informal lobbying and information exchange the Commission tried to formalize and channel the interest intermediation. DG XI increasingly sought well-informed, representative and pan-European interlocutors (Porter 1995b: 17). This was, however, more difficult than for the German and the Dutch government in the context of national policymaking. During the course of 1991 the Commission convened a series of official meetings with representatives of trade and industry, consumers and environmentalists, national and 'independent' experts (Porter and Butt Philip 1993: 18). It turned out, however, that it was quite difficult to find representative actors, especially from business.

The Commission, therefore, approached several umbrella organizations with the intention of creating a discussion platform at the European level which would guarantee representation at all levels and all sectors concerned by a group of interlocutors. Initially DG XI only wanted to approach two cross-national policy networks specifically concerned with packaging, one group comprising 16 trade and industry groups, and another, made up of a range of national packaging research associations. But due to their disorganization and internal disagreements DG XI was compelled to approach UNICE (see Porter 1995b: 18). UNICE established the '*Packaging Communication Network*', comprising some seventy affiliates.

More important than this network was the activity of a small number of particularly pro-active and cross-sectoral industry groups which founded the Packaging Chain Forum (PCF). The PCF comprised 25 European level trade and industry associations which organized themselves also in subgroups representing the different parts of the packaging chain. Another influential group was EUROPEN. It was established by the British INCPEN and the Dutch SVM in 1990. This umbrella organization became increasingly well organized and resourced. The members were 22 multinational companies and five national organizations, namely INCPEN, SVM and their Danish, Swedish and German counterparts. By the end of 1993, EUROPEN claimed to represent 600 companies in all sections of the packaging chain and of all materials. EUROPEN became a member of PCF. Another influential group was the American Chamber of Commerce's EC Committee. Since this group advocated essentially the same points as the PCF: strengthening the single market aspect and abolishing the hierarchy of waste management options, they added weight to the PCF position (Porter 1995b: 21).

For the consumer and environmental interests, DG XI invited representatives from the European environmental and consumer umbrella organizations, the European Environmental Bureau (EEB) and Bureau of European Consumers Unions (BEUC). While the BEUC had staff qualified to deal with packaging waste, the EEB decided that two representatives of its 130 odd member organizations would be best suited for the work: the Dutch and the German branches of Friends of the Earth (Vereniging Milieudefensie and BUND). Besides the two formal meetings between the environmentalists and the Commission services there were regular telephone calls and written correspondence between them and other environmental groups.

The Commission finally succeeded in narrowing down the number of actors involved, partly due to overlapping membership especially on the part of trade and industry. Adding the national experts and the COREPER representatives, there were only fifty officials and representatives who were '*closely involved and influential at all or the majority of the stages of the proposal, a surprisingly small number given the range of actors potentially affected by the proposal'* (Porter and Butt Philip 1993: 18).

The DG XI draft proposal

The first draft proposal (April 1991) developed by DG XI designed solely to increase the level of environmental protection. It contained a hierarchy of waste management options, by which DG XI intended to encourage what it considered to be the most ecologically rational disposal methods: prevention, re-use, recycling, incineration with energy recovery, incineration without energy recovery and landfill. The proposal envisaged two ambitious measures in regard to waste prevention and recycling. Member States should ensure that the amount of packaging waste per person would not exceed the EC average in 1990, which was thought to be 150 kg. This ceiling of the amount of packaging waste would have required a serious reduction by the wealthy northern states with a high consumption of packaging. Member States should also ensure that within five years at least 60 per cent of packaging waste is recycled and another 30 per cent incinerated with energy recovery, while no more than 10 per cent is disposed untreated, the so called 60/30/10 formula (ENDS Report May 1991: 37, Gehring 1997: 19). According to the Commission's own figures these targets amounted to a threefold increase in the share of recycled packaging (see ENDS Report August 1990: 20). The draft proposal also stated that Member States would be obliged to introduce measures to ensure that economic operators and competent authorities cooperated in organising take-back and/or collection systems for waste packaging (ENDS Report May 1991: 37).

The draft proposal was very much informed by the Dutch and the German approach. The 90% collection target and the 60% recycling target were identical to the targets of the Dutch covenant. This similarity is hardly a coincidence since the first draft was written by a Dutch and a Belgian civil servant[16]. The Dutch civil servant was delegated from the Waste Division of the Dutch Ministry of the Environment (Interview VROM 4.12.1995). In addition, the independent experts who were employed by the Commission at an early stage were Dutch and

[16] Hans Erasmus (NL), Willem Vermeer (B)

Belgian consultants who drew particularly on Dutch data. The idea of take-back obligations and/or collection obligations was taken from the German draft Packaging Ordinance. The actors who developed the draft proposal were familiar with the discussions in Germany and the Netherlands. The German and Dutch approaches therefore served as models for the first draft of the European Directive.

As Gehring points out, this original proposal envisaged an exclusively environmental project. *'It would have a positive impact on the single market only if the Member States implemented the harmonized European approach and* voluntarily *(stress by the author) refrained from adopting additional measures on packaging waste'* (Gehring 1997: 348). The focus on the environmental aspects might be surprising since the driving forces of the Europeanization of the issue had been trade effects and the wish for harmonization. The environmental perspective can be explained by the dominant problem perception of the General Directorate of the Environment. The task of DG XI is the formulation of environmental protection policy rather than ensuring the working of the single market. The 'environmental protection' frame is therefore linked to the role of this DG and amplified by the recruitment practice of this DG. When the DG XI was set up in the late 1960s environmentalists rather than internal officials were appointed (Cini 1995).

In the course of the consultations, DG XI received sharp protests from numerous interest groups, in particular from trade and industry. But despite this pressure, the Environment Directorate General retained the main duties of the first draft proposal. This draft then was discussed by the *chefs des cabinets.* This steering body immediately below the College of Commissioners, '*requested that the ambitious environmental project be thoroughly revised in collaboration with other DGs, especially the directorate responsible for single market affairs (DG III)*'(Gehring 1997: 349).

From the DG XI draft proposal to the Commission's proposal

The DG XI draft proposal underwent a couple of changes before it became the official proposal of the Commission. In addition to the leading department, some 15 other sections (most notably DG III, DG IV and DG XXIII) were involved in the internal discussion over this directive. In this phase there was a '*fervent*' (Porter 1994: 26) debate about whether the Directive was more appropriately a single market directive or an environmental one. The weight of trade and industry lobbying became stronger as did the weight of those Member States who emphasized the single market aspects of the future directive. These single market interests were supported by a number of Directorates of the Commission, most notably by DG III. According to Porter (1994: 26) it was the combination of this pressure that forced DG XI to choose Article 100a as the legal base of the Directive. Note that the choice of the legal base is not just a technical question. It reflects the dominant problem perception but also determines the specific decision-making procedure used. Furthermore it shapes the balance of powers and influences the actors involved (Porter and Butt Philip 1993: 16, Gehring 1995). The choice for Article 100a meant that the objective of harmonization was given priority, since Article 130s would allow Member States to adopt stricter measures. The choice for 100a also meant that the cooperation procedure would apply. This gave both the Commission and the European Parliament more power than would be the case under the consultation procedure, under which directives based on Article 130s have to be decided upon (see above). In accordance with the choice of Article 100a, the Commission's initiative included elements which should help to ensure the single market of goods. Those elements reflect Member State's and industries' complaints about the adverse effects of national policies on market integration. The single market dimension took the form of a number of product standards, accompanied by the general obligation to accept that packaging meeting these standards can be marketed throughout the internal market. *'This additional product-related component effectively addressed the difficulties created by the Danish and German schemes that distorted the single market and de facto discriminated against foreign producers by actively promoting one type of packaging (e.g. refillable containers) over another, or discouraging or even prohibiting certain types of packaging (e.g. beverage cans)'* (Gehring 1997: 350).

Besides the inclusion of single market elements, the environmental profile of the directive has partly been weakened. The Commission dropped the heavily criticized ceiling of the amount of packaging waste per capita for political and conceptual reasons. The ceiling caused strong resistance from a number of Member States and almost all representatives of trade and industry. Member States that exceed the European average per capita, might be forced to promote refilling systems so as to reduce the packaging waste down to the average. This potential consequence caused vigorous resistance from the PCF, EUROPEN and other trade and industry interests. But there was also a conceptual reason for the dropping of the ceiling. It was argued that the target, measured in weight, would have the undesired effect of favouring comparatively light composite and plastic packaging over heavier materials, such as glass, even though the former would be more difficult to recycle (Gehring 1997: 349). The Commission also dropped the clear hierarchy of waste management objectives. The DG XI had stipulated in its first drafts that material recycling was environmentally more sound than incineration with energy recovery. This reflected very much the Dutch and German approach. A clear hierarchy of waste management objectives was strongly opposed by most of the trade and industry interests and a number of Member States. Again, business lobbying and political pressure as well as conceptual reasons came together and reinforced each other. The Commission was influenced not only by external interests but also drew its lessons from LCA's that demonstrated instances where an inflexible application of the hierarchy would provide the wrong solution. Therefore the policy objectives were somewhat altered and the hierarchy was made more flexible. The official proposal stated: *'Whereas, as long as life cycle assessments justify no clear hierarchy, re-usable packaging and recoverable packaging waste and, in particular, recyclable packaging waste are to be considered as equally valid methods for reducing the environmental impact of packaging...'* (OJ No C, 263, 12.10.1992, p. 1). The Commission was also influenced by the concept of *valorization* which is essential to the French system managed by *Eco-emballages* (Porter 1994: 25). This concept is also reflected in the British approach of '*recovery*'. Valorization and recovery are based on the assumption that recycling, incineration with energy recovery and composting are equally valid waste management options. In the course of the pre-proposal negotiations the ambitious recycling targets were also modified but remained ambitious. The transitory period was extended from five to ten years. This relaxation reduced the adaptation pressure of lagging Member States and their trade and industry. The targets were also tightened to some extent. While targets initially proposed were global targets that apply to packaging, the new targets applied to separate classes of material. Sixty per cent of each material had to be recycled irrespective of waste material characteristics. This would prevent countries from meeting the obligations by merely recycling the comparatively easy recyclable heavy fractions, glass and metal, while incinerating plastics and composites.

The Commission weakened the environmental profile of the Directive and added single market considerations. These modifications reflected major concerns of trade and industry and number of Member States. Environmentalist groups such as Greenpeace and the Friends of the Earth (FoE) branches from the Netherlands, the UK and Germany took a very critical stance to the Commission's proposal. They argued in favour of Article 130s, a clear hierarchy of waste management options, and objected the drop of the ceiling on the overall amount of packaging waste (see Golding 1992: 7-9) The French FoE group and Southern European environmentalists, however, had opinions on the packaging proposals which differed from those of the official representatives from FoE. The French saw Article 100a as the appropriate legal base for the directive and the FoE Mediterranean task force on packaging (comprising FoE Spain, Italy, France, Portugal, Cyprus and Malta) said that they would only accept the FoE common position in 1992 on condition that its general orientation was not too strong in rejecting the Commission's proposal and that emphasis would be given to better control and monitoring of implementation. These interests reflected the Mediterranean waste situation, in particular the low incidence of implementation of EU legislation, the susceptibility to the import of waste from other parts of the Community because of lower standards, and the widespread illegal dumping of waste (Porter 1995b: 26).

Though the Commission's proposal was weakened, it still aimed at a strong intervention intensity, since 90 per cent of the packaging waste had to be removed from the waste stream, 60 per cent of each packaging material had to be recycled - not a small task, given the estimated average level of some 20 per in the year 1990 and 30 per cent had to be burnt for energy in incinerators that meet '*rigorous*' (The Economist 21.8.1992) EC standards.

The Commission's proposal underwent more substantial changes before it became law. But before the political process following the Commission's proposal is discussed in more detail, the Member States' positions will be discussed.

8.2.2 The Member States' Positions

Member States' positions on adequate European packaging waste policy varied. Prior to the Commission's initiative, the issue of packaging and packaging waste had been discussed in nationally distinct ways in the Member States. This study shows that Germany, the Netherlands and the UK adopted different policies and varied in their regulatory styles. Though each Member State perceived its national interest on the issue differently, it is fair to say that the various Member State positions coalesced broadly speaking into two groups: the 'green'[17] or 'forerunner' countries Denmark, Germany and the Netherlands, and the other countries, the 'laggards', which emphasized the 'single market' aspects of the issue of packaging and packaging waste. The basic cleavage between the 'green' countries and the 'single market' countries became apparent from the very early consultations of the Commission of national experts during 1991. Differences in problem perception crystallized at the choice for the legal base of the Directive. While the 'green' countries preferred Article 130s as the legal base, the United Kingdom, France and the other Member States regarded Article 100a as the adequate legal provision (Porter 1995b: 28-30). Differences among Member States existed on many more points. The need for a clear hierarchy of waste management options was controversial as were the level, degree of detail, and time horizon of recycling and recovery targets. The question of whether individual Member States should be allowed to pursue higher targets than agreed on (opt-ups) was also controversial. The 'green' countries were in favour of the Directive as long as it had high environmental standards and left room for countries to go further.

The national positions were shaped by three related factors. First, the perception of the seriousness of the environmental problem related to packaging waste. This was shaped by the geographical situation and the degree of public acceptance of the various waste management options. Second, the existing national policies and levels of recycling and their national regulatory styles. Finally, their vulnerability to external effects of packaging waste policies of other Member States. The 'green' countries had far-reaching national measures in place and took an active and interventionist approach to the environmental problem associated with packaging waste. They already achieved recycling levels above the EU average and even above the level suggested by the Commission. Accordingly, they emphasized the environmental dimension of the issue of packaging waste also on the European level. They advocated a hierarchy of waste management options with a strong emphasis on prevention. They wanted minimum standards allowing Member States to go further. The issue of packaging waste was high on the political agenda of the 'single market' countries. Accordingly, countries like France and the UK planned or adopted comparatively modest packaging waste policies and pursued a soft approach to industry. Their recycling infrastructure was at an embryonic stage and the domestic collection systems were damaged by cheap exports of collected and separated waste from other countries. They therefore strove for a harmonization of national packaging waste policies. These countries also doubted the scientific justification of a clear hierarchy of waste management options.

Porter's network analysis revealed that Germany, France, the United Kingdom, and the Netherlands were the most active Member States throughout the negotiation process (Porter 1995b: 28). Apart from the Netherlands, this finding resembles the observations in other areas of environmental policy, like air pollution policy (see Héritier 1994). Germany, the United

[17] The label 'green' has the heuristical function to separate two group of actors. It is not the purpose of the author to suggest that these groups are 'green' in the sense, that their policy has a greener effect than other policies in reality. In fact one might argue that the exports of German packaging waste to other countries, damaging their domestic recycling schemes might be undesirable even from an environmental point of view. This meant that a 'single market' solution may be better for environmental reasons, than a "green" solution.

Kingdom and France are the largest economies and the most powerful Member States. In addition, Germany and the United Kingdom probably also have the most different environmental regulatory traditions to defend. Moreover, the French government was, as the British one, exposed to pressure from its domestic export and recycling industry, which feared negative consequences of the German packaging waste policy. The active role of the Netherlands can be explained by the tremendous national efforts to come to a comprehensive national packaging waste policy. Moreover, Minister Alders committed himself in the 1991 Packaging Covenant *'to do his utmost to have community regulations drawn up regarding packaging waste'* (Packaging Covenant 1991).

Before the policy process is described in more detail, the positions of the four most active Member States will be sketched as they were prior to the December 1993 agreement on common position. It will be shown to what extent the Member States' positions were shaped by their existing national policies and regulatory styles, and whether Member States changed their preferences due to learning processes or because they saw it as an opportunity to escape from domestic constraints.

Table 21 Positions in the EU Directive negotiations of D,NL,UK, F

	D	**NL**	**UK**	**F**
Degree of legal codification of harmonization	high	high	high	high
Legal base	130	130	100a	100a
Objective	reduction of packaging waste	reduction of packaging waste	harmonization	harmonization
Clear hierarchy of waste management options	yes	yes	no	no
Level of environmental targets	high	high	low	low
Degree of detail	high	medium	low	low
Instruments for reaching the environmental targets	clear producer responsibility	regulation	national guidelines, loose EC framework	-

The German Position

The German government emphasized the environmental protection objective of the Directive and therefore argued in favour of Article 130s as a legal base. It also supported a clear hierarchy of waste management objectives (prevention, re-use, recycling, energy recovery, landfill) and high minimum targets. Member States should be permitted to go further. The government objected to the Commission's recovery target, because this would imply the incineration of collected packaging material. This option was not allowed according to the German Packaging Ordinance. In accordance with its national regulatory style, the German government wanted a clear definition of producer responsibility and the establishment of clear liability (EC Packaging Report Nov. 1993: 4, Porter 1995b: 29). Hence, the German government stuck to its existing national policy approach. This included even the ban of incineration with energy recovery for collected material, though this provision caused a number of problems within and beyond Germany.

The Dutch Position

The Dutch government shared the German priority for Art 130s as a legal base, the clear hierarchy of waste management options and the possibility for Member States to go further. In contrast to Germany, but in accordance with its own national approach, the Dutch government excluded landfill as a waste management option. It accepted the Commission's proposal of a standstill principle (packaging waste output in 2000 should be no higher than in 1990). This was in line with its national approach. The Netherlands also opted in the Council for re-introduction of the standstill objective after it was skipped from the draft by the Commission (Gehring 1995: 33). It favoured high and global recycling targets, rather than targets per material. The 60 per cent recycling target and the emphasis on an increase in incineration resembled the national targets approach. In line with the Covenant, the government supported long (ten years) time frame (EC Packaging Report Nov. 1993: 18, Porter 1995b: 29). As in the German case, therefore, the position of the Netherlands resembled the existing policy approach.

The UK Position

While the Commission's approach was very much in line with the German and Dutch policy concept, it was far away from the approaches of the laggards. The UK government strongly emphasized the single market aspect of the packaging directive at a hearing organized by the House of Lords: '*...independent action by individual Member States can lead to distortions of the single market. We therefore see the principle concern of the packaging Directive as being to prevent those distortions of the Single Market, thereby enabling Member States to achieve higher levels of environmental protection*' (Eggar, Minister for Energy at the DTI, in House of Lords 1993 (Evidence):1). In line with that statement the British government saw Article 100a as the adequate legal base. The British government objected to a clear hierarchy of waste management options. Recycling and incineration with energy recovery should be treated equally. This opinion was rooted in the traditional reliance on sound scientific evidence in the UK. The government argued that there is no proof for a general environmental advantage of recycling over incineration with energy recovery. The government objected to the 90 per cent target for the removal of used packaging from the waste stream. The UK government saw landfill as a viable waste disposal route, due the favourable geographic situation in the UK and the comparatively high public acceptance of landfill. Given the existing low level of recycling and recovery in the UK and in line with the British traditional regulatory style, the UK wanted a more modest and more flexible approach in regard to recycling and recovery. The government was in favour of lower targets and of intermediate targets to evaluate the Directive after five years, recycling targets should not be higher than incineration targets, and targets should be global and not specified per material. The government's critique of the degree of detail of the targets wanted by the Commission was also in line with the traditional British regulatory style which favours broad provisions and ensuring flexibility. The Directive should place as little a burden as possible on the industry. In regard to environmental instruments the government favoured national guidelines and a loose EC framework (Porter 1995b: 30).

The government objected in particular to the uniform 90% target for recovery (60% recycling, 30% incineration with energy recovery). '*I think we have quite severe reservations about the overall target to begin with, the 90 per cent. We doubt its technical achievability. We certainly have severe reservations about the economics of that target, and then we have a degree, I think, of inherent suspicion of a Directive which splits down the targets in considerable detail. That seems to us to be against, if you like, the spirit of subsidiarity and against achieving the best environmental mix for particular domestic markets, and that surely should be the objective. We should not have a sort of externally imposed framework which is appropriate perhaps for some countries but not for others*' (Eggar, Minister for Energy at the DTI, House of Lords, Evidence 1993: 2). The Minister supported his view with reference to the German approach '*... we hope that the Directive in its final form would first set targets that are challenging but realistic... we believe the Directive should take full account of the needs for markets of material recovered. In other words it is no good just accumulating material if you do not have an end use for that material. We think this would avoid the pitfalls of the German approach...*' (Eggar, Minister for Energy at the DTI, House of Lords. Evidence 1993: 1). As in the case of Germany and the Netherlands, the British government favoured an approach similar to the national one.

The French Position

France took the same basic line of argument as the United Kingdom. Ensuring the Single Market was the overall priority and the government was therefore in favour of Article 100a. It ranked prevention over valorization, i.e. essentially recycling and incineration with energy recovery, and finally landfill, but, like the British, argued that for a clear hierarchy of waste management options more scientific evidence was needed. The government was in favour of intermediate five year targets of 60% recovery and 40% recycling. Like the British government, the French had reservations about national programmes to promote prevention as they might create barriers to trade (Porter 1995b: 30).

8.2.3 The first reading of the Parliament: greening the Commission's draft

The cooperation procedure introduced by the 1986 Single European Act grants the EP more power. Under this procedure a rejection of the EP can only be overruled by the Council by unanimity. Under the co-decision procedure introduced by the Maastricht Treaty, the EP enjoys even greater power since it has the right to veto the Council's common position. Besides this increase in decision-making power, the EP has also been successful in influencing the agenda-setting within the EU (Wallace 1996: 63). Research has shown that the EP have already developed under the cooperation procedure from '*a hapless spectator to an important partner in the EC policy-making dialogue, often working in conjunction with the Commission against the Council*' (Golub 1996: 326). The EP has been particularly powerful in the field of EC environmental policy, the EP has been powerful. It altered a number of major directives with a view to stricter environmental standards, for example over car emissions (Arp 1993). The pro-environment stance of the EP has been related to the cross-national character of many environmental problems. In the case of environmental protection the need for cross-national cooperation is rather evident to the public. Environmental protection therefore is perceived as a legitimate field of European integration. It is therefore no surprise that environmentalists have natural allies in the EP even among non-Green MEP's.

Much of the EP's success has been due to the '*ideological devotion and tireless efforts*' (Golub 1996: 326) of the chairman and the other members of its Environment Committee. Given the institutional change one could assume that the strengthening of the EP would automatically strengthen the environmental dimension of policies adopted under the cooperation and co-decision procedure respectively. This only holds true, however, other things being equal. The neo-institutional argument also suggests that in accordance with the institutional change, interest groups come to rethink their strategy and increasingly use the EP as a point of access to the decision-making process. The case of packaging waste shows that the new found power was recognized by lobbyists. A considerable number of groups lobbied the EP. Porter counted that prior to the EP's first reading, the *rapporteur* on the issue, Vertemati, was contacted by 105 separate organizations and received over 270 different position papers and suggestions for amendments (Porter 1995b: 16). In this case too, trade and industry groups were far more numerous than consumers and environmentalists. An imbalance that has been criticized by some MEP's and added to calls for a lobbyists' code for both the Commission and the Parliament (Porter 1995b: 16).

The business influence at this stage was, however, partly weakened due to a lack of unity. The PCF was united in its opposition to ceilings on the absolute amount of packaging waste. Once the Directive came before the Parliament, however, the internal divisions and tensions of the PCF on other packaging waste issues burst into the open. The submissions to the *rapporteur* indicates according to Porter (1995b: 22), that PCF members put forward a particular sectoral or national point of view without other PCF members immediately being aware of it. It increasingly turned out that the different sub-groups of PCF representing different sections of the packaging chain had different perspectives which made co-ordinated action among such a diverse group of interests increasingly difficult.

Despite the lobbying of trade and industry, the Environment Committee strengthened the environmental profile of the proposal. The report of the Committee included 79 amendments to the Commission Proposal[18]. The committee reintroduced most of those provisions originally favoured by DG XI, but dropped during the preparation of the Commissions proposal. This included the hierarchy of waste management options and binding five-year collection, recovery and recycling targets. The Committee sought to guarantee that Member States were allowed to pursue higher standards and to take into account the special conditions affecting islands.

[18] Each reading in the European Parliament is a two-part process. First, the Environment Committee of the EP issues a report suggesting amendments. Second, this report is adopted in part or in full by the rest of the EP.

The EP accepted most of the Environment Committee's major recommendations despite heavy lobbying of interest groups (Das Parlament 9.7.1993). It agreed on the hierarchy of waste management options and the intermediary targets. The EP expanded the scope of unilateral national action by accepting the amendments which enable Member States to adopt national economic instruments to pursue environmental goals. But it rejected all three amendments aimed at allowing broad national opt-ups (OJ C194, 19.7.1993, Golub 1996: 312, 321).

8.2.4 The Common Position of the Council: single market vs. environment

The Commission produced a new amended proposal (9.9.1993*)* in which the amendments of the EP were only partly taken up. It reintroduced the binding five-year targets but abandoned a hierarchy of waste management options. The Commission did not include the amendments proposed by the EP Environmental Committee, but rejected by the EP's first reading, which would have explicitly allowed widespread national opt-ups. But it adopted provisions which allow Member States to use economic instruments (OJ C 285, 21.10.1993, p.1). The new proposal of the Commission had a weaker environmental profile. It was miles away from the drafts developed by DG XI. The new text anticipated concerns of the majority of Member States, and reflected the Commission's thinking that a political compromise would be necessary in order to get an agreement on the directive.

Member States had been involved in drafting the proposals at a very early stage. National experts were already consulted by the Commission in 1991. COREPER working groups began to examine the packaging dossier in October 1992 prior to the December 1992 environmental council. These meetings continued with differing degrees of frequency and regularity over the following year. Porter noted that the Danish presidency of the first half of 1993 came in for some veiled criticism from those Member States, feeling the effect of German packaging ordinance on their domestic recycling industries, for not pushing the issue hard enough. At this time the Spanish, the French and the British government in particular tried to highlight the urgency of the problem and to ensure that the single market concerns were at the centre in preparing the packaging directive (Porter 1995b: 30). During this time the 'green' countries developed close working relationships. Initially the Danes were not wholly supportive of the German-Dutch position because of their high use of incineration with energy recovery and their initial preference for Article 100a. But in the latter half of 1993 the Danish joined the Dutch-German club when they became aware that the Directive was potentially damaging to their national packaging scheme. They met frequently and on a regular basis. They discussed each others drafts and prepared new proposals together (Interview VROM 4.12.1995). In 1993 and 1994 Germany, the Netherlands and Denmark came to develop a common position on the packaging waste directive at the various COREPER and Council meetings. Porter noted that evidence from the COREPER officials suggests that the interaction in fact mostly happened in national capitals through contacts between experts in respective environmental ministries rather than in Brussels (Porter 1995b: 31). The trans-national cooperation was facilitated by overlap between the membership of

COREPER working groups and meetings between national experts which had earlier taken place. In addition, the officials concerned also met on other occasions to consider other dossiers, something which helped to confirm their positions towards the Directive.

Though the frequency of COREPER meetings increased in May and June, the Danish presidency was not able to reach a compromise. Then the Belgians took over and negotiations in COREPER working groups and full meetings intensified. By the latter half of the year, officials were meeting twice a week (Porter 1995: 30). Though the Belgian sympathies lay with the forerunner countries, as President of the Council Belgium had to stay neutral and it tried hard to find a compromise. In particular the ambitious recycling quota proposed by the Commission were heavily disputed. They were vehemently supported by the 'green' group and were equally vigorously rejected by the other countries (Gehring 1997: 350). The 'laggard' countries favoured lower recovery and recycling targets and the omission of binding recycling targets for each type of material. The situation became quite tense. The EC Packaging Report noted that '*the Belgians are rushing out new compromises like confetti*'" (Nov. 1993: 10). One compromise said that re-use should be counted to the recycling quota. The Commission, fearing adverse trade effects of re-use provisions, even threatened to withdraw the proposal for a Directive (EC Packaging Report Nov. 1993: 10.)

The group around the UK had a majority. The UK took the initiative and proposed dropping any mention of *recycling* but retaining a goal of 50 per cent *recovery* within ten years. This target for recovery was similar to the minimum target the UK government asked its domestic industry to achieve (see chapter nine). This group succeeded in weakening the standards. The common position adopted by the Council included a recovery target of 50-65 per cent to be achieved within five years, and an overall packaging waste recycling target of 25-45 per cent, with a minimum recycling rate of 15 per cent for each type of material. The decision on the level of binding ten-year targets was postponed to an unspecified date after the adoption of the Directive. The common position merely stipulates that the Council will determine these targets *'with a view to substantially increasing the targets'* (OJ, No C 137, 19.5.1994, p. 65). What's more, less developed countries were granted derogation giving them ten years to meet the targets.

There were two reasons for lower recycling targets however. Besides Member States' interests in weaker standards in order to reduce the costs for its industries, the weakening of the targets was also the consequence of the perceived failings of the German Packaging Ordinance. As said earlier, there were not sufficient recycling capacities to meet the high recycling targets set by the Ordinance. The negative lessons drawn from the German policy also contributed to the fact that *maximum* quota's for recovery (65 %) and recycling (45%) were introduced. Weaker standards were also favoured by most parts of trade and industry. At this stage the PCF was divided, however, about the degree of detail of the recycling targets. A number of associations representing sectors which could recycle relatively easily opposed the PCF common position to favour a global recycling target, because they felt that they would have to bear more of the responsibility for reaching the recycling targets than other sectors, since global targets favour industries where recycling is more difficult. The European associations for glass, aluminium, paper, and steel packaging formed the "Group of Packaging

Industries for Equitable Burden Sharing" to lobby the Council, the Commission and the EP to this effect *'thereby undermining the stance taken in particular by the more 'holistic' industry groups representing the whole packaging chain such as EUROPEN, AIM, and ERRA'*(Porter 1995b: 22). By the end of 1993, when the core groups within the PCF, the European Recovery and Recycling Association, Association of Plastic Manufacturers in Europe, European Association of Industries of Branded Products, and EUROPEN proposed a slightly modified approach to the lobbying over the recycling targets and their scope, several groups, including the Alliance for Beverage Cartons and the Environment, the European trader association, and the glass and steel packaging associations felt unable to support the proposed changes for different reasons, though (EC Packaging Report March 1994: 5, Porter 1995b: 23).

The 'green' countries '*struggled hard*' (Gehring 1997: 351) and eventually successfully for an exemption of the maximum recycling. Member States are allowed to exceed this so called cap provision, if a Member State has sufficiently high capacity to process collected material, and the measure does not distort the internal market or hinder compliance by other Member States with the Directive. The latter condition reflected the lessons learned about the effects of the German system on other national recycling schemes. As elaborated on above, countries like the UK have argued that national collection schemes cannot develop when the recycling industry use cheap foreign material from used packaging. The 'green' countries also succeeded in including provisions that allowed them to stimulate the prevention and re-use of packaging. This did not establish positive obligations but may have the effect of expanding opportunities for future domestic action. The single market component of the Directive remained virtually undisputed. Germany and the Netherlands did not attempt to erase it from the directive. Implicitly, however, the exemptions granted to forerunners limited the impact of the single market component (Gehring 1997: 350). The common position was agreed upon in the Environment Council on 15 December 1993, but was formally adopted as late as 4 March 1994. It was decided by the qualified majority of the Council. Though the Netherlands, Germany and Denmark succeeded in getting exemptions, they voted against the common position. Moreover, these countries then lobbied the EP Environment Committee to push for amendments including higher standards (EC Packaging Report March 1994: 2, EWWE 18.3.1994: 2)

8.2.5 The second reading of the Parliament: not green any more

The amended proposal of the European Commission indicated that the environmental bias of the Environment Committee was no longer supported by the Commission. Nevertheless, the Environmental Committee tried again during its second reading to reintroduce some of the provisions originally favoured by DG XI, and a few which it had put forward in its first reading. But the Environment Committee's second reading *'was a token gesture compared to its earlier efforts'* (Golub 1996: 322). The Committee agreed on 38 amendments. The provisions included the removal of any restrictions on maximum targets as well as an increase in the global recycling requirements from 45 to 50%, and for specific materials from 15 per cent to 25 per cent. The Committee also reintroduced the hierarchy of waste management options.

In the second reading of the EP on 4 May 1993 each of these proposed substantial amendments failed to gain sufficient support. The Parliament agreed only on 19 less substantial amendments. As a consequence, the Council's position remained virtually unchanged. The reasons for this U turn of the EP will be discussed in section 8.3.2.

8.2.6 The conciliation procedure: persuading Belgium

The Commission agreed to all amendments made by the EP. One might assume that a common position of the Council which was almost entirely accepted by the Parliament would easily get the assent of the necessary majority of Member States. But things got yet more complicated. At the last minute Belgian objections blocked a common position in the Council. The Belgians were dissatisfied with the wording of one amendment by the Parliament dealing with national economic instruments. The Belgians feared that this amendment would jeopardise Belgium's ecotax. The Belgian government had promised the introduction of national ecotaxes to the Belgian Green Party, in order to get their votes for a constitutional reform in Belgium.[19] (EWWE 17.6.1994: 1). A Belgian defection would have deprived the proposal of the 54 votes necessary for a qualified majority. The problem was compounded when Luxembourg threatened to resist the proposal for similar reasons. Since the attempt in the COREPER to persuade the Belgians to step back from their objections failed, changes were necessary to resolve the deadlock. The proposal therefore passed into the conciliation procedure as provided for by the Treaty. It was the first environmental proposal ever decided upon at this stage. An equal number of MEP's and members of the Council met to '*hammer out a compromise*' (Golub 1996: 324). A couple of MEP's supported by the 'green coalition' of the Council, tried unsuccessfully to reopen the debate on various amendments which were defeated in the second reading (EWWE 7.10.1994). In the course of the meetings the Council accepted the EP amendments. The wording of the section on economic instruments designed to reach the recycling targets was revised, so as to get the assent of Belgium. Britain which had also threatened to withhold support, was equally satisfied that the proposal would not allow national economic instruments which distorted the market, nor would it herald the future widespread imposition of Community eco-taxes on Member States (Golub 1996: 324).

On 20 December 1994, a qualified majority of Member States voted in favour of the Directive. Denmark, Germany and the Netherlands voted against the Directive. The Directive comprised the major elements of the 1993 Common Position of the Council, along with the vague section on national economic instruments and a commitment to revisit the entire issue of waste again in ten years in view of substantially increased recycling targets.

---

[19] The Belgian case shows that intergovernmental and national policymaking is strongly interrelated, and this even beyond the issue at stake, given the conclusion of cross-issue package deals on the national level

8.2.7 The content of the Directive on Packaging and Packaging Waste[20]

The Directive aims to harmonize national measures concerning the management of packaging and packaging waste while providing a high level of environmental protection (Article 1.1). The Directive states that measures aimed at prevention of the production of waste have first priority. But the Directive does not include a clear ranking of waste management options. The Directive stipulates only that, *'the directive lays down measures aimed at re-use, recycling, and other forms of recovering waste in order to reduce the final disposal of waste'* (Article 1.2). Member States shall ensure that prevention measures beyond the scope of Directive are implemented (Article 4) and they may encourage re-use systems of packaging in conformity with the Treaty (Article 5). Member States shall take the necessary measures to attain 50 to 65 per cent recovery[21] and 25-45 per cent recycling, with a minimum of 15% for each material, within five years (Article 6.1.) The decision on ten years target has been postponed (Article 6.1.c, 6.3.b.). Member States shall, where appropriate, encourage the use of recycled material (6.2.). Greece, Ireland and Portugal get a temporal derogation from complying with the above mentioned targets (Article 6.5). Member States have to take the necessary measures to ensure that return/and or collection and recovery systems are established in order to meet the Directive's objectives (Article 7). Member States are allowed to adopt economic instruments to promote the implementation of the Directive's objectives (Article 15). Packaging sold in Europe has to meet essential requirements after three years from the date of the coming into force of the Directive (Article 9). The Commission shall promote European standards related to the essential requirements (Article 10). Member States shall not impede the placing on their market of packaging which satisfies the provisions of the directive (Article 18). The directive has to be implemented by laws, regulations and administrative provisions before 30 June 1996 (Article 22). Member States have to notify draft measures intended to transpose the Directive into national law. Notifications must comply with the procedure set forth in Directive 83/189 on technical harmonization (see for details chapter 9).

8.3 The role of actors revisited

With the Europeanization of the packaging waste issue, policy-making entered a new arena. New actors became involved and other rules applied. This part of the chapter will review the role of the different actors in the process towards the Directive as well as their preferences and their influence.

---

[20] The Directive on Packaging and Packaging Waste (94/62/EC) replaced the 1985 Directive on containers for liquids for human consumption.

[21] Weightpercentage of the overall waste stream

8.3.1 Towards Convergence? The failure of the Commission

The first proposal of the Commission would have meant a convergence of national Member State policies. Uniform targets for collection, recycling and incineration with energy recovery would have been mandatory for all Member States. This would have implied a catch up of the laggard countries to the levels envisaged by the German and the Dutch national scheme. The degree of convergence would have been quite high since no opt-ups and derogation were provided. The central elements of the Commission's proposal did not survive the Council decision-making, however. As Golub claims *'Even if one assumes a certain amount of gamesmanship, whereby the Commission habitually puts forward extremely ambitious proposals and targets which it knows will be sacrificed during subsequent negotiations with the Council, the weakness of the Commission as agenda-setter is striking'* (Golub 1996: 325). The Commission's second proposal was less informed by a green profile than by the wish to get the directive via the Council. Several scholars have argued that the Commission's sole right of initiate legislative proposals and its active attitude as policy entrepreneur provides it with considerable control over the timing and content of the first proposal of a new directive. The case of packaging waste shows, however, that this does not mean that the Commission can also substantially shape the outcome of the policy process.

8.3.2 The U-turn of the European Parliament

The position of the European Parliament in its first reading would have implied a significant upward convergence of national packaging waste policies, forcing the 'laggard' countries to achieve the standards of the front-runners. The degree of convergence might have been less high, than in the case of the Commission's proposal since the European Parliament amendments provide for opt-ups for front-runner and derogation for small islands. The degree of convergence implied by the EP position in the first reading could not be retained during the second reading. The European Parliament did not use its new powers granted by the co-decision procedure to counteract the common position of the Council which implied less convergence (broader range of targets, opt-ups and derogation) and this on a lower level (see below). There were several reasons for this U-turn. In the first reading the amendments only needed a simple majority of the members present in the parliament. In the second reading, however, any modification of the Council's common position needed the absolute majority of all MEPs, that is 260 of 518. Due to the coming EP elections many MEP's campaigned in their home countries and were not present at the decision. Not all of those present voted. The MEP's had to vote for each amendment separately. The Minutes of Proceedings (94/C205/45-47) reveals that the total number of votes lies between 265 to 329 depending on the amendment concerned. Many parliamentarians did not follow party line but national positions. French and Southern socialists voted against the amendments of their socialist colleague Vertemati. They were concerned about criticism in the run-up to the elections (EWWE 6.5.1994: 2). Still, for a number of amendments the difference between success and failure was a matter of a few votes. In the case of the amendment requiring an increase of the general (from 45 to 50 %) and the material specific (from 15 to 25%)

recycling quota, the difference was 22 votes. Moreover, the Greens disliked the watered down proposal of the Environment Committee and put forward a substitute. They might have supported a 'greener' proposal but preferred no action to bad action.

So, why was the second report so weak? First of all the EP Environment Committee was no longer supported by the Commission. While the EP and the Commission were often seen as natural coalition partners, this coalition faded away as soon as the Commission decided to drop elements of its original opinion even before the first official draft. The Commission had refused to adopt amendments favoured by the Environment Committee or the Parliament which would have strengthened the environmental profile of the directive. Second, the *rapporteur* Vertemati was not as "green" as many of his colleagues in the Environment Committee (EWWE 15.4.1996: 6). Third, the lobbying of trade and industry became stronger.[22]

8.3.3 Pressure from trade and industry

The Europeanization of the packaging waste issue provided interest groups with new points of access. It provided the opportunity for those industries which suffered from interventionist approaches, such as the German system, to weaken its impacts on their business. The European decision-making process is, however, quite challenging to interest groups due to the many potential points of access. To influence the internal Commission debate alone, lobbyists had to direct their attention to several DG's. In addition, the heterogeneous Parliament and the Council had to be addressed. Trade and industry had met this challenge quite successfully. Business effectively influenced the wording of the first drafts of the ordinance. Dropping of the ceiling on packaging waste generation, the hierarchy of waste management options, and the ambitious five year recycling and recovery targets in the first draft can be related to industrial lobbying (Porter 1994:21). In addition, the relatively weak environmental profile of the second EP Environment Committee report and the rejection of the most ambitious provisions still included, was at least partly due to the lobbying of trade and industry. Trade and industry itself was a quite heterogeneous group of actors. Who was influential and why?

Within the Packaging Waste Forum (PCF), there were about 8 leading groups (ERRA, APME, AIM, CIAA, EUROPEN, BCME, FEAD and SEFEL) which were particularly active and influential. Apart from FEAD and SEFEL, they were not traditional Euro-groups, that is federations of national associations, but cross-national networks composed partly or wholly of multinational companies. ERRA and BCME for instance, both established when the EU debate on the directive started, comprise 30 and 5 companies respectively. As Porter argues '*because of the size and the economic weight of their members, these organizations can still claim to be representative of a large amount of economic activity and because of their streamlined structure, they are very responsive and able to provide information with speed*'

[22] The story would have taken another direction if an undivided Parliament has developed a 'green' alternative. The unanimous vote in the Council required to overturn amendments would most probably have been prevented by Germany, Denmark or the Netherlands, who lobbied the Environment Committee to take a 'greener' position.

(Porter 1995b: 21). These groups did not suffer from the use of vetoes of national associations[23]. Other groups such as AIM (1967), APME (1976), CEFIC, and EUROPEN had hybrid structures, combining direct membership of multi-national companies with their traditional national associations and a sophisticated committee structure. This structure enabled them to monitor a wide range of issues.

The influence of PCF members on the Commission can be related to the information dependency of the policymakers in Brussels. ERRA, although essentially an industrial association, became a valued source of data because of its *'original research and extremely professional approach'* (Porter 1995b: 21). The association of plastic manufacturers (APME) was able to support its arguments with the research provided by the Plastic Waste Management Institute. The brand producer association (AIM) was also able to generate much research information (Porter 1995b: 21). The transmission of trade and industry interests to the Commission was facilitated by the overlapping membership of a number of individuals. Key representatives of the hard core of the PCF often held prominent positions in multi-nationals such as Procter & Gamble, Coca-Cola or Tetra Pak. These people were also sometimes members of national trade associations. *'As a result, a relatively small core of people met regularly and liased over this issue in a cross-national policy network...It is this smaller group of people which at times came closer to constituting a 'policy community' than any other group and indeed, they met regularly with Commission officials on an informal basis'* (Porter 1995b: 22).

8.3.4 Weak environmental and consumer groups

In contrast to trade and industry, there was a limited number of societal groups that have lobbied over the packaging directive. Environmental and consumer groups lacked the resources available to their counterparts. They must therefore be selective in the issues to which they devote a significant amount of time and effort. The environmentalist networks came into existence after the Commission had asked the EEB to provide data. They lagged, therefore, behind trade and industry in their mobilization. Friends of the Earth was the most active environmental group. '*In fact, the history of the cross-national policy network which developed from the environmental side is in effect a history of cooperation between various national branches of Friends of the Earth and much more a story of decentralized local action and campaigns than European level lobbying effort*' (Porter 1995b: 24). Environmental groups organized themselves in the Sustainable Packaging Action Network (SPAN). This happened at a conference in Amsterdam that was supported financially by the Dutch Ministry of the Environment and DG XI (ENDS Report December 1992: 23, ÖB 20.1.1993). Within SPAN, the FoE branches, especially those in the Netherlands, Germany and the UK, have taken a lead in drawing up policy positions, co-ordinating activities and representing environmental positions to the Commission and the Parliament. FoE exchanged information quite regularly not only with environmental groups and

[23] As Porter reported, the waste management association, FEAD, had just such a problem in agreeing a position due to the stance adopted by its German members (Porter 1995b: 20).

Green MEPs, it also spoke every fortnight with BEUC. The link with consumer organizations over this issue was apparent in other related networks (Porter 1995b: 26). Though joint press releases were issued, there were also points of difference between the consumer groups who favoured single market (100a) legislation and their Northern FoE counterparts who favoured environmental legislation based on Article 130 (Porter 1995b: 27). The access of environmental groups at the early stages of the policy process was not significantly better than that of trade and industry. *'Contrary to some business views, DG XI did not appear to have a particularly cosy relationship with environmental groups but dealt with them like any other interested, representative and informed group'*. When there was some influence, it was at the national grass-root level and to some extent in the EP (Porter 1995b: 25). The lack of influence can be partly explained by the institutional politics. The EP's support of Article 100a, rather than pushing for Article 130, was related to the greater powers it enjoys under this procedure.

8.3.5 Member States ensuring policy autonomy

The Member States weakened those components of the Directive that would induce strong adaptation pressure on their national policies. The Directive contains lower environmental targets than in the forerunner countries, who therefore threatened to continue regardless. These targets were, however, too tough for many others, who achieved temporary opt out. This undermined the harmonization objective necessary for the single market to operate desired by majority as a priority. Given the trade effects of the existing regimes as well as Article 100a as the legal base of the new Directive, the lack of harmonization is surprising. How can it be explained that the Packaging Directive left ample leeway to Member States for both the implementation of the Directive and the maintenance and introduction of national policies? The majority of Member States formally had the opportunity to reject the demands for exemptions asked by Germany, the Netherlands and Denmark, which would have induced more pressure towards convergence. But '*this is not how it works*' (Interview DoE, 12.7.1996). There is a procedural consensus between the delegates of Member States that certain courses of action are out of question. The range of what is allowed by this consensus is narrowed. Kerremans argues that this is reinforced by what is called the effect of cooptation in EU decision-making. '*When the same actors participate successively to different processes, it becomes rational to take into consideration not only the possible effects of choices on each process on its own, but also the possible consequences of these outcomes on future processes*'(Kerremans 1996: 233). In addition to this rational choice argument, the structural embeddedness of the Member States has to be taken into account. European decision-making is complex in particular in the case of co-decision procedure. All institutions concerned have ample opportunities to block each other. Member States depend on each other to reach decisions, to conduct policies (Kerremans 1996: 233). Therefore informal relations and informal negotiations gain importance in order to design a compromise. These negotiations are guided by normative principles such as fairness, reciprocity, subsidiarity and '*Gemeinschaftstreue*' (Eising, forthcoming). The main consequence is that there are neither distinct losers nor winners. Instead, all the participants are sharers, all of them gain advantages and all of them have to concede.

The degree of national autonomy preserved by the Member States should not be exaggerated, however. Though most Member States might be happy with the outcome of the negotiation process *at the time of the decision*, it limits the options of Member States by preventing backsliding, by setting compliance deadlines and targets which otherwise might have been repealed or amended by the national executives. Member States such as Britain are now forced to carry out national plans. Moreover as Golub emphasized: *'(t)his loss of national executive autonomy becomes even more apparent when new national executives are unhappily bound by EC obligations entered into by previous administrations. In the current case, a new government in any of the Member States will have to comply with mandatory EC recycling targets with which it may or may not agree'* (Golub 1996: 334). The following part will define the scope of national autonomy in more detail.

8.4 The European Packaging Directive and national discretionary space

It has been argued that the Member States quite successfully defended their national autonomy in regard to packaging waste. This part of the chapter will look more into relevant legal aspects. What is the potential influence of the European Directive of Packaging Waste on national policies and regulatory styles? Is it likely that the Directive leads to harmonization of national approaches and therewith to convergence? Are effective legal means provided by the Directive to slow down the front-runners and to push the laggards? What is the national leeway concerning the form in which the Directive has to be implemented? The answers to these questions require an analysis along three dimensions. First, the potential convergence pressure entailed in the Directive itself. Second, the potential for digression by front-runner countries provided for by the opening clause of the legal base of the Directive Article 100a (4). Third, potential convergence pressure induced by the legal requirements for the implementation of the packaging directive. While the first two dimensions concern the strictness of standards, the third dimension primarily deals with the degree of formalization.

8.4.1 Pressure to convergence by the content of the Directive

Prevention and refill

The Directive contains provisions for additional national measures aiming at packaging waste prevention and the promotion of re-use systems. The 'green' countries fought hard to get these provisions included. The conditions under which Member States can introduce such measures is rather unclear, however. Article 4 states that '*Member States shall* ensure, (emphasis M.H.), *that in* addition (emphasis, M.H.) *to the measures to prevent the formation of packaging waste taken in accordance with Article 9* (essential requirements, M.H.), *other preventive measures are implemented...if appropriate with economic operators'* The directive explicitly states that the measures should '*take advantage of the many initiatives within Member States as regards prevention'*. Member States are not free in the design of these measures, however. Article 4 stipulates that the measures have to comply with the Directive's objectives as defined in Article 1 (1). Hence, the measures shall not threaten the functioning of the market and should not erect obstacles to trade or distort or restrict competition within the Community. Member States are also allowed to '*encourage*' re-use systems of packaging, which can be re-used in an environmentally sound manner (Article 5). In this case, the Directive is not referring to Art 1(1) but stipulates more generally that these measures have to be *'in conformity with the treaty'*. The domestic scope for national policies aiming at prevention and the encouragement of re-use is not very clear. It is not the aim of the study, however, to delineate the *potential* impact in great detail. Rather the *factual* impact on the national policy process and outcomes will be analysed. This will be done in the chapter nine.

Recycling and recovery

In regard to recycling and recovery, the Directive itself prescribes a range of targets to be met. Member States are bound by the deadline of five years, but they may choose the level of targets they want to achieve within the margin of 50 to 65 per cent for recovery and 25 to 45 per cent of recycling. Member States are also free to choose their own mix of packaging materials to be recycled, as long as at least 15% of each waste material is included. The national variation in policy targets is not confined to the margins, however. Portugal, Ireland and Greece are granted temporal derogation from meeting the targets. Moreover, Member States are allowed to maintain or introduce programmes aiming at higher recycling and recovery targets. Again, the front-runner countries fought hard to get this provision included and - yet another parallel - the conditions under which these measures are allowed are rather unclear. The Directive stipulates that these measures have to avoid distortions of the internal market, do not hinder compliance by other Member States with the Directive, do not constitute an arbitrary means of discrimination or a disguised restriction on trade between Member States.

Golub has argued that the targets reflect the victory of the 'single market' countries over the 'green' countries. This claim is based on the assumption that conditions for the opt-ups are very restricted (Golub 1996). This would imply a slow down for the front-runner countries, and would therefore contribute to greater convergence. It is doubtful, however, whether the opt-ups for front-runner countries are as limited as suggested by Golub. Golub argues that '*by emphasizing national self-sufficiency in waste recycling, as well as the importance of each state being able to meet agreed targets, the directive provides Britain, France and other states with legal grounds on which to block such exports...the ECJ would probably have considered such restrictions to discriminations on trade and therefore contrary to Article 30'* (Golub 1996: 23). Under these conditions the maximum targets would mean that the front-runner Member States have to increase their national recycling capacities or would have to increase incineration or landfill of packaging waste. Yet the Directive is saying only that Member States have to '*provide...appropriate capacities for recycling and recovery*' (Article 6.6), whether this must be on their own territory is unclear (Douma 1995: 12, ENDS Report Dec. 1993). If packaging waste is welcome in other countries, the capacity might also be abroad. Whether the Directive provides sufficient legal grounds to block waste shipments is also unclear. The 1993 Council Regulation on waste shipments includes a green list of items which benefit from the free movement of goods. This list includes all packaging relevant materials (No 259/93, OJ No L30, 6.2.1993, p.1.). Hence, Member States like Germany can easily argue that collected and separated plastic and paper packaging waste fall under these categories and cannot be banned.

Golub also claims that the targets reflect the least common denominator of the Member States: *'Thus, in almost all cases, the directive actually approximated recycling rates which already existed or were already planned throughout the Community. In the remaining few instances, it allowed substantial derogation which postponed any required changes to state practice. Thus there is little evidence that 'the grubbier majority will under the directive have to come up with plans to do much more', and considerable evidence that the directive represented the lowest common denominator'* (Golub 1996: 329, including a quotation of The Economist 18.12.1993). And Gehring (1997) maintains that *'(t)he majority lowered the recovery and recycling targets to a level that did not require serious adaptations of existing programmes in most Member States'* (Gehring 1997: 351).

I would argue that these scholars underestimate the adaptation pressure for national governments. As a matter of fact, probably most of the representatives at COREPER and the Council who negotiated in Brussels did so. Given the low level of recycling and recovery in a number of countries (in particular Southern Europe) an upward pressure on environmental standards is almost inevitable. This holds true also for those countries that benefit from derogation. The exemption is only temporary and just weakens the adaptation pressure.

Moreover, the existing level of recycling and recovery in general and especially in the 'laggard' countries might be exaggerated. As elaborated on earlier in this study all figures concerning waste generation and management should be treated with care. The figures are often not valid, in particular for countries with less elaborate statistical capacities. It may well be that the figures which are tabled by institutions like the OECD will not stand a more rigid EU monitoring system.

Furthermore, the 50-65 per cent recovery target is a tricky one. According to PRG figures 32 per cent of packaging was recovered in the UK of the early 1990s (PRG 1994). This country has only a small capacity for incineration with energy recovery. It may increase to burn some 8 per cent of the overall packaging waste. This is quite costly, however, given the comparatively strict EU standards for incinerators. Even when this capacity can be built up, it still means that the remaining recovery has to be done by recycling. Hence, rather than the 25 per cent envisaged in the Directive, the UK government has to ensure that 42 per cent of the overall packaging has to be recycled to meet the minimum target of 50 per cent.

Finally, the very fact that European requirements match national plans does not logically indicate that there was no EU induced adaptation pressure. As the UK case will show, national plans were adopted precisely because of European developments, or to be more concrete in anticipation of the EU targets[24].

---

[24] Even if existing national levels match European targets it would not mean that the national plans and policy are sufficient. They may be sufficient in substance but not in form. As I will argue in section 7.4.3, the measures for the implementation of the European directive have to meet certain legal requirements.

Instruments and organization of waste management

The Directive leaves leeway to Member States concerning the concrete instruments and the organization of waste management. Take-back responsibilities for trade and industry were mentioned in the first internal drafts, but were dropped. According to the Environment Commissioner it was on line with the principle of subsidiarity that Member States are to decide for themselves how return and management systems should be organized and funded (EC Packaging Report Nov. 1993). The Directive requires only that the collection and recovery systems a) have to ensure that the objectives of the Directive are met; b) shall be open to the participation of the economic operators of the sectors concerned and to the participation of the competent public authorities; c) shall apply also to imported products under non-discriminatory conditions, and (d) shall avoid barriers to trade or distortions of competition in conformity with the treaty (Article 7 (1)). Article 15 stipulates that the Council adopts economic instruments to promote the implementation of the objectives set by the Directive. The Article also details under which conditions Member States may adopt economic instruments in the absence of supra-national action. '*Member States may, in accordance with the principles governing Community environmental policy, inter alia, the polluter-pays principle, and the obligations arising out of the Treaty, adopt measures to implement those objectives*'. This rather vague wording is not surprising, given this article's character of compromise.

Conclusion

In conclusion, one can say that the exact degree of pressure towards convergence of national policies caused by the Directive is difficult to estimate. The Directive established a new, but highly complex and unclear, institutional basis for future national action aiming at ambitious prevention, re-use, recycling and recovery of packaging waste. It is possible that front-runners may not slow down. This suggests less convergence. But 'laggards' might be pushed more than is apparent at first glance. This points to more convergence.

8.4.2 Article 100a (4) as an escape from convergence?

Article 100a (4) provides another opportunity for front-runner countries to pursue their own approaches and thereby weaken the degree of harmonization and convergence. This Article stipulates, that '*(i)f, after, the adoption of a harmonization measure by the Council acting by a qualified majority, a Member State deems it necessary to apply national provisions ... relating to protection of the environment, it shall notify the Commission of these provisions*'. Hence, Member States are allowed to opt out of the internal market rule. But only when the national measure does not constitute an arbitrary means of discrimination or a disguised restriction on trade between Member States (Krämer 1997: 182).

As a matter of fact, when the Netherlands, Germany and Denmark voted against the Council's common position in 1993, they issued a joint declaration in which they announced that they would probably use Article 100a (4) to maintain their ambitious programmes

(Commissie Verpakkingen 1994b: 63).Yet the choice for Article 100a as legal base rather than 130s has restricted the national discretionary space. Article 130t states that '*protective measures adopted pursuant to Article 130s shall not prevent any Member State from maintaining or introducing more stringent protective measures. Such measures must be compatible with the Treaty'*.

The national discretionary space is smaller in Article 100a (4) as compared to Article 130t for two reasons. First, according to the majority of legal scholars, the term "apply" only allows the *maintenance* of national measures, but does not include the introduction of new ones (Geradin 1993: 176, Krämer 1997: 182). A number of scholars, in particular German authors but also Flynn (referred to in Geradin 1993: 176), hold, however, that the term also means the introduction of new measures (Krämer 1997: Footnote 8). In contrast, Member States are allowed under Article 130t to *introduce* new measures under certain conditions[25].

The second factor that restricts the national leeway for measures based on Article 100a (4) as compared to Article 130t is that it is uncertain, how strict measures can be that Member States are allowed to maintain. In the first case in which the Commission accepted the reference to Article 100a (4) by a Member State, a total ban on PCP by Germany - rather than a restriction in use as stated in the respective directive, the Court of Justice annulled the decision for formal reasons (C-41/93 [German ban on PCB], ECR 1994: 1829). Since Article 100a explicitly deals with harmonization, while Article 130s deals with environmental protection and explicitly allows for more stringent protective measures, I would argue that the scope for stricter standards is larger under the latter provision, even though it is stated, that these measures must be compatible with the Treaty, hence must -among other things - not hide barriers to free trade (see Geradin 1993: 175).

In conclusion one can say that the conditions under which the opening clause 100a (4) does apply, are far from being clear (Douma 1995: 112). It even not clear which countries may refer to Article 100a (4). It has been argued that only those Member States are allowed to maintain national measures who have voted against the respective Directive. Whether this is the case is subject to legal debate (Douma 1995: 112). It is worthwhile to note, however, that in the case of the Netherlands the vote against the Directive was partly driven by the need to ensure the opportunity to refer to Article 100a (4) (Interview VROM 4.12.1995, De Volkskrant 14.5.1994).[26]

Hence, the opening clause Art 100a (4), as well as the Directive itself, provides a rather tricky institutional basis for future national action. Countries trying to protect existing legislation and to introduce new ones will meet unclear conditions.

---

[25] The Treaty of Amsterdam clarified the situation. The new Article 95 that replaced the old Article 100a, allows Member States to introduce new measures under certain - very strict - conditions.

[26] This may also be the case for Germany. This may be indicated by the fact that the responsible official of the German minister of Economic Affairs even pretended that Germany did not vote against the Directive (Interview BMWi 10.12.1996), This point may also indicate the confidence the Minister of Economic Affairs places in the directive in contrast to the Ministry of the Environment.

8.4.3 Legal requirements for the implementation of Directives

The Directive on Packaging and Packaging Waste has to be implemented by law, regulations and administrative provisions. Article 189 of the Treaty states that Member-States are free to choose their own forms and methods to implement a directive (Article 189). The room to manoeuvre for national governments in regard to the degree of the legal codification of these forms and methods is restricted, however, by European law and its interpretation by the European Court of Justice. The legal arguments are important for the Dutch and the British in particular, given the predominant voluntary approach in the Netherlands, and the preference for self-regulation in the UK.

The European Court of Justice has defined legal requirements for the form in which directives have to be implemented (C-59/89, [TA Luft (lead)], 1991, ECR p. I-2607, C-361/88, [TA Luft (sulphurdioxide)], 1991, p. I-2567). These requirements are based on the principles of legal security and legal protection. These principles demand that the implementation measure is legally binding and clear. The latter means that the concerned parties know the rights and duties which the measure assigns to them (Heukels 1993: 65, van de Gronden 1997: 178-179). For these reasons most directives are implemented by measures based on public law.

Also, according to ECJ case law, directives have to come to 'full effect' (Steyger 1993: 180). This implies certain obligations for Member States to make directives enforceable. Member States are held liable when the desired result is not achieved. The ECJ has shown that it accepts very little by way of defence if a directive is not properly implemented. The Member State concerned cannot evade its responsibility by blaming private actors, courts or decentralized governmental institutions. This means, for instance, that reliance on the self-regulatory capacity of industry is risky. Member States cannot be sure that the directive is properly implemented (Steyger 1993: 176). This makes national government more cautious in choosing other implementation instruments than command and control measures based on public law.

In short, EU law induces a strong degree of legal codification of national packaging waste policies. This points towards a convergence of formalized policies which might clash in some cases with traditional national regulatory styles. This might have caused more domestic trouble, especially in the UK, than the national negotiators in Brussels were probably aware of.

8.5 Summary and outlook to domestic processes

The German packaging waste policy endangered the functioning of the internal market. Under the Danish Bottle Case jurisdiction it was difficult, however, to challenge the German approach. This was one of the reasons why the Commission chose the way of positive integration to harmonize national packaging waste policies. The other motive was that the Commission saw a window of opportunity to revitalize its own efforts to develop a European packaging waste policy. The Europeanization of the packaging waste changed the structure of the political game. New procedures have to be followed, new actors became important. The European Commission set the agenda and printed its mark on the first drafts. Due to internal conflicts, however, it did not follow a coherent strategy. The content of the final Directive was not very much influenced by the Commission. The European Parliament initially backed the proposal of the Environment Directorate of the Commission and opposed the less environmentally-oriented position of the majority Member State governments. Finally, however, the EP gave its assent to the Common Position of the Council. The Member States were able to protect their own preferences in regard to the level of environmental protection. Though the majority wanted a convergence towards a medium level of regulation, they granted exceptions for both the front-runners and the laggards. Even under conditions of majority voting Member States hesitated to overrule minorities when substantial interests were at stake. The result was a compromise full of vague formulations. The national leeway for the maintenance or introduction of national measures is not clearly circumscribed. This may lead to a stronger role of the Commission and the ECJ. Member States may find the Commission and eventually the Court rejecting new measures in the packaging waste sector that most probably had been in conformity with European law prior to the adoption of the Directive. A company believing itself to be adversely affected by the domestic packaging waste policy of a Member State, may now choose an appropriate case and test whether the Directive has modified the legal situation as compared to the status based upon the Danish bottles case jurisdiction (see Gehring 1997: 351-352).

How did the Netherlands, the UK and Germany implement the Directive? How did the three countries fill in their discretionary space in regard to policy targets and instruments? Did Germany weaken its strict packaging waste policy? Did the Dutch Covenant survive? How did the British government cope with the double challenge to introduce binding regulation and to increase its recycling efforts? The following chapters will show the impact of the European Directive of Packaging and Packaging Waste on the United Kingdom (9), the Netherlands (10.) and Germany (11). It will also reveal that other developments within and beyond these countries impinged upon the national packaging waste policy development.

The national packaging waste policy debates were shaped not only by the perceived adaptation pressure the Directive would induce on the domestic policies. They were also shaped by the intended and unintended consequences of the national packaging waste programmes adopted in the early 1990s. The relative importance of European and domestic factors varied, however, between the three countries.

The European Packaging Directive had a profound impact on the British policymaking process, even though the British government wanted the Directive and voted - unlike the Netherlands and Germany - pro-Directive. The adaptation pressure started already with the Commission's official draft proposal 1992. It was clear to government and many companies that the non-statutory recycling plans of local authorities, the small-scale recycling credit schemes, and the lukewarm commitments of British trade and industry would not be sufficient to meet the targets of a European Directive, even though the precise substance of this Directive was not yet clear in those days. The government was aware of the need for stricter standards and more responsibilities for trade and industry and therefore increased its efforts to force business to establish a sophisticated industry-led solution. What followed was a lengthy, controversial and sometimes chaotic process which developed through different stages, each associated with a specific industry network, created to meet the challenges of the Minister of the Environment, Gummer. This process, characterized by cumbersome intra-industry and industry-government interaction finally resulted in the *Producer Responsibility Obligations (Packaging Waste) Regulation* which came into force on 6 March 1997 (Statutory Instrument 1997, No. 648).

In contrast to the UK discussion, Dutch policy was informed not only by developments on the European level and in other member states but also on the domestic level. The implementation of the Packaging Covenant was perceived as quite successful by the government, trade and industry, and many independent observers. Most intermediary targets for prevention and recycling could be met. The early developments on the European level were welcomed by the government. When it turned out, however, that the recycling levels achieved on the domestic level could be undermined by supranational policies, and that the rather effective voluntary agreement would be replaced by more legalistic and "bureaucratic" measures, the Directive came to be seen with mixed feelings by government, trade and industry, and the wider public. As in the British case, the legal implementation process of the Directive took longer than originally expected and allowed for by the Directive. It mainly took place in two interrelated arenas: first, intra-governmental negotiations between the Minister of the Environment and the Minister of Economic Affairs resulting in a draft regulation; second, government-industry negotiations with the view to continue the covenant approach, albeit based on public law. These negotiations resulted in the Packaging and Packaging Waste Regulation (Staatscourant 1997, no. 125, 4.7. 1997) which came into force on 1 August 1997 and the Packaging Covenant II (Staatscourant 1997, No. 247, 15.12.1997), which came into force on 16 December 1997.

The German discussion initially focused very much on domestic developments, namely the implementation of the German Packaging Ordinance, in particular the working of the recycling system for sales packaging organized by the Duales System Deutschland (DSD). The implementation caused tremendous problems and the DSD was almost bankrupt. Paradoxically, the problems had a lot to do with the success of the collection system. Much more sales packaging was collected than envisaged by the government. Industry lacked the capacities to recycle all the collected waste. As a result, waste was exported and/or illegally dumped. In addition, the DSD ran into a serious financial crisis. Costs were higher than expected, partly due to subsidies paid for the recycling of waste and partly due to free-rider attitudes of German companies. The first draft of an Amended Packaging Waste Ordinance, issued in 1993, was therefore less informed by anticipation of the European Packaging Directive than by the attempt to stabilize the Dual System. Besides the problems of the DSD, the German 72 per cent quota for refillables was the main issue. As the share of refillables threatened to fall below the quota, domestic and foreign companies feared that retailers would start to discriminate against one-way drink packaging. They were afraid of barriers to trade and the EU Commission repeatedly asked the German government to relax the quota.

Table 22 Implementation of the Packaging Directive: an overview

Year	**EU**	**UK**	**NL**	**D**
1992	24.8 Commission's Official Proposal for a Directive			
1993		27.7 Government Challenge to Industry (PRG group)		December 1993 BMU Draft on Amended Ordinance
1994	4.3.Common Position of the Council 20.12 Adoption of Directive	Feb. Draft PRG Plan Nov Final PRG Plan		4.2 Public Hearing
1995		May Consultation Paper on Producer Responsibility Dec. Agreement on Financing Principles	22.12 Draft Regulation	Dec. EU Commission Letter of Complaint on Refillable Quota
1996	1 July: Deadline for transposition into national law	June Draft Regulation		28.1. New BMU Draft 28.2 Public Hearing Sept. New BMU Draft 6.11 Government Draft adopted 12.12 Parliament approves Draft
1997		6.3 Final Regulation	29.6 Final regulation, and agreement on new covenant 15.12.Covenant come into force	25.4 *Bundesrat* rejects Government Draft 21.5 New Government Draft
1998				27.3 *Bundesrat* postpones decision 8.5 *Bundesrat* postpones decision 28.5 *Bundesrat* agrees on Amended Ordinance

However, the government was determined to maintain them. A certain relaxation was included in the second draft of the Amended Ordinance but was opposed by the majority of the *Bundesrat.* The German situation therefore reached a deadlock which was broken only after the *Bundesrat* agreed on the government's draft on 29 May 1998. Table 22 provides a synopsis of the major stages of the packaging waste policy developments in the three countries.

One factor which delayed the implementation process and increased the complexity in Germany and the Netherlands was the notification procedure, which had to be in line with the procedure as set forth in Directive 83/189/EEC (O.J. No. L, 26.4.1983, 109). This Directive, adopted in 1983, laid down a procedure for the provision of information in the field of technical standards and regulations. Technical standards and regulations could result in barriers to trade, and the Directive aimed at providing the Commission and other Member States with information about such policy programmes before they were adopted. This notification procedure applied also to the packaging and packaging waste measures aimed to transpose the Directive into national legislation and virtually all other related measures. Member States had to send their drafts for notification to Brussels and a three month standstill period had to be taken into account. When the Commission and/or other Member States delivered detailed opinions within these three months, to the effect that the measure envisaged should be amended in order to eliminate or reduce any barriers to trade, the adoption of the drafts had to be postponed by six months from the date of notification.

In the remainder of the second empirical part the national packaging waste policy developments parallel to and after the European development of the Packaging Waste Directive will be reconstructed for each country separately (chapters 9-11). Unlike the first empirical part, this part will start with the UK because this country faced the highest adaptation pressure and the government perceived this at a very early stage. The policies adopted to implement the European Packaging Directive will be compared to the policies of the early 1990s. The differences and similarities, hence the continuities and changes, will be described along the policy dimensions developed in the theoretical part. This section identifies the forces and mechanisms that either resulted in changes, or prevented changes in the countries under study. A combined cross-national and longitudinal comparison will be made in chapter twelve. Here, the extent of convergence and persistent diversity and the underlying forces and mechanisms will be discussed. The chapter will also provide a conclusion in which the findings of the case study are viewed in a broader theoretical and empirical context.

9. The inadvertent emergence of British packaging waste policy

9.1 German rubbish and the European Commission's initiative

The British packaging waste policies were affected by two external developments: The imports of cheap packaging waste materials for recycling; and the EC Commission's initiative for a European packaging directive.

Cheap plastic and paper/cardboard packaging waste increasingly entered the British market in the early 1990s. This was a consequence of the German packaging waste policy, in particular its lack of domestic recycling capacities (see chapter 11). Price differences between British and German packaging waste materials were quite substantial. In the second half of 1993, for example, German waste paper - a large source of fibre for cardboard boxes - sold 10 to 15 per cent less than British mixed waste paper, even when transport costs were taken into account (ENDS Report Nov. 1993: 12). Plastic provides an even more extreme example, German HDPE plastic waste material was 30 per cent cheaper than virgin material, while the same fraction in the UK was 20-25 per cent more expensive than virgin material (ENDS Report Oct. 1993: 225: 14). The British recycling companies welcomed the cheap German imports and argued that only the mix of cheap German material and more expensive British material would make economic sense. The government and other companies, however, perceived the imports from Germany as a threat to the British collection schemes which were still in an embryonic phase. They argued that it would make no sense to collect used packaging when the recycling firms would prefer German packaging waste.

The government reacted in three ways to this threat. First, as already elaborated on in chapter eight, the government demanded on the European level either an infringement procedure against the German provisions or the harmonization of national packaging waste laws. Second, the British government talked bilaterally with the German government. Germany had to take substantial action soon unless it wanted to be responsible for the *'most serious setback to European recycling efforts within recent memory'* (quoted in ENDS Report Oct. 1993: 14). The British government called on Germany to amend its Packaging Ordinance that less packaging should be collected or more incineration allowed. Töpfer responded by announcing that he would stop the export of plastic packaging waste to EU countries at the end of 1993 because sufficient capacity to recycle Germany's waste plastic was coming on stream both in Germany and in non-EC countries (ENDS Report Oct. 1993: 14). Third, the government asked industry for short-term action to safeguard the UK's plastic and paper and board recycling infrastructure. Since this challenge to industry was combined with a more long-term challenge to industry to establish a system capable of dealing with the demands associated with the European packaging directive, this point will be described below.

The second development external to the UK which strongly influenced the packaging waste policy agenda in the UK was the Commission's proposal for a European packaging directive presented on 24 August 1992. Though the British government generally welcomed the objective to harmonize national packaging waste policies, it strongly opposed the targets envisaged by the EU Commission, i.e. 90 per cent removal from waste stream, a minimum of 60 per cent recycling of each material, and a maximum of 30 per cent incineration with energy recovery and a maximum of 10 per cent landfill. The government also criticized the predilection for recycling over incineration with energy recovery, and saw no sense in recycling without markets for recycled materials.

The British government thought that the ambitious targets set by the Commission could probably be lowered in the course of negotiations with and among the European Parliament and the European Council. But the British collection, sorting and recycling infrastructure was still at an embryonic stage. This was partly due to a delay in the implementation of stricter landfill standards and control which was meant to increase landfill costs and tip the balance towards recycling (Stupples 1993: 61). In addition, the government's recycling credit scheme was perceived even by government itself as under-funded and it was clear to the government that the lukewarm commitments by the British trade associations, such as the COPAC Plan, the BRC Guidance Notes, and the Packaging Standard Code of Practise would not be enough to meet the objectives of a coming European packaging directive. Local authorities reported an average of less than 5 per cent of municipal waste recycling in 1992 and only a minority intended to meet the 25 per cent recycling target set by government (Coopers & Lybrand 1993).

As in the case of many other environmental directives, the impact of the European Packaging Directive on British national policies was quite significant. This was not due simply to the low level of recycling in the UK. Probably even more important was the different regulatory style prevalent in the UK, and the kind of policies required by - or appropriate to suit - the Directive. What followed was a cumbersome and sometimes chaotic process that took much longer than expected, and resulted in a fragile consensus between government and the majority of industry on the lowest possible level of standards. The traditional policy approach, namely bilateral interest intermediation between government and companies, and the predilection for self-regulation and tailor-made solutions for companies, clashed with the kind of standards required by the packaging directive.

The European Packaging Directive influenced domestic policymaking in Britain long before the formal adoption of the directive in December 1994. There were two reasons for this: First, the government wanted to anticipate the impact of the directive on British policy strategy and industry behaviour and second, and probably more important, the British government wanted to present to the European Council a practical alternative to the legalistic and interventionist approaches planned by the Commission and adopted by some other member-states, in particular Germany and Denmark. As a British industrialist said: '*If you want to join in the debate to talk about football tactics, then you better play a bit of football, because otherwise people will say - look, go and talk about cricket*'(Interview British Industrialist 24.6 1996). The British case made clear that EU policymaking and national implementation were not sequential and empirically separated processes, but took place in parallel. The anticipation and implementation process of the European Directive can be divided into five stages, whereby the first two stages were prior to the adoption of the European Directive.

1) The challenge of the Minister for the Environment Gummer to industry (July 1993);
2) The short-term action ensuring British collection and recycling infrastructure and the development of a business plan by the Producer Responsibility Industry Group (PRG, draft presented February 1994, final version November 1994);
3) The attempt by the VALPAK-Working Representative Advisory Group (V-WRAG) to distribute the costs of an industry scheme (November 1994- November 1995);
4) The draft regulation to implement the European Packaging Directive (June 1996);
5) The final regulation to implement the European Packaging Directive (March 1997).

9.2 Gummer's challenge to industry

The process started in July 1993, a few months before the European Council had to adopt a common opinion on the directive. The British government told industry to come up with a waste recovery and recycling plan by Christmas (DoE press release 27.7.1993). In September 1993, Gummer, the Minister for the Environment, invited representatives of 27 leading companies and challenged them: a) to adopt immediate action to safeguard the collection and recycling infrastructure threatened by subsidized foreign imports; b) to develop a staged plan which built-up progressively to recovery levels between 50-75 per cent of the overall packaging waste by the year 2000; c) to create a demand for recycled materials where appropriate standards were met; d) to establish an effective organization, spanning all the relevant business sectors, which could draw up the plan and put it into action; and e) a commitment by industry that it would meet the costs necessary to fund a new collection and processing capacity and a mechanism for raising the necessary finance. The government combined its challenge with a threat: *'If the industries concerned cannot satisfy us by Christmas, we will need to move towards a legislative approach to mandate producer responsibility'* (DoE press release 27.7.1993). Industry had to brief the Minister for planned action in November 1993, so that the Minister could present the British approach at the Environmental Council in Brussels for the negotiations on a Common Position.

The challenge to industry was different from earlier exercises in two important aspects. First, the challenge was more concrete and more elaborated. Second, rather than challenging industry in general, the government talked to a select group of leading British companies. Representatives of trade associations such as INCPEN or COPAC were not invited. This step made clear that the government was not satisfied with the self-regulatory measures proposed by the trade associations.

The threat by the government to impose unilateral action was not believed by industry, however. Summarizing the weakness of the government's threat, an industrialist described the perceptions of many of his colleagues as follows: *'It was a government with a small majority. A government who did not perceive the issue as being nearly high profile enough. A country which has never had a green political representative. It isn't on Greenpeace's agenda, it is not on Friends of the Earth's agenda, so what happens if we just go slowly and don't do a lot...'* (Interview British Industrialist, 24.6.1996).

Moreover, the companies invited by Gummer did not represent all parts of the packaging chain equally. The companies were chosen in an *ad hoc* manner. Gummer wanted the heavy players in and he referred back to his personal network of contacts, which he built up during his time as Minister at the Ministry of Agriculture, Fisheries and Food (Interview DTI 15.5.1996, DoE 12.7.1996). Therefore, the food industry and the big retailers were well represented. In addition, he invited the very big petro-chemical companies like Shell, ICI and BP, which produce plastic for packaging manufacturers. They had good contacts with the Department of Trade and Industry. The packaging manufacturers did not have as close links to the Ministry of Agriculture and the DTI as the others had. This might explain why only two packaging manufacturers were invited. No producers of tinplate, aluminium or glass packaging sat around the table

The objectives of Gummer's approach to the packaging waste problem differed from those of Germany, the Netherlands, and the European Commission's proposal in that the British government concentrated on recovery as the primary waste management option. Recovery included material recycling and incineration with waste-to-energy. Neither were recycling targets set as in the German, Dutch and the Commission's case, nor were targets set for absolute waste reduction, as in the Netherlands. The government's approach was therefore criticized by environmentalists. They objected that no targets and no timetables were set for the avoidance and recycling of waste (Stupples [FoE] 1993: 62).

9.3 The PRG Plan: regulate us, please!

The companies invited by Gummer formed an *ad hoc* group: the Producer Responsibility Industry Group (PRG). Their senior executives had overall control. The work was guided by a steering group of 15 senior executives led by a managing director of Procter & Gamble.

Table 23 Members of the (UK) Producer Responsibility Industry Group

Raw Material Supplier	BP Chemicals, ICI, Shell Chemicals (all plastic)
Packaging Manufacturer	Bowater (later renamed: REXAM, paper and board, plastic), D S Smith (paper and board)
Packer/Filler	Allied Lyons (drink), Grand Metropolitan (drink), Guinness Brewing Worldwide (drink), Coca-Cola Great Britain & Ireland (drink), Nestlé UK (food), Northern Food, Procter & Gamble (diverse), RHM (food), Unigate PLC (food), Unilever (food *et al.*), United Biscuits (food)
Retailer	Argyll Group, ASDA Group, Burton Group, The Boots Company, Kingfisher, Marks & Spencer, J. Sainsbury, Tesco Stores
Other	Forte (hotel and leisure)

When Gummer asked the companies for short-term action safeguarding UK's paper and board and plastic recycling and processing infrastructure, he added that he would take the interim plan as an indication of progress on the longer-term objectives, and legislation would follow rapidly if the industry failed to deliver (ENDS Report Oct. 1993: 13). The problems of the PRG to meet this challenge shed light on the insufficient self-regulatory capacity of British industry when cross-sectional issues are concerned. At a meeting with Gummer on 12 November 1993, the PRG presented its plan for intermediate action. As far as the paper and cardboard industry was concerned, the PRG itself admitted that it had not responded fully to the paper industry's need for immediate support (ENDS Report Nov. 1993: 12). They did not guarantee any protections for the paper and board recycling industry. The PRG's announced that collection levels would be maintained provided the cost was equal to or less than other disposal options. With this statement, however, PRG made clear that it would do no more than what would happen anyway. The ENDS Report commented: '*Where cardboard recycling is economic it will take place, but where it costs more than landfill or incineration it will not. It is difficult to see how this will alter companies' normal response to market force.*' (ENDS Report Nov. 1993: 13). It appeared that the most important issue under negotiation between retailing interests, strongly represented in the PRG, and waste paper interests, not well represented at PRG, was how a voluntary funding mechanism should be applied and which level of funding would be needed. Conflicts on this point caused the delay and were not solved. It is likely that the major retailers played for time. They might well be content to see this issue unresolved so that the government would be forced to introduce legislation to force all packaging users to pay their share (ENDS Report Oct. 1993, 225). Fortunately for the British situation, an upturn in the economy occurred which had a positive effect on the British cardboard recycling business (ENDS Report January 1994: 15).

As far as plastics were concerned, the announcement of PRG sounded more persuasive: PRG promised to safeguard over 90 per cent of the plastics recycling infrastructure for the next six months. The group stated that recycling of plastic film would definitely continue, that funding for Recoup - the industry recycling organization for plastic bottles- had also been secured for another year and that reprocessors of collected bottles had also agreed to maintain prices for the next six months at 1 October 1993 (ENDS Report Nov. 1993: 12, Dec. 1993: 13). Some weeks later, however, Recoup said that its funding was not secure, and that not all reprocessors had guaranteed to maintain their prices in order to sustain collection. The PRG's promise of full protection for plastic bottle recycling was made without any such commitment from plastics recyclers themselves. No guarantees had been offered - only a general expression of positive intent (ENDS Report Dec. 1993: 13).

Despite the serious shortcomings of the PRG's approach to help the ailing paper and board and plastic recycling industry, the Environment Secretary said he was *'very pleased'* with PRG's response (ENDS Report November 1993 (226): 13). This comment should be interpreted against the ongoing discussion on the European level. Gummer needed a successful voluntary approach to cite in the continuing negotiations on the EU Packaging Directive. DoE officials were briefed by PRG on its long-term plan so that Gummer could hold up his approach as an example for the rest of the Community to follow (ENDS Report Nov. 1993: 12). Gummer told the PRG three weeks before the Environment Council in Brussels *'I will not succeed unless I can point clearly to a successful voluntary scheme in the UK'* (quoted in ENDS Report Nov. 1993: 13). Friends of the Earth got the point when they argued that Gummer's '*political credibility is relying on a voluntary initiative*' (quoted in ENDS Report Nov. 1993: 13). The negotiation in Brussels put the Environment Minister under pressure to accept any reasonable plan from industry even if it fell short of truly effective action.

The PRG was unable to meet the deadline set by Gummer. A draft business plan was presented in February 1994 and the final report was published - after a period of consultation and revision- as late as November 1994. The plan stipulated an increased rate of recovery of packaging waste to 58 per cent by the year 2000, including 50 per cent for material recycling and 8 per cent incineration with energy recovery. Packaging waste reduction and increasing re-use was not on the PRG's agenda. And there were neither timetables nor concrete commitments announced (ENDS Report Nov. 1994: 11). It is worth noting that though trade associations such as INCPEN always argued that the usage of packaging decreases, the PRG based its plan on an *increase* in the amount of packaging entering the market from the present 7.3 million tonnes per year to some 8 million tonnes in the year 2000. PRG promised that industry would support 'buy recycled' campaigns to stimulate the market. However, the plan also said that the achievement of recovery targets will depend on the development of reliable sources of collected and sorted packaging waste, both in terms of volume and quality, to justify investment in reprocessing capacity. There is, however, a dilemma involved which can be illustrated by the situation of plastic bottle recycling, where precisely the lack of collection schemes hampered the development of end use markets (ENDS Report Nov. 1994: 13).

In the final plan PRG had reduced the planning funding for VALPAK's first year from 100 million to just 40 million pounds. And the funding required in the year 2000 has been cut from an early indication of 300-500 million pounds to 100 million pounds. The reduction was due partly to the improved economics of paper and board recycling over the previous twelve months. The situation improved since the wood pulp price doubled to 700 dollars a tonne in summer/autumn 1994, partly as a consequence of growing environmental restrictions on forest products and the threat of a strike in British Columbia, one of the world's biggest suppliers (FT 7.10.1994). It was also argued, however, that reduced cost estimation had also to do with the attempt of PRG to allay concern within the industry. Moreover, the PRG took the advice of waste management business rather late and incomplete on board. This added to estimations of cost levels which, according to some industry sources, were completely unrealistic and would not be borne out in practice (ENDS Report Dec. 1994: 12). A packaging consultant estimated the long term costs to be 270 - 550 million pounds per year - much closer to PRG's original figures. The consultant said that the organizers of an industry-led scheme should not be '*lulled into a false sense of security and fail to make provisions for price support mechanisms which can be drawn upon when the need can be proven. If overseas competitors have the benefit of secure prices and supplies by virtue of a managed market subsidized by their customers and by the consumer, UK companies can be expected to lose ground. And once the infrastructure has been lost it will be difficult to restore it*'(quoted in ENDS Report Nov. 1994: 11).

The PRG was unable to decide at which point in the packaging chain it should raise the funding necessary for the new organization. It made several suggestions. None of them would mean a serious financial burden on the retailers. This reflected the unevenness of interest representation in the PRG. The packaging chain was deeply split on this question. Even the advice of an independent expert, an eminent banker from the City and former chairman of the Monopolies and Merger Commission, appointed by PRG and the government, was not able to solve the conflict (European Chemical News 1.8.1994, Interview British Industrialist 24.6.1996). It would take another year before a fragile consensus was found on this issue (see below).

The PRG announced the establishment of a formal industry organization, provisionally called VALPAK, to oversee the implementation of the plan. The question was, however, who could administer such a permanent scheme? Unlike in the Dutch case, there was no strong trade association in the UK that represented large parts of the whole packaging chain. The broadest group was the consortium of the packaging chain COPAC, a group of seven trade associations. However, COPAC's failure to produce an acceptable plan in 1992 put it out of favour with the government (ENDS Report Sept. 1993: 16). As a consequence of the lack of organizational infrastructure, a totally new organization had to be set up. In its draft, the PRG group itself was rather pessimistic about its authority to put the business plan into practice and to ensure compliance from industry. They wrote in the draft: '*Experience with earlier attempts to develop packaging recovery voluntarily suggests that some players would opt out from any purely voluntary involvement. This would jeopardise its accomplishment and place an undue burden on the conscientious participants*' (PRG 1994: 5). Therefore the PRG asked for the legal backing of their strategy: '*The plan can only move ahead with a commitment by government to provide the legislative backing to enforce compliance by all members of the packaging chain*'(PRG 1994: 5). A situation emerged where industry asked: 'Regulate us, please!' and the government was reluctant to do so. It took another half year and repeated pressure by the PRG group before government accepted that regulation was needed. '*We will tell them that we can deliver the plan - if they deliver the legislation*', said a PRG member shortly before the government accepted the statutory backing of the plan (quoted in ENDS Report Sept. 1994: 13).

9.4 V-WRAG or who pays?

'The feeling that everyone is passing the packaging levy parcel and hoping the music won't stop' (Director of V-WRAG)

After having submitted their plan, the PRG disappeared and made place for a new organization, the VALPAK - Working Responsibility Advisory Group (V-WRAG). There were two reasons for this. First, both PRG and the government were rather optimistic about the achieved results. The government claimed that the work was near completion (DoE news release 28.9.94) and the Members of PRG, all of them managing directors or board members of their companies, said that they had done their job, and did not want to put any more time into the project. Second, the PRG and government believed that a broader organization was needed, including all parts of the packaging chain. The packaging industry, in particular, had heavily criticized the PRG and its plan as the following quote illustrates: *'Unfortunately, the group (PRG) has failed to broaden its membership to reflect the importance of packaging manufacturers in properly developing such a plan. However, it has suggested that these same unrepresented packaging manufacturers should pay any additional costs of increasing recovery and recycling rates. In addition, the group's estimates of the amount of these costs may well prove to be wildly inaccurate. And now it is reported that the group is developing a plan to implement its ideas through a top-heavy bureaucratic organization to be called VALPAK, whose structure will further disadvantage packaging manufacturers. Surely the time has come for Mr Gummer to call a halt to this process which he, himself, initiated. He should insist that the group reconstitutes itself to be more broadly representative of the entire packaging chain before finalising proposals which are of such long-term, strategic significance to this leading industry'* (Linpac Plastics International, Letter to the FT, FT 23.9.94).

V-WRAG, the new group, consisted of representatives from some 50 companies, including interests from all relevant packaging materials. Its chairman was a senior figure, the outgoing president of the British Chamber of Commerce (FT 10.10.1994) and its managing director had 25 years experience as a manager in a packaging manufacturing company. His appointment symbolized the better representation of packaging manufacturer interests than in the PRG.

In the meantime, the European Packaging Directive had been adopted. V-WRAG was therefore challenged to set up a collection, recovery and recycling system that could meet the European targets for recovery (50-65 per cent) and recycling (25-45 per cent).

The main task of V-WRAG was to distribute the costs associated with the future organization VALPAK across the packaging chain. Industry within and outside V-WRAG was deeply split on the question of where a levy was to be raised. Generally speaking there were three camps. The raw material and packaging producers, argued that the levy should apply at the retailer stage. They argued that the levy amounted to a small proportion of retailer's margin, and the retailers were closest to the consumer and hence it would most likely passed to them. The retailers were in favour of a levy at the stage of the packaging manufacture, since they are the producers of packaging. A third group headed by large packers and fillers such as Unilever, Procter & Gamble and Coca Cola, wanted a shared approach (ENDS REPORT July 1994: 15).

In April 1995, V-WRAG put forward four proposals to the Department of the Environment. From the perspective of a representative of a large packing and filling company the main difficulty with the policy process at this stage was that *'each of these proposals was a very narrow sector proposal. It didn't start by saying "What is the best approach for the UK as the whole?; i.e. how will we share the pain?" Instead, it was "What is the best approach for my business sector, regardless of anyone else". You had all this nonsense of people going to extraordinary lengths to argue why they had nothing to do with it'* (Interview British Industrialist 24.6.1996).

After extensive consultation within industry and between government and industry, V-WRAG, INCPEN and several trade associations developed a shared producer responsibility approach. V-WRAG claimed that 'the plan has the unanimous support of the 50 V-WRAG members' (V-WRAG information sheet 30.8.95), and that it built on the common ground between all sectors of the industry (V-WRAG information sheet 13.6.95). This industry plan was based on the traditional British environmental policy style. It was modelled after the Health and Safety at Work Act. Within the framework of an umbrella act, guidelines describing best practice were to be developed in regard to household packaging waste. In addition, it was proposed that individual companies had to negotiate individually the specific recovery and recycling responsibilities of the packaging waste arising on their premises with the Environment Agency. It was estimated that all business sectors contributed equally, i.e. 25 per cent to the total costs (ENDS Report September 1995: 248, Interview British Industrialist 24.6.1996).

The government, however, was not prepared to accept this compromise, even though it appreciated the stated consensus. '*We made amendments, because the plan did not give the guarantee that the targets will be met. They are not subject to infringement procedures*'(Interview DoE 12.7.1996). The DoE argued that it remained unclear *'how the legal obligation for individual businesses would be set out in a way which enables the government to implement the EC Directive on Packaging and Packaging Waste* (DoE Paper 1995b: 3)· The government also said that *'It would not be acceptable to leave this target (which could apply to 350,000 businesses...) to the discretion of the businesses or to be settled in negotiations (i.e. on a case by case basis) with the Environment Agency. To be enforceable and legally effective as a means of implementing the EC Directive, the relevant targets must be included in national legislation so that any business contemplating individual route compliance knows in advance what is required'* (DoE 1995b: 3).

V-WRAG was very disappointed about the government's reaction and said that it would resist any imposition of arbitrary government targets for recovery or recycling of packaging waste. *'Targets will be inaccurate if they are based on packaging waste figures that are not available. V-WRAG has developed codes of practice and other details. These would obviate the need for arbitrary government targets'* (V-WRAG information sheet 30 August 1996). Hence, the European approach to set obligatory and quantified targets that Member States are obliged to meet, clashed with the British style of codes of practices and bilateral negotiation between individual companies and implementation agencies within the broad legal framework. This clash of administrative styles had a negative impact on government-industry relations.

It also had a negative impact on intra-industry negotiations. Though trade and industry was united in its opposition to the government's approach, it was divided on the best reaction to it. In the period between August and December there were *'a lot of under-the-table negotiations. People were looking for a compromise. There were some 200 different shared approaches'* (Interview DoE 12.7. 1996). In the autumn of 1995, the conflicts within the packaging chain reached a peak. An industrialist said: *'The even-handedness of some people involved in this exercise was questionable. At the end there were changes in staff. And the tension became so great that the whole process was extremely close to breaking up.... it has to be said that there were probably a limited number of companies who were trying to make the process go nowhere, who believed that their business should have nothing to do with the plans. There was a destructive element in the process, leading to forms of behaviour which are not acceptable, like, providing untrue information to the press. People also agreed to certain things and then ignored what they had agreed to. I mean, it became a very unpleasant environment. The V-WRAG process was a dreadful process*' (Interview British Industrialist 24.6.1996).

V-WRAG was unable to present a financing option that had a broad majority across the packaging chain. In contrast to the Netherlands (SVM) and Germany (BDI, DIHT), there was no organization capable of aggregating cross-sectoral and cross-material interests and to make credible commitments to the government. Gummer had to ask an independent senior businessman to broker a deal. On 15 December 1995 Gummer met with representatives of 32 leading companies, some of them the most senior figures in British Industry. The V-WRAG management was not invited. The group 'thrashed out' a sort of compromise, at least for the first two years. The 'Parker Plan' distributed the obligations as follows: raw material suppliers 5.5 per cent, converters 14.5 per cent, packers/fillers 35 per cent and retailers 45 per cent (Material Recycling Week 12.4.1996: 14). After this compromise, Gummer appointed a new group, the Packaging Advisory Committee (PAC), to help him draft the regulations needed to implement the European Packaging Directive. This group was made up of eight senior officials of leading companies representing all parts of the packaging chain.

9.5 The draft Regulation

After three years of '*gruelling*' (FT 19.6.96) negotiations with industry, government presented its approach to implement the European Directive. In contrast to the 58 per cent recovery target initially promised by the PRG group, the final output of the negotiation was the lowest recovery and recycling target allowed for in the Directive: 50 per cent recovery, 25 per cent overall recycling and 15 per cent recycling of each material - all by 2001 (DoE 1996a). The government's draft regulation also set interim targets for 1998: 40 per cent recovery and 8 per cent recycling. In contrast to the draft version, it was not only companies that used less than 50 tonnes packaging or packaging materials which were exempted but also companies with audited accounts of less than 1 million pounds (until 2001 companies with less than 5 million pounds).

Each of the other companies, estimates run from 25,000 to 50,000, had an individual legal obligation to contribute to the 50 per cent recovery and 15 per cent recycling of each of the materials they used. The distribution of responsibilities across the packaging chain was calculated according to the proportion of each sector's activity and business turnover directly concerned with packaging. The result of this calculation differed only slightly from the compromise on financing principles reached on 15 December 1995. According to the new formula, the allocation of the responsibilities was as follows: raw material suppliers 6 per cent, manufacturers of packaging 11 per cent, packers/fillers 36 per cent and retailers 47 per cent. That meant for example that a packer/filler had to prove that he took responsibility for 36 per cent of the 50 per cent recovery target and the 15 per cent recycling target, for the packaging he used. The overall costs were estimated at 250 million to 500 million pounds by the year 2001. That is five times more expensive than Gummer's initial estimate of 50 million to 100 million pounds (FT 19.6.1996), but still less expensive than the German system which cost 1.7 billion pounds annually. As in the Dutch or German situation (see chapters 9.3 and 9.4), individual companies could join a collective system. The legal obligation would then rest with the collective scheme rather than the individual businesses as long as the appropriate recycling and recovery levels were achieved. It was planned, however, that the British scheme allows for rival recovery schemes, in order to ensure competition.

Gummer was very proud of the results. According to him, Britain's shared producer responsibility system - the only one in Europe hammered out with industry - will ultimately prove the fairest and most cost-effective way of complying with this Directive (FT 19.6.96). *'No-one wants the bureaucratic system in Germany and the mixture of blackmail and obligations in Holland'* (quoted in FT 12.7.1996). The draft was not based on a large consensus among industry, however. The chairman of the PAC said that, *'There are still serious areas of concern in these proposals and there will be changes'* (FT 12.7.1996). The packaging manufacturers in particular were still critical. The Chairman of the British Fibreboard Packaging Association said that the UK government's draft regulation *'just will not work'*. He also said that the regulation *'will result in heavy regulations that cannot be met, huge administration costs, disadvantages for the UK manufacturing industry and problems for the packaging industry in increasing recycling targets'* (Packaging Week, quoted in EU Packaging Report July 1996).

9.6 The Producer Responsibility Regulation

It took another eight months before Parliament passed the Producer Responsibility Regulations at 4 March 1997. This happened just three months before the existing Parliament rose for the last time before the coming election. The main provisions of the regulation did not deviate from the draft version. The distribution of financial responsibilities were still allocated according to the '6,11,36,47' formula. The overall targets for recovery and recycling were slightly increased to 52 per cent and 26 per cent respectively, as well as the material specific targets (16 per cent per material). The short term targets were slightly weakened. This reflected the concern of industry, that the short term targets were unrealistic. But they built up more progressively, in order to allow for a more smooth adaptation to the levels required in the year 2001.

Table 24 Recycling and Recovery Targets UK Scheme

	Draft Regulation 1996		**Final Regulation 1997**	
	Recovery	**Recycling**	**Recovery**	**Recycling**
1998	40	8	38	7
1999	40	8	38	7
2000	40	8	43	11
from 2001 on	50	15	52	16

Companies need either certificates to prove that they have met their obligations, or they can join a collective scheme. The companies and schemes have to register and to furnish a certificate of compliance to the appropriate agency. They have to furnish information in connection with competition scrutiny and allow agencies to enter and inspect their companies. It is an offence to contravene any of these responsibilities, and liable to a fine.

In January 1998 a system of packaging recovery notes has been introduced. These notes can be issued by packaging reprocessors such as paper mills, glass bottle makers, or incineration firms in exchange for cash. The reprocessors must be accredited by the Environment Agency. Obligated companies or their compliance schemes must either obtain these notes from accredited processors or prove that they have recovered waste at non-accredited reprocesssors (ENDS Report, Feb. 1998: 17).

The first deadline for the registration of individual firms or collective schemes was 31 August 1997. In July 1997, only 12 companies had registered individually. Most firms will join a collective compliance scheme. In early 1998 ten compliance schemes had gained approval from the Office of Fair Trading. *VALPAK* is the biggest. It had some 750 members in July 1997, including large packer/fillers and retailers, e.g. Coca Cola & Schweppes Beverage; Procter & Gamble, Anheuser-Bush, Nestlé, Booker and Sainsbury. It will probably be responsible for some 70 per cent of the total recycling and recovery obligations (ENDS Report Feb. 1998: 18). A number of large food retailers, Asda, Sommerfield and Co-operative Retail Service organized a scheme called *Wastepack*. The federation of the dairy industry announced a separate scheme for companies from the dairy sector, *Difpak*. The waste giant Biffa announced a scheme called *Biffpack*. *Cleanaway, Waste Link* and *Onyx* are organized by other waste management firms. Companies in the paper sector organized in the recycling initiative REPAK have launched a system called *Paperpack.* Eight major glass container manufacturers organize *Glaspak*. UK Waste another large waste company and Marks & Spencers plan a separate system for the high street retailer and its suppliers, to be called *Recycle UK* (ENDS Report Feb. 1998: 18, European Packaging & Waste Law, July/August 1997).

9.7 The future of British packaging waste policies

The labour party criticized the regulation when it was in the opposition. The new Labour Environment Minister Meacher announced that he wanted to change the regulation. The Minister argued that the retailers did not take a fair share of their responsibilities, because they concentrate their recycling efforts on back-door waste (transport packaging), which is easier to collect and to recycle than sales packaging. In a statement on 5 June 1997 the Minister presented a two-staged approach to changing the regulation. In the first stage he wanted to look at the sharing of responsibilities (6,11,36,47) and the exemption for small businesses (under 50 tonnes a year). He argued concerning the latter that '*those affected here might be those who manufacture packaging materials and find that importers of significant volumes of packaging or packaging material can escape obligations*'(quoted in European Packaging & Waste Law, July/August 1997: 23). In a second stage, a more fundamental review is planned with a view to requiring all obligated businesses to take part in household packaging waste recovery on an equitable basis and strengthen the role of local authorities and voluntary bodies (European Packaging & Waste Law, July/August 1997: 23).

It is unlikely that the UK will meet the minimum targets of the Directive. Apart from delays in the practical implementation there is another reason for that: In 1997 approximately 28 per cent of the packaging waste was recovered (ENDS Report Feb. 1998; 17).This is nowhere near to the 52 per cent recovery target. Moreover, there is a lack of waste incineration facilities which recover sufficient energy, most of the recovery has to be done by material recycling rather than energy recovery. It is estimated that 42 percentage points out of the 52 per cent targets have to be achieved by material recycling. This is substantially more than the 26 per cent recycling target set by government or the 25 per cent minimum target set by the EU directive.

9.8 Continuities and changes in British packaging waste policy

Adaptation to the requirements of the European Packaging Directive was the major driving force behind the process of policy change in the UK. The packaging waste announcements in the 1990 White Paper and the provisions in the 1990 Environmental Protection Act were an expression of an unrivalled peak in the pro-environmental attitude of the public. This environmentalism was carried into the government by the relatively pro-environment Minister Patten. In the subsequent years, however, economic issues became again far more important than environmental ones. And it is very likely that the government would have accepted the vague business plan of COPAC and the code of practices. It was only due to external events, the consequences of the German and other national schemes and the environmental provisions of the European Packaging Directive, that the British government pressed for more action of industry.

Table 25 Continuities and changes in British Packaging Waste Policy

	White Book, 1990, Codes of Practice 1992	Producer Responsibility Regulation 1997
FORMALIZATION	**low**	**much more formalized**
degree of legal codification	Government and industry statements, not legally binding	statutory instrument, based on public law
STRICTNESS OF STANDARDS	**low**	**slightly stricter**
packaging reduction targets	no	no
re-use target	no	no
material recycling targets for one-way packaging (levels already achieved)	half of recyclable household waste, i.e. some 25%, incl. Composting (1990= > 5%)	26% (overall) at least 15% for each material
recovery target	no	52%
incineration target for packaging waste	no explicit targets	(included in recovery target)
landfill for packaging waste	no explicit targets	no explicit targets
time horizon	2000	by 2001
prescription of implementation	general vague target, no prescription of implementation	more detailed
flexibility	complete flexibility	less flexible
ALLOCATION OF RESPONSIBILITIES	**public and private**	**more privatized**
collection, sorting and recycling	public and private pilot schemes	public and private (all sectors of the packaging chain)
licensing, monitoring, assessment	-	Environment Agency (EA)
dispute settlement/ enforcement	-	Administrative courts, EA
MODES OF INTEREST INTEGRATION	**pluralists**	**pluralist but selective**
intensity of integration	low	more intensive
mode of interest integration	pluralist, companies and trade associations	pluralist, companies
ORIENTATION TOWARDS TARGET GROUP	**laissez-faire**	**more coercive**
general orientation	laissez -faire	consensus-oriented
character of incentives	stimulating: recycling credits	repressive, obligations

The Route Towards Formalization

The adaptation pressure for the British government was particularly high in regard to the legal codification of its policy. This aspect will therefore be discussed in more detail. Forced to develop an alternative to the legalistic approaches in Germany and Denmark which were in the EU agenda in 1992 and 1993, the government asked industry to develop a scheme on a voluntary base. There were at least two reasons for that. First, self-regulation by industry was in line with the general ideology of the government. In contrast to the Netherlands or Germany, the UK government pursued a market-led approach to sustainable development. It believed that economic growth was the precondition for environmental protection and that the free market would help to promote sustainable development. In line with the liberal idea of a minimized role of the state in society, a voluntary approach was to be developed: An industry-run scheme that puts the lowest possible costs and administrative burdens on industry while having broad support among industry. A second reason was that the government wanted to place the political responsibilities on business. The government had nothing to win and much to loose. When the minimum targets of the Directive are achieved, the political gains are not very high. When the targets are not achieved the government will be blamed. The ENDS Report wrote: *'A purely voluntary approach has minimized government work, and, perhaps more importantly, absolved it of any responsibility if the plan failed'* (ENDS Report Sept. 1994: 13).

Industry was not able, however, to develop a scheme based on a broad industry consensus as was the case with the 1991 Covenant in the Netherlands. The British government was therefore not able to present a persuasive alternative to more legalistic approaches in Brussels. There are three interrelated reasons why British industry could not come up with a voluntary scheme. First, the government's threat to intervene was not believed by industry. Second, the government did not provide a clear context for intra-industry negotiations, and finally, the British industrial community lacked the capacity to aggregate interests within and across sectors.

The UK government repeatedly threatened to take unilateral action against industry, if industry was not able to come up with a business plan. A threat, however, is effective only when it is believed and this was not the case in the UK. The weakness of the threat was admitted by an official at the Department of Trade and Industry· *'I think the government is generally unwilling to introduce new legislation. And there are also political reasons for that. There is the general political trend, which is against extra regulations, extra bureaucracy...'*(Interview DTI 15.5.196). This became clear when industry asked government to regulate, but the government was very reluctant to do so.

Moreover, the British policy process was not very well structured. The British government had no clear hierarchy of waste management options and there were no clear targets. The 1993 Gummer challenge only fixed a range (50-75%) for packaging waste recovery. It was also unclear for a long time, whether government would accept industry's call for legal backing. The unstructured context of the government's challenge provided no incentive for industry to cooperate and left the packaging chain with the difficult task of developing both the specific targets and the distribution of costs. Throughout the lengthy implementation process, industry was working in something of a vacuum and found it difficult to firm up its proposals. The lack of context for the implementation of the European Directive can be related to the prevalent idea of British environmental policy. As said earlier, British government usually resists setting national and quantified targets because it argues that those targets are seldom based on sound scientific evidence. '*Facts not fantasy*' matter (HM Government 1990). Environmental regulation has always been broad, and targets were set for each company individually in negotiation between the agency and the company concerned.

The political context became more clear, however, when the EU Packaging Directive was adopted. The Directive set quantified minimum targets for recovery and recycling. It was still difficult, however, for industry to develop a plan. British industry was still much more passive and reluctant to take responsibilities than its Dutch counterpart. Packaging manufacturers and their suppliers on the one end of the packaging chain and the retailers and distributors on the other end took a minimalist approach and argued that the other end of the packaging chain had to foot the bill. A more cooperative stance was taken by the large packers and fillers including Procter & Gamble, Unilever, Coca-Cola, and Guinness. They have been the driving industry force towards a workable shared approach and took central positions in the industry networks. These companies operate in a world market. They strongly supported the harmonization objective of the EU Packaging Directive. They experienced government intervention in other countries which resulted in strong market interventions such as the ban of certain types of packaging in Denmark or the German Packaging Ordinance. They knew that they would be affected by regulations in any case, and tried to make the best of it (British Industrialist 24.6.1996).

The most powerful players in the game were the retailers. In contrast to their Dutch counterparts, however, they did not take a cooperative stance. As said earlier, this general unwillingness of large parts of British industry has its roots in the British individualistic industrial culture (see chapter 7.6).

When the Directive was adopted it became almost inevitable for the British packaging waste policy to be based on public law (see chapter 8.4 for the legal arguments) and the government finally accepted that. The government, however, resisted the industry's call for a broad framework regulation and individual targets to be settled between individual companies and the Environment Agency. The government argued that the European institution would probably not accept such a broad regulation. It is rather unclear, however, whether a framework regulation would be objected by the EU Commission or the ECJ. A second reason was that the UK government used the EU as a scapegoat to escape from the need to administer a scheme in which ten thousands of firms asked for tailor made solutions. The government required clear responsibilities. This resulted in the '6,11,36,47' formula. This development makes clear that the UK government was prepared to deviate from its traditional environmental policy approach, i.e. broad framework regulation and discretionary authority for implementation agents, in order to adopt an effective packaging waste policy with a minimum of bureaucracy for the Environment Agency. The statutory instrument was based on Section 93 to 95 of the 1995 Environment Act. This Act was primarily adopted to create a national cross-medium environment agency. The development of this Act was independent from the British packaging waste policy. Sections 93 to 95, however, were included in order to create the legal base for the adoption of statutory producer responsibility instruments.

Strictness of standards

The government still took a very minimalist approach to the problem of packaging waste, because environmental policies were not very high on the political agenda in the mid 1990s. There are no targets for packaging or packaging waste reduction or for the use of re-usable packaging. The British recovery and recycling targets reflect the minimum targets set by the European Packaging Directive. Without the European Directive, however, there would be no quantified recovery and recycling targets for packaging waste in the UK at all. The minimum targets are however, relatively ambitious in the British context. It is estimated that the current recovery rate has to be doubled. The heads of the material organizations founded in the context of VALPAK estimate the current material recycling levels as follows: glass: 22.5 per cent; paper and cardboard: about 45 per cent (most of it commercial packaging); aluminium: probably close to 20 per cent, steel: 13 to 14 per cent; plastic: 6 per cent (almost entirely from commercial sites) (Materials Recycling Week 4.8.1995: 16-19). It remains to be seen whether the UK is able to achieve the targets. As said earlier, because of the lack of adequate burn-to-energy facilities, most of the recovery has to be done by material recycling. The 52 per cent recovery target means that some 44 per cent of the overall packaging waste has to be recovered by material recycling.

The system which is likely to emerge exhibits a certain degree of flexibility. Organizations such as VALPAK function as intermediary between the companies who carry the obligation and the waste recovery and recycling industry. The latter has itself organized in a number of waste material organizations. Packaging material organizations announced to offer so-called packaging recovery notes (PRN), that prove the recovery and recycling of a certain amount of used packaging. These PRN's can be purchased by individual firms or by collective schemes. There will be a financial incentive to recover and recycle packaging waste as cheaply as possible. It is very likely that most companies will take their responsibilities by paying for the recycling and recovery of commercial and industrial packaging. In contrast to the German system they are not obliged to recycle or recover sales packaging.

Allocation of responsibilities

Initially the local authorities played an important part in the collection, sorting, recycling and recovery of waste. With the Producer Responsibility Regulation, trade and industry became more important. The European Packaging Directive did not determine the allocation of responsibilities between private and public actors. The shift towards more responsibility of the private sector can therefore not be explained by European requirements. As stated already, the privatization of responsibilities was partly due to blame avoidance. It also fitted into the general belief of a minimized role of the state in industry. Putting recycling and recovery responsibilities on the public sector would not be consistent with this belief. Moreover, an increase in recycling credits to stimulate recycling would mean more public spending, which is also against the belief in the self-functioning of markets. The local authorities still play an important role in the collection, sorting and recycling of household packaging waste. The government learned from pilot projects that a nation-wide private packaging waste system that functions independently and parallel to the local waste management system is neither economically efficient nor necessary to meet the comparatively low recovery and recycling targets envisaged in the UK. There will be no dual collection system in the UK. The producer responsibility obligations create a market for collected and separated packaging waste, which is mediated by packaging recovery notes or functional equivalents. This means that the incentive to collect and separate packaging waste increases. It remains to be seen to what extent local authorities will be able to compete either alone or in partnership with private actors in these markets against private waste management firms.

The formalization of the British packaging waste policy by statutory measures implies the need for licensing, control and enforcement. These tasks are allocated to the new national Environment Agency. The Environment Agency was established by the 1995 Environment

Act. This institutional innovation has nothing to do with the European Packaging Directive. It is the result of the perceived need to overcome the decentralized and sectorally fragmented implementation infrastructure in the UK (Carter and Lowe 1995).

Modes of interest integration

The mode of interest integration has also changed over time. While in the early 1990s trade associations played a visible role, individual companies became more important later on. The involvement of trade associations was rather untypical for the British context. In contrast to Germany and the Netherlands, in the United Kingdom the most important form of business-state contact is the direct one between company and government. The government prioritizes such forms of contact over associative intermediation. The government has to rely on individual companies, because British trade associations are rather weak (see chapters 6 and 7 for details). That government did not focus its attention on individual companies in the early 1990s indicates the lack of importance of the issue of packaging waste and the minimal approach to deal with it. Accordingly, the growing need for a national packaging waste policy, induced by external pressure, led the government to direct relations with major companies. While in the Netherlands and Germany the modus of interest integration is more characterized by continuous contact with a small number of representatives of high aggregated interests, in the United Kingdom different *ad hoc* networks were created which were more or less selective. The PRG group was an *ad hoc* creation. The rationale for the selection was provided by former contacts of the Minister for the Environment, when he was Minister for Agriculture. The 26 PRG members were high ranked, they represented leading British companies, but did not represent all sectors of the packaging chain equally. Its successor V-WRAG was made of representatives of some 50 companies. V-WRAG presented the different sector and packaging material interests more equally. The problem was, however, that the group was too large to reach compromises. Moreover, most of its members were not high ranking but delegates. They had no leeway to negotiate a compromise (Interview British Industrialist 24.6.1996). The Packaging Advisory Committee (PAC) was the third industry group created. The government learned from the weaknesses of both networks that a business group was needed that a) was not too large; b) had to represent all sectors of the packaging chain adequately; and c) was to be made up of high ranking figures. Accordingly, the PAC consists of eight senior managers, two of each sector from the packaging chain. In addition to these core groups of business, the government integrated other interests through the formalized way of consultation papers. Consultation is required by Section 93.2 of the 1995 Environment Act. One consultation paper dealt with the four ways of financing an industry-led scheme, put forward by industry. And the other consultation procedure was set out for the draft Regulation (DoE 1995, 1996a)

Orientation towards target group

The orientation towards the target group changed throughout the process. This again has to do with pressures imposed by the European Packaging Directive. In line with the general policy tradition, the government pursued a consensus-oriented approach for a long time. It wanted an industry solution that had the commitment of all parts of the packaging chain. In line with

European requirements and informed by efficiency reasons, however, the government resisted the industry's proposal for a framework regulation. Moreover, the coming deadline for the implementation of the Packaging Directive forced government to hammer out a compromise in regard to the distribution of costs. The weakening of the intermediate targets in the course of the consultation procedure indicates that government took a pro-industry stance and tried to make it as convenient for industry as possible. Still, the new regime is based on a command and control approach, exhibiting relatively detailed standards. It remains to be seen, how strictly the implementation agency will control and enforce the producer responsibility obligations.

9.9 Summary

The British packaging waste policy became more formalized, involved stricter standards, and placed more responsibilities on business. The policy process was characterized by the *ad hoc* creation of business groups made up of companies rather than trade associations. In the course of the policy process government took an increasingly impositional stance towards industry. The driving force behind the British packaging waste policy development was the adaptation pressure deriving from the European Packaging Directive. Where the Directive left leeway, the British approach was informed by the government's belief in a limited public sector and clear and uniform responsibilities for industry in order to ensure the efficient implementation by the Environment Agency. The new approach of government is not in line with the traditional environmental policy approach. It remains to be seen whether the new approach will be effectively implemented.

10. The Netherlands: From negotiation to hierarchy?

10.1 Domestic developments: Implementing the Packaging Covenant

In contrast to the UK, the Dutch political debate did not concentrate on new policy initiatives to meet the standards of the European Packaging Directive, but on the implementation of the 1991 Packaging Covenant. The implementation of the covenant was perceived by government, trade and industry, and more independent actors as relatively successful. The quantity of packaging material coming to the market was stabilized. Business had taken measures to optimize the use of packaging, superfluous packaging was often omitted, environmental unfriendly materials were replaced by environmental friendly ones, and the quantity of materials per packaging unit had been reduced (Commissie Verpakkingen, 1995).

Table 26 Amount of packaging entering the Dutch market (1986=100)

Type of Material	1991	1992	1993	1994	1995
plastic	123	128	129	131	131
paper/cardboard	122	124	120	122	124
glass	99	99	97	96	93
tinplate	113	118	110	110	113
aluminium	108	116	104	100	115
total	113	116	113	114	114

Source: Commissie Verpakkingen 1996: 18

The intermediate goal to increase the separate collection of white, green, and brown glass to 50 per cent in 1994, was not achieved. But the Packaging Committee wrote that, based on the figures provided by the branch association of glass, the target would be met in the first half of 1995. The obligatory interim goal for recycling of 40 per cent for 1995 was already reached in 1994 (Commissie Verpakkingen 1995:5). The dynamic of the last years indicated that the 60 per cent target for the year 2000 could be achieved, though any further increase in recycling would probably involve relatively more efforts and costs. In particular, because until now industry had concentrated on packaging waste from commercial sites.

Table 27 Development of Dutch recycling quota of used one-way packaging

Type of Material	1993	1994	1995
plastic	9,3 (1,5)	10,0 (1,4)	11 (2)
paper/cardboard	56,7 (35,1)	55,7 (33,4)	62 (36)
glass	65,5 (67,6)	71,7 (74,1)	74 (76)
tinplate	37,3 (13,0)	33,9 (13,3)	42 (14)
aluminium	22,2 (-)	10,5 (-)	15 (-)
total (1986 ca 25)	47,4 (38,2)	46,2 (38,3)	50 (39)

Source: Commissie Verpakkingen1995: 23,29; 1996: 19, in brackets: household packaging waste

Two, more critical, remarks relating to this evaluation must be mentioned. First, although the interim goals were met, it is not likely that the obligatory prevention result will eventually be reached. In order to reach the level of packaging produced in 1986, the annual amount of packaging entering the market would need to be reduced by roughly one sixth in the year 2000 as compared to the early 1990s. This seemed rather unlikely, given annual economic growth rates of around three per cent. More recent figures already indicate that a further increase, rather than a reduction of the total amount of packaging is likely: the total amount of packaging has increased by roughly 3 per cent in 1996 in comparison to 1995 (Commissie Verpakkingen 1997: 7).

Second, it is difficult to establish a causal relationship between the Packaging Covenant and packaging and packaging waste developments. It is quite likely that other factors are also important for reducing packaging growth and increasing recycling. Hence packaging choices have been influenced not only by the covenant. Pure economic cost considerations and technological developments have led to the permanent optimization of packaging. In addition, many brand producers and retailers were anxious to establish and maintain a good environmental image by choosing 'environment friendly' and easy recyclable packaging (Interview British Industrialist 24.6.1996). Furthermore regulations in other countries such as Germany also played a role in the development and brought environmental friendly packaging to the market. Still, a certain role of the Covenant is acknowledged by all actors. The environmental effectiveness of the Covenant can be explained by the commitment of industry. Aalders (not the Minister of the Environment) argued that the principle of *pacta sunt servanda* is strong in business (Aalders 1993: 82). The commitment of industry can also be explained by the same factors which contributed to the conclusion of the agreement, most notably the credible threat of unilaterally imposed interventionist measures (see chapter 7.2). It is worth noting that many observers related the effectiveness of the Covenant to the generation and diffusion of knowledge about environmental effects and possibilities for optimization in the course of the strategic discussion (van Kempen 1992: 194), and the life cycle analysis.

The Environment Minister also fulfilled his obligations to create a framework for increased incineration and the separate collection of packaging waste materials. A ban on landfill for certain materials -including packaging came into force on 1 October 1995. Since there were not enough incineration capacities, however, provinces could ask for temporar exemptions from the ban (AOO 1995). It was reasoned, however, that by the year 2000 sufficient incineration capacities would be in place. The government, the provinces and the local authorities did develop also programmes for the separate collection of glass, paper and cardboard (AOO 1996, see also 3.4).

10.2 The European threat to the Dutch covenant

In the course of the discussions on the European Directive the Netherlands constantly tried to translate the goals of its own Packaging Covenant to the European level. As stated in chapter eight, the Dutch government succeeded in achieving exemption clauses (Interview VROM, 4.12.1995). Member States are allowed under certain conditions to follow a more ambitious policy than the one set out in the European Packaging Directive, including national action towards packaging prevention, the encouragement of re-use or aiming at higher recycling quotas. It was possible in principle to aim for the reduction and recycling targets of the Dutch Packaging Covenant (Douma 1995 Interview VROM 4.12.1995). Even if it were not possible to pursue a far-reaching policy in compliance with the Directive, the Dutch government could refer to Article 100a (4) of the Treaty. The reference to this escape clause could be accepted by the Commission and the ECJ, because the Netherlands voted against the Directive and the industry signed the Covenant voluntarily (Douma 1995: 112, see also chapter 8.4).

The maintenance of high standards in the Netherlands was not the most relevant problem. More important was whether the degree of legal codification of the covenant was sufficient to meet the European legal requirements. Earlier, it was indicated that there were virtually no possibilities for implementing a European Directive through voluntary agreements, when the Directive does not explicitly allow for this. The majority of legal scholars argue that covenants are therefore not sufficient to implement European Directive. Covenants lack binding character and provide insufficient legal security (Douma 1995: 111, Jans 1994: 131, Sevenster 1992).

The implementation process of the European Packaging Directive was lengthy and more complex than the development of the 1991 Covenant. The deadline set out in the Directive could not be reached due to cumbersome inter-ministerial bargaining, the need to notify the national regulation and the new covenant to the Commission, and government-industry negotiations on a new covenant. In contrast to the British case, however, the process was less chaotic and more structured. Highly aggregated business interests and a generally cooperative stance on the part of industry, made it easier for the Dutch government to negotiate a new packaging waste policy than was the case of the UK. Still, the Dutch packaging waste policy debate was also rather complex. References were made to three different programmes. The European Packaging Directive, the draft Regulation and the draft Covenant. The complexity is indicated by the fact that even specialized journalists sometimes got lost, e.g. confused the Regulation with the Covenant, or made wrong statements about the targets.

The Dutch policy process can be divided into three phases, whereby phases two and three were more or less parallel. The first phase was the preparation of the draft Regulation (Dec.1995). In the second phase the draft was transformed into a regulation (July 1997). This phase took one-and-a-half years. Parallel to this process the Dutch government negotiated a new covenant with industry, which had to be notified to the European Commission. The new Covenant came into force in 1998. For analytical reasons, first the two phases resulting in the Regulation of Packaging and Packaging Waste will be discussed. The chapter then traces the process towards the new covenant.

10.3 The draft Regulation on Packaging Waste: Inter-ministerial bargaining

Immediately after the adoption of the European Directive, the Dutch government began the transformation procedure. In comparison with the policy process that resulted in the Packaging Covenant, the context had changed significantly. The content of the covenant was the result of a voluntary agreement between VROM and the SVM. VROM, the SVM and those parties who signed the private law contract committed themselves to action. Though the SVM represented a large part of the packaging chain, it did not represent all companies. And even though other Ministries had been consulted during the negotiation process, VROM was the main Governmental actor.

The Ministry of the Environment made clear from the outset, that the European Directive was to be implemented by legal and administrative provisions. A covenant under private law was not perceived as an alternative option (Interview VROM 4.12.1995). Legal and administrative provisions for the implementation of EU Directives have to be developed with all relevant departments. Since the European Packaging Directive dealt not only with environmental protection but also with economic harmonization, the Department of Economic Affairs (Economische Zaken, EZ) became a central actor in the network, though VROM was the leading department. Hence, the framework of the Directive's approach, linking environmental and the economic issues, and the institutional rule that all departments concerned have to take part in the transformation of the Directive, altered the policy network. There were also other ministries involved including the Ministry of Public Health and the Ministry of Agriculture. But they had no substantial influence on the content of the regulation (Interview VROM 4.22.1995). The more central position of EZ, however, had a substantial effect on the policy process and on the shape of the new Dutch packaging regulation. It provided a better point of access for business interests than the Ministry of the Environment.

In general, the government had two options for transforming the Directive into national law. It was possible to implement the Directive through an Order in Council (a*lgemene maatregel van bestuur*, AMvB) or a Ministerial Order (*ministeriële regeling*), (Environment Act, *Wet Milieubeheer,* Article 21.6 section 6, see also Bekkers *et al.* 1995: 403-404). Both legal instruments are generally binding. Each alternative, however, is accompanied by a specific procedure which provides different points of access for interest groups. The Dutch government chose for the Ministerial Order. The main reason for this legal option was, that the procedure of the Order in Council would be too time consuming. It must be decided by the government (*Ministerraad*). The Council of State - the most important advisory committee - must give its opinion. And both chambers of parliament must have the opportunity to give statements

(Environment Act 21.6. section 4). It would have been difficult to meet the deadline of 30 June 1996 (Interview VROM 4.12.1995, see also Bekkers *et al.* 1995: 411-412). The Dutch constitution does not prescribe a specific procedure for Ministerial Orders. Consultation of parliament or other actors is not required. This makes it the swiftest way to implement directives (Bekkers *et al.* 1995: 404). The choice for a Ministerial Order was facilitated by the fact that the Regulation was basically a one to one translation of the Directive. If the Ministries had gone beyond the scope of the Directive an Order in Council would have been more appropriate. Given the high political salience of the issue, however, the government decided to publish a draft in the Official Journal (Staatscourant) and to inform parliament and provide the opportunity to discuss the Regulation there (VROM Interview 4.12.1995). The choice for a Ministerial Order - which implies a lack of democratic control - was subsequently criticized by those who felt negatively affected by the Regulation - in particular representatives of small and medium sized industries (Raad voor het Midden en Kleinbedrijf, Staatscourant 26.1.1996, 3), as well as by a Member of Parliament (Lansink 1996: 77). But this critique subsided.

The first draft of the new regulation was produced by VROM. At this stage VROM did not want a one-on-one translation but intended to exploit the national discretionary space as much as possible in order to achieve a high level of environmental protection. The Ministry intended to translate the central elements of the 1991 Covenant into the new regulation. Therefore, the draft comprised the recycling goals of the covenant (60 per cent). VROM argued that the Directive explicitly allowed for higher quota than those set out in the European Packaging Directive. In order to achieve a high level of protection, VROM was even ready to rely on Art 100a (4) of the Treaty. Apart from the Packaging Covenant the German Duales System also served as a model. The first drafts comprised, like the German Packaging Ordinance, a take-back responsibility for the retailers. This was done to put pressure on the packaging chain to establish a private system. Next, similar to the German system, the first concept placed the responsibility for the collecting and sorting of packaging waste on the packaging chain. This element was also in line with the general strategy of VROM to achieve overall product responsibility by industry (Interview VROM 4.12.1995). In short, the first draft of VROM aimed at a translation of the targets of the covenant into a provision of public law. Furthermore, central elements of the German system were included in the concept.

As outlined above, EZ became the second central actor in the policy network. The approach of EZ was less oriented to the environmental protection objective of the Directive than to the harmonization goals. Therefore EZ did not agree on the 60 per cent mandatory recycling target favoured by VROM. EZ wanted a one-on-one translation of the Directive into the Ministerial Order, in order to avoid market distortions. EZ argued that any further-reaching measures could be included in a new covenant. With this position they went along with the industry's Foundation of Packaging and the Environment (SVM). Furthermore, EZ opposed the new responsibilities for the industry included in the draft of VROM, i.e. take-back responsibilities for retailers, responsibilities for collecting and sorting of waste for the packaging chain. It was argued that no additional burden should be placed on industry. This could interrupt the ongoing processes in the Dutch industrial community which were induced by the Covenant. These differences between VROM and EZ resulted in countless consultations. As a consequence, a significant delay in the policy process occurred. The discussion between EZ and VROM was very intense. These two actors held the central positions in the network. It was very much an inter-ministerial process, though on behalf of the industry, the SVM had regular contact with the Ministries (communication with VROM 7.5.1998). The environmental movement and consumer interests were not involved. The Departments saw no need to consult them. And the groups themselves saw no reason either. At this stage, the process was seen as a bureaucratic exercise (Interview Milieudefensie 20.11.1995). That the negotiations between VROM and EZ were difficult is indicated by the breach of an informal rule by EZ. This Ministry made an internal concept of VROM known to the public. This breach of confidentiality distorted the negotiation process. Through this 'leak' VROM was exposed to critics from industry at an early stage of the programme formulation process. The main debate between VROM and SVM concentrated on the question of the extent to which elements of the covenant, in particular the targets, should be integrated in the Ministerial Order.

The draft Regulation was published in the official journal of the Dutch government on 22 December 1995 (Staatscourant 22.12.1995). At the same time it was sent to the European Commission for comments. At this point the government was positive that the regulation could be adopted on 1 July 1997 (AOO 1996). The draft revealed that VROM could not succeed with its draft on three points: higher targets, take-back responsibilities for the retailers, and collection and sorting responsibilities for the packaging chain. The targets were identical with the maximum targets set out in the Directive, that is 65 per cent recovery, recycling and incineration with energy production, and 45 per cent material recycling. At least 15 per cent of each material was to be recycled. Note that, apart from the minimum target for the recycling of plastic waste, all targets were already met by the packaging chain in 1994. The draft placed the legal responsibility for take-back, recovery and recycling on the packers/fillers and the importers of packaged goods. Every company of this group had to meet the recovery and recycling targets individually and had to set up a data base to monitor the progress and to report annually. It was possible, however, that the packaging chain established an organization which set up its own system. Every company that joined the collective system could be exempted from the individual obligations. The new system could get the legal form of a covenant. The concentration of the legal responsibility on a single part of the packaging chain was defended with the argument that an identifiable group was needed to make the regulation enforceable (VROM 1997b: 4). Though the packers and fillers were primarily responsible, the draft also emphasized that the raw material producers had to support the producers and importers of packaged goods by meeting the targets. It is worthwhile to note that almost all big retailers such as Albert Heijn are also packers and fillers in the sense of the draft Regulation, because they commission others to produce goods under their own name, the so-called *'huismerken'* (own brands). They therefore had in fact take-back and recycling responsibilities. Only small retailers were excluded from this obligation (VROM Interview 4.12.1995). Incineration with energy production could be done by local authorities paid for by industry. The local authorities were responsible for the separate collection of paper/cardboard, and glass. The draft stipulated that the Regulation would be evaluated in 1998. In the event that the provisions of the regulation were not met or the covenant did not fulfil the expectations of government, more far-reaching measures could be adopted.

The failure of VROM to put through higher targets and more responsibilities for trade and industry has partly to do with the European Directive that weakened the position of VROM concerning higher targets, since EZ could argue that besides environmental protection the Directive aimed at harmonization. Higher targets than stipulated in the Directive would contradict the harmonization objective. More business responsibilities were also prevented by the strong opposition of EZ, but also by the packaging chain, in particular the retailers. The statutory business association of the retailers ('Bedrijfsschap detailhandel') and the SME's association ('MKB'), which represent among others small retailers, lobbied very hard to prevent take-back and separate collection responsibilities. They used their traditional good contacts with the Ministry of Economic Affairs. They succeeded in preventing responsibilities for last minute packaging (also called service packaging), that is packaging being added to the product at the point of sale, for example, paper bags for bread in a bakery, or - to give a typical Dutch example - the box for French fries. Even though they are packers/fillers in this sense, the responsibilities in this case were placed on the producers of the packaging. The government legitimized this exemptions for reasons of efficiency and a reduction of bureaucracy (Staatscourant 4.7.1997: 15). This might be a second reason for this exemption.

10.4 Towards the Packaging and Packaging Waste Regulation

For many companies the draft Regulation was a surprise. Five years earlier, business representatives agreed upon a covenant with the Ministry of the Environment after a laborious process and now suddenly a Directive came out of Brussels. Importers and producers were confronted with detailed legal obligations without being involved in the policy preparations. An industry newspaper's comment on these developments was: '*After long years of effort to push back the quantity of waste in the Packaging Covenant together, industry receives a huge blow. A blow to the back of the head.*' (Forum 16.1.1995). And a parliamentary member criticized: '*We have 60% to 70% of the waste under control. So we are well on the way. It now seems that we consider bureaucracy as more important than the environment.*' (NRC 31.1.1996). The Dutch Union of Wholesalers and the Peak Association of the small and medium sized industries declared the regulation impossible to implement and spoke of an additional financial burden of 3.5 million guilders for the 300,000 producers and importers involved (Financieel Dagblad 13.3.1996, NRC 31.1.1996). That would turn out to be five times as expensive as the much criticized, costly German packaging waste collection and recycling system. This estimate is undoubtedly exaggerated, but does illustrate the frustration in certain circles of industry. This discontent is not so much a consequence of the content of the new Directive, but rather a pointer to the manner of its introduction, from the top down, without intensive preceding consultation with industry's actors apart from the SVM. As the MEP Lansink said: '*Het Nederlandse bedrijfsleven is kennelijk zo gewend aan zelfordening en convenanten, dat regelgeving van bovenaf op voorhand argwaan oproept'* (Lansink 1996: 77).

The discussion in the Environment Committee of parliament took place on 17 April 1996. The majority supported the draft Regulation. Only the representative of the liberal party (VVD) criticized the individual responsibilities for business and the bureaucracy involved. Together with the other large parties he strongly supported the idea of a second covenant as an alternative means for compliance (Financieel Dagblad 18.4.1996).

Against the background of the broad parliamentary support for a new covenant, the government gave the covenant provision a more prominent place in the draft - from Article 9 to Article 2, and facilitated the procedure of joining a covenant. While the old draft stipulated a time-consuming and bureaucratic procedure for attaining exemption from individual obligations, according to section 3.5 of the Dutch General Administrative Act, it was now sufficient to declare that one would join a covenant. This modification had the additional effect of decreasing the paper work of VROM (communication VROM 7.5.1998). The prospect of a new covenant dampened the industry' critique on the Ministerial Regulation (Verpakkingsmanagement 6/1996: 32).

As late as 18 June 1996, the draft was notified to the Commission. Since a notification implied a three months standstill period, the Dutch government was no longer able to meet the deadline for the implementation of the Directive. The Regulation could not come into force on 18 September 1996, however. The Commission raised objections to the Dutch regulation in the form of a detailed opinion. This triggered a second standstill period until 18 December 1996. The Commission criticized the fact that recycling and re-use are defined differently than they are in the Directive. Even though differences in definition can have considerable consequences, this was not the case in regard to the definitions of re-use and recycling. For 're-use' for example the Directive spoke in terms of a '*minimum number*' of trips, while the Dutch version said '*more than one*'. The Commission also criticized the Dutch definition of 'recovery' which included only recycling and incineration with energy recovery. This would be too limited. The Dutch government, however, argued that in the Netherlands there were no other alternatives. The term would reflect reality (European Packaging & Waste Law Oct. 1996: 10). In the official reaction to the Commissions' detailed opinion, the Dutch government said that it is difficult to respond to the objections because the opinion was not detailed at all. The government also said that it was surprising that the Commission had postponed the implementation of the regulation by making detailed opinion. The Dutch government argued that the discrepancies in terminology established by the Commission would not affect the free movement of goods within the framework of the internal market. It also said that Article 189 would not require that definitions used in a Directive be transposed into national legislation verbatim. (European Packaging & Packaging Waste Law Feb. 1997: 3).

The Commission did not impose any more obstacles to the enactment of Packaging and Packaging Waste Regulation. Its adoption was further delayed for domestic reasons, however. The debate on a new covenant progressed more slowly than initially expected by the government (see next section), and a time gap of more than thirteen weeks between the adoption of both programmes would cause practical problems. The Packaging Regulation said that producers and importers had to notify the minister, within thirteen weeks of this Regulation becoming applicable to them, how they intend to fulfil their obligation. Without a covenant being enacted at that time, the companies would have been forced to announce that they would comply individually. Since the new covenant also had to be notified to the Commission, the Dutch government had to take into account another three months of standstill. In December, the government therefore announced that the Regulation would come into force 1 April 1997. Another delay in the negotiation with industry resulted in further delay to the Packaging Regulation. It was adopted by the government as late as 29 June 1997 (Staatscourant 4.7.1997) and came into force on 1 August 1997. At this point government and industry had reached an agreement on a new covenant.

Neither the influence of interest groups nor the reaction of parliamentarians or the European Commission produced substantial changes the draft Packaging and Packaging Waste Regulation. The central elements remained the same: 65 per cent recovery, 45 per cent overall recycling, and at least 15 per cent per material. The allocation of responsibilities did not change, nor did the possibilities for escaping individual obligations by joining a collective system. Only minor changes took place. Generally speaking the strictness of standards decreased slightly. First, due to delay in the implementation process the timetables were relaxed. The obligations for industry came into force on 1 August 1998 instead of 1 July 1997. This was also the deadline for the first company report including the measures taken, the results and the quality of cooperation with other parts of the packaging chain. The targets agreed on in a covenant have to be met by 30 June 2001 instead of the originally intended 31 December 2000. Second, the regulation responded to the critique of too much bureaucracy. It replaced the annual reporting requirement for companies which take individual responsibilities. These companies have to report only every three years. To stress the idea that not only producers and importers of packaged goods have to take responsibility, the Regulation requires companies to report whether there are problems with the cooperation with other parts of the packaging chain. The government wants to use this information for its evaluation of the regulation by 31.12.1999. In order to gain flexibility in enforcement of the regulation, the Ministry may attach rules or restrictions to approval of the notifications by which companies have to prove that they comply with the Regulation. The Minister may also stipulate that the approval shall only apply for a limited period.

10.5 Towards a New Covenant: 300 or 300,000 does not matter?

The announcement in the draft regulation that a new covenant could be negotiated with industry was embraced not only by all large Dutch political parties, but also large parts of industry. As already said, the alternative - individual legal obligations - caused some agitation. Industry criticized the fact that due to the Directive from Brussels, the consensual solution that worked effectively, had to be replaced by detailed regulations and bureaucratic obligations for individual companies. The European Directive also had an impact on the institutionalized Dutch system of packaging interest inter-mediation. The Foundation of Packaging and the Environment was 'the spider in the web'. Over the last twenty years, the SVM had successfully performed the double function required in a corporatist arrangement. On the one hand, the SVM represented the interests of the packaging chain concerning the issues of packaging and the environment *vis-à-vis* the government and the public. On the other hand, when agreements were made, the SVM was quite effective in assisting implementation. In the course of the implementation of the European Directive, however, the SVM had to be replaced by a more complex system, since, due to the draft Packaging and Packaging Waste Regulation *all* companies of the packaging chain were affected.

That a new covenant was needed was already part of the SVM lobbying in the course of the development of the first draft regulation in 1995. Officially the SVM declared its support for a new covenant in early 1996. The SVM argued in favour of a new covenant because individual obligations would be too bureaucratic and the restriction of the primary legal responsibility to the packer/fillers and importers would lead to a passive attitude in other parts of the packaging chain and would endanger the whole system, which depended on cross-sectoral cooperation (Milieutechnologie 23.2.1996: 1).

The parliamentary support of the Packaging and Packaging Waste Regulation resulted in the initiative of the Peak Association of the Dutch industry associations (VNO-NCW) to discuss the basic structure of a new covenant with the SVM on behalf of the packer/fillers (Lansink 1996b: 69). Industry formed the Packaging Platform for negotiating a new covenant. This platform consisted not only of the SVM and the VNO-NCW but included also the umbrella association of the small and medium sized industries (MKB). Initially this change in the packaging policy network caused some problems. The SVM represented the more active part of the packaging chain. The driving forces were large retailers and wholesalers such as Albert Heijn, Vroom&Dreesman and companies with brand products such as Campina Melkunie, Douwe Egberts/Sara Lee or Unilever (Interview VMK 9.9.1997). These companies in particular benefit from a good environmental image. Now, more or less the whole of Dutch industry was represented in the network, that is 300,000 instead of 300 companies (NRC 8.7.1997). The association of SME companies had been particularly critical about the Dutch packaging waste policy, and only very few of its members had joined the first covenant.

Since most of the SVM members were also members of the VNO-NCW, the allocation of competencies within industry became unclear, in particular the role of the SVM. Its managing director argued that '*driehonderd of driehonderdduizend bedrijven, dat maakt niet uit*' (quoted in Pak Aan 3/1996: 20). Notwithstanding this, the SVM had to establish contacts with new sectors, such as construction and service. The intra-industrial co-ordination problems however, never reached the dimensions of the British situation. They were eventually solved or at least no longer came to the surface. The SVM remained a powerful actor. It was again the most important focal point of the government-industry interaction. The Packaging Platform met once or twice a month. Some 50 representatives of the three associations were involved. The meetings were chaired by a managing director of the large packers/fillers (Campina Melkunie), which is an affiliate of the SVM and member of VNO-NCW. He chaired the discussions on behalf of the latter organization (Interview VMK 9.9.1997). Industry was in favour of a covenant approach, because it feared the bureaucracy and rigidity implied by the Packaging and Packaging Waste Regulation. On behalf of government VROM negotiated with industry. The role of EZ was rather restricted in the process, it was informed about the progress, but did not take part very actively (communication VROM 7.5.1998). VROM, like industry, was interested in continuing the covenant approach, because it could not succeed in making higher targets generally binding in the Packaging and Packaging Waste Regulation. The Covenant negotiations provided a second chance to increase the environmental standards.

There were three major controversial points in the industry-government interactions. These were essentially the same issues, which were already discussed at length in the context of the first covenant. The first contended issue dealt with packaging and packaging waste prevention. The second concerned the contributions of the different materials to the overall recycling target, and related to this the allocation of responsibilities between industry and local authorities. The third issue concerned the promotion of re-use.

Prevention

The government wanted a quantified prevention target as in the 1991 Covenant. Industry opposed a quantified target very strongly. It argued that any ceiling on the amount of packaging would mean a ceiling on economic growth. As said earlier, industry argued for instance, that due to economic growth and the decreasing average in the size of households, it was increasingly unlikely that the total amount of packaging entering the Dutch market could be reduced to the level of 1986 (Herweijer 1996: 33). Industry was in favour of a general prevention principle. This principle was called the ALARA principle, or As Low As Reasonably Achievable. That is: 'restrict volume and weight to the minimum amount necessary to fulfil the required level of safety, hygiene and consumer acceptance'. The Environment Ministry was sceptical about this approach, fearing that it was to difficult to control. An official at the Ministry said, '*the industry's prevention principle sounds good, but we want to know what it precisely will mean and how it will be put into concrete practice*' (Communication with VROM 30.11.1996, EU Packaging Report Sept. 1996: 15). This conflict was finally solved by a compromise. A quantified, albeit conditional, packaging prevention target was included in the *Integration Packaging Covenant*, signed by VROM,VNO/NCW, SVM, and MKB. The quantity of packaging to be newly placed on the market in the year 2001 has to be reduced by at least 10 per cent in relation to the quantity of packaging in the year 1986, corrected for both the trend in GDP since 1986 and the use of secondary materials insofar as this results in heavier packaging than is the case in using primary raw materials. The ALARA principle has been included in a *Sub-Covenant Producers/Importers* signed by VROM and by producers and importers of packaged goods, either directly or through representatives. An instrument clarifying the ALARA principle, was to be developed by the Technical University of Delft (Verpakkingsmanagement 1/1997: 9). Hence, for a second time a major conflict between government and industry was solved by involving academics from University. Furthermore, these were the same administrative scientists who - then associated with the Erasmus University - developed the process standard for the LCA's and who chaired the steering committee of the implementation of the LCA's.

A provision stipulating an absolute ceiling on the amount of *packaging* could not be achieved. The Integration Covenant stated, as had the 1991 Packaging Covenant, that the amount of packaging *waste* to be incinerated or to be landfilled is restricted to 940,000 kilotonnes, however. This implies that any increase in the amount of one-way packaging has to be countered by an increase in packaging recycling.

Recycling

In the discussion between VROM and the Packaging Platform, the Ministry made clear that it wanted a higher recycling target in the Covenant than stated in the Regulation. As a compensation, the administrative burden for members of a covenant would be lighter and the flexibility higher than in the case of individual obligations. Already in 1995, the government envisaged a recycling target of 65 per cent for the year 2001. This would be five percentage points more than stipulated in the 1991 Covenant for the year 2000 and 20 percentage points more than the maximum target set out in the European Packaging Directive. The main representatives of trade and industry had no strong objections against this general recycling target. The question was, however, how much the different materials have to contribute to this target and related to this was the question of who paid for the separate collection of materials for recycling from household sites.

Under the Waste Collection Programme adopted in early 1995, local authorities were already responsible for the separate collection of paper, cardboard and glass. In the same year the Association of Local Authorities (VNG) and associations of packaging raw material suppliers and packaging industry concluded declarations of intent about the separate collection and recycling of these materials with private organizations. These declarations of intent have been transformed into sub-covenants. Each of these covenants has an appendix which stipulates the required quality of the material separately collected and offered. In the Sub-covenant Paper/Fibre the VNG ensures the separate collection of 85% of waste paper and cardboard, and its offer to those waste paper traders that affiliate with the Organization for Paper Recycling. These companies guarantee the take-up and recycling. In regard to glass packaging, the VNG guarantees that 90 per cent of glass packaging will be collected separately and offered to a company which is a member of the glass recycling organization. These companies guarantee the take-up and recycling of this glass.

The local authorities have no specified role in regard to metal packaging. Here the Metal Recycling Federation, the Organization for the Promotion of Metal Packaging Recycling and VROM undertake to ensure that industrial metal packaging is kept separate. VROM undertakes to help creating an infrastructure for household metal packaging. The Metal Recycling Federation guarantees to take up the collected material and to ensure 80 per cent recycling of all metal packaging.

The most controversial waste stream was plastic packaging. In contrast to the other materials there was no prospect of economic benefits from recycling, due to high recycling costs and the lack of end markets. Industry initially made a promise to recycle 35 per cent of all packaging waste. The government wanted industry to stick to this promise, but industry was reluctant to restate their announcement, partly because in those days only some 11 per cent of plastic packaging waste was recycled. Interests of about half of all plastic manufacturers were represented by the Association of Environment and Plastic Packaging (*Vereniging Milieubeheer Kunststofverpakkingen*, VMK). The VMK had federative relations with the National Plastic Federation (NKF), i.e. all VMK members were members of NKFD but not vice versa. The VMK was founded as a reaction to the 1991 Covenant, because companies felt that plastic interests were not adequately represented (Interview VMK 9.9.1997). Due to the lack of plastic interests integration in the 1991 Covenant negotiations, an ambitious target had been included: 50 per cent recycling of all high grade plastic bottles. This target was not achieved, however, and the plastic industry strongly opposed high recycling targets. Finally VROM and VMK committed themselves in the Sub-covenant Plastic to help to separate used plastic packaging. The VMK guaranteed the material recycling of 27 per cent and committed itself to try to achieve 35 per cent.

The government also wanted a separate covenant for drink packaging made from composites (Tetra pak). But it could not succeed in that, partly because the EU Packaging Directive does not demand separate action for deposits. It was decided that these composites fall under the Sub-Covenant for paper (Verpakkingsmanagement 1/1997: 9).

Reuse

The 1991 Covenant and the subsequent LCA's and MEA's dealt with the extension of re-use to new product groups. But the LCA's did not result in an extension (see 5.3.6.). The issue of the extension of re-use to new products disappeared from the political agenda. As in Germany, however, the share of one-way drink packaging started to increase slowly. In late 1996 rumours arose that Coca-Cola wanted to introduce a one-way bottle. This put the issue back on the political agenda (Staatscourant 19.12.1996). Later it was revealed that these rumours were wrong: they were spread by the company itself to test the reaction of Dutch consumers (Verpakkingsmanagement 12/1997: 8). Partly due to these developments the *Consumentenbond, Milieudefensie* and *Stichting Natuur en Milieu* offered a report to the government in which they ask for the inclusion of provisions in the new covenant for stimulating an increase in re-usable packaging (1996).

The Ministry of VROM was still in favour of refillables, at least in the sectors where they are already established. It was not prepared to accept an uncontrolled decrease in refillables. The new Covenant therefore addressed the issue of re-use. In contrast to the 1991 Packaging Covenant the new provisions aim neither at the extension of re-use to other product groups nor to at an increase in those categories where the refillable systems were already. As in the case of Germany, the provisions were aimed at a maintenance of the status quo level, to be more concrete: a minimum of two percentage points below the status quo level. In the Sub-Covenant Producers/Importers the companies undertook, for the products for which refillable packaging is currently dominant, i.e. beer, water and soft drinks, to replace refillable packaging by one-way packaging only under certain conditions. These conditions are specified in a protocol, which was attached to the covenant and became an integral part of it.

The protocol stipulates that one-way packaging may be introduced only when either the share of refillables measured at the time of the enactment of the Covenant does not decrease by two per cent or when life cycle analyses demonstrate that the introduction has less or at most the same impact on the environment as the existing refillable system. The LCA's have to be carried out by an independent research institute and will be checked by another independent institute to be designated after consultation with the Minister.

Hence, the government agreed on a covenant scheme which actually consists of a number of covenants each signed by a specific set of actors. There is a general covenant, the so called Integration Covenant, and a number of sub-covenants: one addressing the producers and importers of packaged goods, and the others dealing with specific materials including glass, paper/cardboard, metal, and plastic. The new covenant scheme is much more detailed (and more intricate) than the 1991 Packaging Covenant, already indicated by the sheer volume of the document. While the English translation of the 1991 Packaging Covenant covers less than 14 pages, the English translation of the new covenant, excluding the explanatory notes, runs to 59 pages. All covenants include the definitions and scope of the respective covenant, the obligations of the parties that signed the covenant, provisions in regard to monitoring, reporting, duration and the legal base. All covenants have the force of an agreement under civil law. They will terminate on 31 December 2002. With the exception of the articles on monitoring and reporting, however, the covenants substantively terminate on 31 December 2001.

The Integration Packaging Covenant stipulates not only the general packaging prevention target and the packaging waste ceiling, but also the material recycling target: at least 65 per cent. This target applies in so far as the requisite packaging waste is offered properly segregated in accordance with the conditions agreed in the sub-covenants on material recycling.

The Sub-Covenant Producers/Importers not only entailed the ALARA principle and the provisions in regard to re-use. In this covenant, producers and importers committed themselves to segregate the discarded packaging in their companies as much as is reasonably possible; to ensure that the secondary raw materials are used in accordance with market principles; to report annually on the measures taken and to endorse the contents of the Integration Packaging Covenant.

The covenants exempted companies who use less than 50,000 kilo packaging per year from reporting requirements. This was some 90 per cent of the 300,000 companies which were potential candidates for the covenant. The overall costs of the covenant scheme were estimated to be substantially lower, as was the case with the German DSD scheme. Industries costs were estimated by the chairman of VNO-NCW at some 10 per cent of the 1,5 billion guilders the German system costs, when adjusted to the Dutch situation. The government did not quantify the estimated costs. In the explanatory memorandum to the covenants it states that the costs for collection are borne by local authorities because they perform this service at the lowest possible social costs. The costs for separate collection would be set against lower incineration costs and by the proceeds from the separate collected streams, so that on balance disposal costs for local authorities would not rise. It also said that companies already paid for the disposal of their own waste. Moreover, new regulations based on the Environmental Management Act would incorporate requirements for the separation of industrial waste. In connection with the ban on disposal at landfill sites and rising incineration rates it would become attractive to separate the packaging waste and present it separately for recycling. It is admitted in the explanatory notes that recycling of paper is not always economically viable. Any so called chain-deficit had to be borne by industry rather than by the local authorities, through a disposal fund organized by the Industry Organization for Paper Recycling. It is also likely that plastic recycling has to be subsidized, but the explanatory notes did not allude to that. The administrative costs for industry were estimated at 20 to 24 million guilders a year.

After the Covenant was accepted by government and industry it was sent to parliament and notified to the EU Commission. The parliamentary committee concerned discussed the Covenant on 24 September 1997. The representatives of the four large parties PvdA, VVD, D66 and CDA were very positive about the covenant, though they had a number of questions, in particular regarding the complicated provision concerning packaging prevention and the control of compliance. Only the representative of the left Green Party (*Groen Links*) was dissatisfied with the Covenant. She complained that the provision regarding prevention was incomprehensible and that the trend towards more one-way packaging was not effectively restricted (Verpakkingsmanagement 10/1997: 5).

The adoption of the Covenant was delayed because the Regulation in which the Covenant was embedded was notified later by VROM than initially announced. The Dutch government did not receive reasoned opinions and was rather surprised about that, given the provision about refillables, an issue which caused many problems in Germany (see chapter 9.4). There were two reasons for the Commission not challenging the Covenant. First, DG XI (Environment) was positive about voluntary agreements in general and covenants in particular (CEC 1996), and it did not want to frustrate the Dutch approach. Other DG's, in particular DG III (Industry), were generally critical about refillable provisions, but ultimately accepted the Dutch solution because it did not imply a fixed quota, but employed a more flexible approach based on LCA (communication with VROM 7.5.1998).

The Covenant came into force in December 1997. This was six weeks after the packers/fillers and importers had to notify the Minister about how they wanted to comply with the Regulation. The Ministry was, however, pragmatic and did not demand the individual compliance route. It was sufficient that the concerned companies had declared by 1 November 1997 that they would join the new covenant.

The SVM, the MKB and the VNO-NCW were very satisfied with the Covenant (Verpakkingsmanagement 1/1998). The chairman of VNO-NCW praised the agreement calling it the green polder model (De Financiële Telegraaf 8.7.1997). It was believed that almost all companies, some 90 per cent, were to join the new covenant (Verpakkingsmanagement 1/1997: 9). In order to facilitate the monitoring and reporting obligations, these companies will cluster together in some 250 groups. This clustering will happen mostly within the same branch. Trade associations will play a key facilitation and co-ordination role (Pak Aan 3/1997).

Societal interest groups, i.e. *Stichting Natuur en Milieu, Consumentenbond,* Greenpeace and *Milieudefensie*, were not as active as in the 1980s. When the covenant was finally adopted they prepared a joint declaration in which they stated the critical opinion that the packaging prevention provision actually means that the amount of packaging may increase by 10 per cent; that re-use is not encouraged, and that the maintenance of re-use systems does not apply for milk and other dairy products. They asked industry to undertake more far-reaching measures, and they requested from government strict control and additional policies, if it became apparent that the targets would not be achieved (Verpakkingsmanagement 1/1998).

In contrast to the British case, it is out of the question, that the Netherlands will exceed the minimum targets set out in the European Packaging Directive. The Netherlands already exceeds the maximum recycling target. It is doubtful, however, whether industry will achieve the targets set out in the covenants. In particular the restriction to 940,000 kilotonnes allowed to be landfilled or incinerated will probably not be met. It would mean a packaging waste reduction of some 30 per cent as compared to 1996. Even if a recycling level of 65 per cent were to be achieved, the amount of packaging entering the market may not exceed 2,7 million tonnes. This is the level reported by the RIVM for 1996 (Commissie Verpakkingen 1997: 12). It is unlikely that industry will be able to stabilize the amount of new packaging given economic and socio-demographic trends. If the industry does not succeed in that, an increase in recycling has to occur above the 65 per cent, to an extent that would compensate the failure of packaging prevention. This seems rather unlikely even if one takes into account that almost all companies joined the covenant, at least as long as there is no manipulation of packaging and waste statistics, substantial innovations in recycling technologys, or a wider definition of what can be considered as recycling. The failure to meet this target will be partly due to the government's lack of will to control industry. Though the Environment Minister promised parliament that the agreement would be tightly controlled, it is rather unlikely that this will be the case. The Chairman of the Packaging Platform already warned against inquisitorial action. He said that it should be clear *'dat bij de controle van individuele bedrijven het uitgangspunt is dat er geen heksenjacht mag ontstaan en dat bureaucratie achterwege dient te blijven'* (Kleibeuker, quoted in Verpakkingsmanagement 1/1997: 9).

10.6 Continuities and changes in Dutch packaging waste policy

The packaging waste policy in the 1990s was embedded in a broader economic and political context which was not in favour of interventionist environmental policy making. The political debate was very much informed by issues of globalization and national competitiveness. The government coalition D66, PvdA and VVD (1994-1998) made unemployment and economic growth the central political objectives. Deregulation and market incentives were seen as the main means to achieve these aims (Geelhoed 1997). The relative increase of economic issues went hand in hand with a decrease in environmental policy issues. The idea of ecological modernization changed its meaning. The new target for packaging waste prevention nicely illustrated the new definition of ecological modernization of the Dutch government: when the economy grows, the amount of packaging may grow. The waste issue in general became less of a priority. Less waste than officially expected had to be disposed of and new incinerators increased waste disposal capacities (AOO 1995).

The issue of packaging waste decreased in importance. Environmental groups as *Milieudefensie* or *Stichting Natuur en Milieu* did not give high priority to this issue any more (De Volkskrant 29.11.1997), with the exception of the campaign at the end of 1996. One reason for this was the depoliticization of the issue of re-use. Consumers were uncertain about the environmental superiority of multiple-use packaging over one-way packaging, and were therefore difficult to mobilize for environmental protest, such as boycotts. This uncertainty was raised by contradictory outcomes of life cycle analyses. The LCA's are probably also one of the reasons why there were no motions by MP's any more, demanding an increase of mandatory deposit systems. Hence, four factors supporting a strict packaging waste policy, i.e. a policy paradigm emphasizing the reconciliation of economic growth and environmental protection, lacking waste disposal capacities, societal pressure and political pressure, decreased in importance. How did the Dutch packaging waste policy change against this background?

Table 28 Continuities and changes in Dutch packaging waste policy

Dimension	**Packaging Covenant 1991**	**Packaging and Packaging Waste Regulation 1997**	**Packaging Covenant II**
FORMALIZATION	**medium**	**higher**	**higher**
degree of legal codification	Covenant: voluntary agreement binding under private law for signatories	Regulation based on Public Law	Voluntary Agreement enabled by Regulation based on Public Law
ALLOCATION OF RESPONSIBILITIES	**public and private**	**still public and private, but public enforcement**	**still public and private**
separate collection	public and private, including pilot schemes	public and private	public and private
recycling and recovery	private	private, primarily packer/fillers and importers of packaged goods	private, whole packaging chain
licensing	-	VROM	VROM
monitoring and assessment	public-private committee	public	public-private committee
dispute settlement/ enforcement	arbitration/civil law courts	administrative courts	arbitration/civil law courts

Dimension	Packaging Covenant 1991	Packaging and Packaging Waste Regulation 1997	Packaging Covenant II
STRICTNESS OF STANDARDS	**medium/high**	**less strict**	**still medium/high**
packaging reduction targets	stabilization below the level of 1986 (i.e. reduction of ca 15 per cent in comp. with 1991)	only best practice, no quantified targets	10 per cent lower than 1986, corrected by economic growth
packaging waste reduction target	max. 940,000 kilo tonnes	-	max. 940,000 kilo tonnes
re-use target	increase of re-use under certain conditions	only best practice, no quantified targets	no targets, but switch to one packaging constrained conditions
recycling (levels already achieved)	60% (1986 =25%)	45 %, 15% per material (1995=50%)	65% (1995=50%) 27-90% depending on material
recovery	not explicit, but implicit virtually 100 per cent	65%	not explicit, but virtually 100 per cent
incineration	increase to max. 40% = 940,000 kilotonnes		max. 940,000 kilotonnes
landfill	phase out	phase out	phase out
time horizon	in the year 2000	1.8.1998	in the year 2001
prescription of implementation	modest degree of detail and precision	more detailed and more precision	more detailed and more precision
flexibility	self-regulation within broad framework, change of targets under certain conditions	self-regulation within stricter legal framework	self regulation within broad legal framework
MODES OF INTEREST INTEGRATION	**corporatist**	**still corporatist**	**still corporatist**
intensity of integration	high	high	high
mode of interest integration	corporatist single issue cross sectional foundation	still corporatist but more pluralized	still corporatist but more pluralized
ORIENTATION TOWARDS TARGET GROUP	**mediating**	**still mediating**	**still mediating**
general orientation	mediating	mediating	mediating
character of incentives	repressive and stimulating: deposits and subsidies, public infrastructure	repressive and stimulating:mandatory obligations, infra-structure	repressive and stimulating

Formalization

As in the case of the UK, the Dutch government formalized its policy approach. This change in style was, however, less controversial than in Britain. The Dutch government, in particular the Ministry of the Environment was not completely despondent about the European Directive, even though it voted unlike the UK against the Directive. The 1991 Packaging Covenant was an agreement under private law, that only applies to the voluntarily signing companies (some 300).

The Ministry points out that with the new regulation *all* producers and importers are covered, and, particularly that the many small companies involved now must also comply. A general binding rule may very well be more convenient, to prevent free-rider behaviour of those that have not signed the covenant (Interview VROM 4.12.1995). Through this the effectiveness of the packaging policy can be increased. Preventing free-riding was also the main reason why the binding character of the new regulation was valued positively by those companies that signed the covenant. Free-riding companies could distort competition: this was to be prevented by the new regulation (Interview SVM 9.11.95). In the words of the managing director of the SVM: *'Hoe zwakker de wetgevingscontrole en hoe meer free riders, hoe geringer de kans dat een tweede convenant tot stand komt'* (Managing Director SVM, quoted in Verpakkingsmanagement 6/1996: 32). Moreover, it was unlikely that the packaging prevention target of the 1991 Covenant would be achieved. In other words, the European Packaging Waste Directive provided the Dutch government and its partners with an opportunity to escape the agreements they had made without losing face (see below). The broad industry support for a new packaging covenant can be explained by a smaller environmental burden and more flexibility for those companies who were primarily affected by the ministerial regulation, i.e. the packers/fillers, the importers of packaging goods, and the large retailers who commission the production of goods. This involvement of packaging material and packaging producers can be explained by the economic benefits of recycling. Since in the case of paper and cardboard, glass, and metal public authorities carry responsibilities for separate collection, they have a rather cheap supply of secondary raw material. In the case of plastic, the plastic material organization VMK took responsibilities on behalf of their members in order to prevent higher targets than the modest 27 per cent or more interventionist instruments. The German system functioned particularly in this respect as a warning (Interview VMK 9.9.1997). Note that the 1991 Dutch covenant scheme did not prevent the amount of plastic packaging rising by 30 per cent as compared to 1986, while, under the German 'green dot' scheme, plastic became a more expensive material, and accordingly its use steadily decreased in the 1990s (DSD 1996).

Strictness of standards

The Packaging and Packaging Waste Regulation set lower standards than the 1991 Packaging Covenant. In line with the European Packaging Directive there were neither quantified packaging reduction targets nor concrete measures to increase the re-use of packaging. The Packaging and Packaging Waste Regulation stipulates the maximum recycling and recovery targets set out in the European Packaging Directive (recovery 65 per cent, recycling of all material 45 per cent and 15 per cent each material). The 1991 Packaging Covenant implied a recovery rate of virtually 100 per cent by the year 2000, because landfill is banned, at least 60 per cent had to be recycled, and all Dutch incinerators have burn-to-energy facilities. The 45 per cent recycling target set out in the Packaging and Packaging Waste Directive is 15 percentage points lower than 1991 Packaging Covenant target. Note that the maximum recycling targets set out in the EU Packaging Directive for the year 2001 were already met in the Netherlands in 1993. In fact only the 15 per cent minimum recycling targets for plastic and tinplate were not already met in 1995. The lowering of standards cannot be explained by adaptation pressure induced by the European Packaging Directive, however. As outlined above, it was legally possible to pursue higher standards than set in the Directive. The weaker standards are the outcome of a 'package' deal basically involving the Ministry of EZ and the SVM on one side and VROM on the other side. The deal was to translate the Directive one-on-one into national regulations, and to pursue any higher standards in a new covenant.

More interesting than the comparison between the 1991 Packaging Covenant and the 1997 Packaging and Packaging Waste Regulation is therefore the comparison between the 1991 Covenant and the 1997 Covenant, also because virtually all affected companies joined the new Covenant. A comparison between these programmes reveal that standards for *packaging* prevention have been substantially relaxed, that the standard for packaging *waste* reduction has been maintained and the targets for recycling have been slightly increased. These targets are related like '*kommunizierende Röhren*', which make them still ambitious.

The most substantial change was made in regard to the total of packaging which was allowed to enter the market. The 1991 Covenant ruled that this amount had to increase to the level of 1986, some 2,35 million tonnes. When the 1991 Covenant was agreed, the amount had already risen by some 15 per cent. Hence, industry had to reduce the amount of packaging they put on the market. Industry was able to stabilize the amount of packaging, but increasingly argued that it was not realistic to reduce packaging in absolute terms against economic growth. Hence, the European Directive gave VROM and industry the possibilities to re-negotiate the Covenant, and to change this target without losing face. The new packaging target relieved industry from much of the pressure to reduce packaging at source. Rather than decreasing the absolute amount of packaging, industry is now allowed to increase the amount substantially, because the prevention quota is now conditioned on economic growth. According to calculations of the Ministry of VROM, the absolute amount of packaging may increase to 2,96 million tonnes, given economic growth of 40 per cent in this period (VROM 1997c). According to the logic of the covenant, this means that an increase in the amount of packaging from 2,35 million to 2,96 million tonnes is a prevention of 10 per cent. With this prevention target, the Dutch government did not follow the logic of its general environmental policy which informed the first covenant, namely that economic growth can go hand in hand with an absolute decrease in pressure on the environment.

There is another target, however, which is likely to discipline industry to some extent. The ceiling to the absolute amount of *packaging waste,* which may be landfilled or incinerated. This still amounts to 940 kilotonnes. Hence, industry is only allowed to produce 2,96 million tonnes of packaging for the Dutch market, if it recycled 68 per cent. If industry only recycles the envisaged 65 per cent, then the total amount of packaging is restricted to 2,69 million tonnes. This amount of packaging is not only one-sixth more than allowed under the old covenant, but it would also mean a stabilization of the current level, which might be difficult, given economic growth and socio-economic developments that points to more packaging.

The standard for recycling has been increased from 60 per cent to 65 per cent. The time frame has been extended from 2000 to 2001. This standard is stricter than the old one, if one takes into account a further growth in packaging. Since the standard is relative, any increase in the absolute amount of packaging implies an increase in the absolute amount of recycled material to keep the percentage constant. But as is the case with all the other new targets, these standards are weaker in the sense that the targets are now placed on more shoulders. Rather than companies representing 60 per cent of the turnover in packaging, now some 90 per cent are represented.

On balance, the standards decreased. Industry succeeded in negotiating a covenant which relieved it from strong pressure towards packaging prevention. The slight decrease in the strictness of standards can be seen against the above mentioned general political background emphasizing economic rather than environmental objectives. Economic concerns were mediated in the political process by the Ministry of Economic Affairs, which for institutional reasons, achieved a more important position in the policy network. As a representative of the plastic material association VMK said '*The access to EZ is easier because they understand that too much regulation is not good*' (Interview VMK 9.9.1997). On the side of industry economic considerations were very much emphasized by the MKB and the VNO-NCW who gained a more important role on the industry side. The Environment Ministry lost part of its power to countervailing forces. Not least, because parliamentary pressure and societal pressure towards less packaging and more re-use decreased in importance, and the lack of waste disposal opportunities was pictured less dramatically. The packaging prevention target of the 1991 Covenant would never have been achieved. The Packaging Directive therefore provided industry with a chance to negotiate a more favourable covenant.

The degree in detail of standards adopted also reflect instrumental learning processes. In regard to packaging prevention, the quota not only took into account economic growth, but also additional weight induced by the replacement of one-way packaging by re-use packaging. The government learned from the LCA's that refillables are usually heavier than one-way packaging. A focus on prevention, usually measured in weight, might frustrate the introduction of re-usable packaging. The government therefore accepted an increase in the weight of packaging when it was caused by the replacement of one-way packaging by refillables. In regard to re-use the government learned that one-way packaging is not always more environmentally damaging than re-use packaging. The government therefore allowed that under certain conditions the industry can move from re-use packaging to one-way packaging.

The new targets are very complex. When they were discussed in the concerned Parliamentary Committee some members admitted, that they did not understand them. Other members asked for additional information. Intended or not, the covenant increased the complexity, and this will probably result in a depoliticization of the issue. What in the United States is called the 'Dan Rather test' applies here: *'Reforms are less likely to generate a public outcry if television reporters cannot explain the implications of the new policies in fifteen seconds or less'* (Pierson 1995: 21).

Allocation of responsibilities

The Packaging and Packaging Waste Regulation and the new Covenant clarifies the allocation of collection, sorting and recycling responsibilities between public and private actors. Basically local authorities have always been responsible for the collection of household waste. The Packaging and Packaging Waste Regulation states that the duty lies on industry to take back packaging waste released from households from a point determined by the local authority (Art.5). Provinces shall incorporate in their provincial environmental ordinances a requirement that with effect from 1 August 1998 local authorities shall provide for the *separate* collection of glass and paper and cardboard. The agreements between industry and local authorities concerning glass, paper and cardboard has been formalized in Sub-Covenants. There is no financial responsibility for local authorities to collect plastic and metal separately. Producers and importers may reach agreements with the local authorities on how the other packaging materials will be separately collected (Regulation Art. 10). In practice metal and plastic from household sites will be collected shortly before, or in the case of metal after, incineration. Separate collection of plastic and metal packaging will be done on a large scale only from industry and commercial sites. A new Order in Council based on the Environment Management Act will demand from industry the separate collection of these materials.

This allocation of responsibilities can be explained by two factors. First, the general consensus-oriented approach and the stronger role of EZ in the policy network prevented the situation where industry had to take responsibility for the separate collection household packaging waste. Secondly, the government drew lessons from its own pilot projects and the experience of the German DSD. Pilot projects in the early 1990s revealed that the separate collection of plastic and metal from households is very expensive and not necessary to reach high overall recycling quota. That high costs are involved in the separate collection of household waste has been confirmed by the high investment and running costs of the German DSD which is based on separate collection. Accordingly, the government also learned that concentration on packaging from commercial sites is more cost effective. The requirement for the separate collection of these materials from industrial sites has to be seen against this background.

The formalization of the Dutch policy approach into public law implies that state agencies and administrative courts gain enforcement responsibilities. In regard to the Covenant, the government and industry stick to the old monitoring and assessment system of public-private committee. The membership has been increased from five to nine, mainly to reflect the broader range of interests affected. The provision of dispute settlement and enforcement remained unchanged as compared to the first covenant. The regulation is still binding under private law. And dispute settlement is governed by the Dutch arbitration regime. No actor saw a need to change these provisions partly because they have mainly a symbolic function. All parties are interested in a consensus-oriented approach. In this context these provisions were seen as measures of last resort, which are not very relevant for the practise of the implementation of the Covenant.

Modes of interest integration

In the preparation of the first Covenant, strategic discussions including a broad range of interests had been organized by VROM. They had the function of creating and diffusing information about the packaging waste problem and the possible solutions. Such a network was not created again. The main reason for this was that the new programme was not something substantially new, but a reaction to the European Packaging Directive and the further development of the 1991 Packaging Covenant. Rather than consultation, the function of negotiation had to be performed. The environmental and consumer groups were not needed for that, because they did not control necessary resources. Moreover, the environmentalists themselves did not have the packaging issue high on their agenda any more.

The choice for a general binding rule for all packers and fillers, importers of packaged goods and retailers who have their own products, implied a broadening of the target group. Rather than bilateral negotiations between VROM and the SVM, now industry organized itself in the packaging platform, where representatives of VNO-NCW and MKB also participated. Still the network remained corporatist. A small number of key figures from trade associations negotiated with the government. The SVM remained a central actor in the network. It had already stabilized relationships with VROM. And VROM saw the SVM as a reliable partner, due to its pro-active approach in the course of the implementation of the Covenant (Interview Expert 11.9.1997). Initially there were some irritations about the role of the SVM in relation to the role of the MKB and the VNO-NCW. These problems could be accommodated however. The possibility of high interest aggregation in the Netherlands as compared to the UK has to be seen against the character of the intra-industry relations in the Netherlands. The persons involved in the packaging waste policy have known each other for many years: they form a specialized network. They met regularly, not only in the course of the strategic discussions or the process of life cycle analysis, but also at all sorts of conferences. Large symposia are organized almost every month. '*Now, one knows each other better, understands the position of the other bette*' (Interview VMK 9.9.1997). Intra-industry relations are also tightened by overlapping membership. The chairperson of the packaging platform, for example, was there on behalf of the VNO-NCV, but his company is an active member of the SVM. It is also important to note that the SVM, the MKB, and the VNO-NCW are located in the very same building. More generally speaking, representatives of companies affected by packaging policies are often members of each other's boards. One should not forget in this context, that intra-industrial conflicts about packaging waste are often small as compared to other issues such as wage negotiations with trade unions. This means that companies try very hard not to raise conflicts between companies about packaging issues, to a point where they might hamper intra-industry unity in wage negotiations with trade unions, or other important topics (Expert Interview 11.9.1997). This aspect became particularly relevant when the VNO-NCW, the peak employer association, became part of the packaging waste network.

Orientation towards target group

The Dutch government continued its consensus-oriented approach. Initially attempts of the Ministry of VROM to include tough responsibilities on industry were prevented by the Ministry of Economic Affairs. Very close personal relations between representatives of VROM and representatives of the SVM had evolved over the last 10 years. The managing director of the SVM and the head of the waste division of VROM met each other frequently on an informal basis. This relation, which was made possible in part by the continuity of people involved, is close to what can be described as regulatory capture, at least if one takes a traditional pluralistic perspective. Though it might be exaggerated that the SVM wrote the entire Covenant, as one interviewee claimed, it is certainly the case that industry put its mark on the Dutch packaging waste policy. The interaction between VROM and the SVM was facilitated by an administrative scientist from the TU Delft, who - paid by the SVM - functioned as an independent broker (Interview Expert 11.9.1997). This academic was also heavily involved in the development and the implementation of the LCA process. He is also part of the group which has to concretize the ALARA principle.

10.7 Summary

The Netherlands did not move from negotiated agreements to hierarchical control. The Dutch government succeeded in maintaining its covenant approach, though in a more formalized way. Unlike the 1991 Covenant, the new one is embedded in a generally binding regulation. The standards for packaging reduction were weakened but remained strict in regard to packaging waste recycling. The policy process became more complex. The interests of virtually the whole of Dutch industry had to be integrated, while industry participants of the 1991 Covenant were mainly confined to the pro-active industries. Industry was, however, capable of aggregating the diverse interests, so that government could broker a compromise with a small group of actors who were capable of making comparatively credible commitments. The Packaging Directive was the main reason that the Dutch packaging waste policy formalized. The need for a generally binding agreement led to a pluralization of interests to be considered in the policy process. High recycling standards could be maintained. As a concession to economic interests the packaging reduction standards were, however, substantially weakened. Economic interests could effectively influence the shape of the policy content not only through the Ministry of the Environment, but also via the Ministry of Economic Affairs. This Ministry gained a more central position in the network, due to the need to transpose the Directive into a generally binding regulation. Hence, the need to adapt to positive integration changed the Dutch policy network and therefore indirectly influenced the strictness of standards. The Directive provided an opportunity for the Environment Ministry and the SVM to re-negotiate the over-ambitious 1991 Covenant. Without the Directive it would have been more visible to the public that the targets of the 1991 Covenant would not have been met by industry. The contrast to the British case in regard to the integration of interests is striking. Dutch government and industry succeeded in finding a compromise that has comparatively broad support among industry. What would be labelled from an (Anglo-Saxon) pluralist perspective as regulatory capture, is celebrated in the Netherlands as the green polder model.

11. Defending the German model

In the years following the adoption of the German Packaging Ordinance, the political debate in Germany concentrated not on the Commission's initiative for a European packaging directive but almost entirely on domestic developments. The political and economic actors as well as the wider public became aware of the consequences of the German Packaging Ordinance and the business organization that had to co-ordinate the sales packaging recycling, the Duales System Deutschland. Only a year after the last elements of the Packaging Ordinance had come into force, the Minister of Environmental Affairs presented a draft for an Amended Packaging Ordinance. The objective of the Amended Ordinance was to stabilize the DSD, which ran into serious financial problems and was at one stage almost bankrupt. The stabilization of the DSD was also an important topic after the adoption of the European Directive in 1994.

In contrast to the Dutch and British cases, the European Packaging Directive did not challenge the degree of formalization of the national policy approach. The public law codification of the German packaging waste programme matched the legal requirements for the implementation of the European Packaging Directive. As in the case of the Netherlands, however, the German standards were higher than those envisaged by the Directive. The German mandatory recycling targets for the second half of 1995 - ranging from 64 to 72 per cent depending on the material - clearly exceeded the Directive's general recycling target of 45 to be achieved by the year 2001.

Moreover, the inclusion of a recovery target in the Directive was in contradiction to the German approach. Recovery included incineration with energy recovery. But incineration was banned in Germany for packaging waste that fell under the DSD regime. The European Packaging Ordinance also did not entail concrete measures in regard to re-use. It only said that member states can encourage re-use. The question was whether the German refillable quota could be regarded as a measure of encouragement, or something more interventionist. The refillable quota was heavily contested by governments of other Member States and a great number of companies and business associations. The Commission started an infringement procedure, but hesitated to bring Germany to court. The German government was very determined to maintain the refillable quota and included it in a slightly relaxed form in the draft of the Amended Ordinance. The German parliament agreed on the draft. But more than eighteen month later it was still unclear whether the *Bundesrat* would agree to the relaxation of the quota. While the Commission might argue that the quota is still an illegal barrier to trade, the *Bundesrat* wanted to maintain the initial refillable quota or even increase the standards for re-use. This question increasingly became a topic of partisan politics in Germany. This has to be seen against the background of the impending elections in Germany.

The deadlock was finally solved when the *Bundesrat* agreed on the government's draft and included only minor amendments which were accepted by the government and the German parliament.

The German packaging policy process can be divided into three themes. The crisis of the DSD and the government's attempt to solve this crisis through amending the 1991 Packaging Ordinance. This phase was characterized by domestic developments (9.4.1, 9.4.2). Second, the German battle against the EU Commission as well as domestic and foreign companies to maintain the refillable quota (9.4.3). And third the government-*Bundesrat* bargaining to come to an Amended Packaging Ordinance. This theme includes the latest developments of the two other themes (9.4.4, 9.4.5).

11.1 The crisis of the DSD

The implementation of the German Packaging Ordinance was a highly controversial process and attracted much media publicity inside and outside Germany. The critical attention focused on the Duales System Deutschland (DSD), the private organization which had to ensure that the collection, sorting and recycling targets for sales packaging stated in the Packaging Ordinance would be achieved. The DSD had to establish a nation-wide collection scheme for sales packaging, independently, but in mutual agreement with, the local authorities. The DSD had to prove for each German Land individually that it fulfilled the requirements of the Ordinance. Without this approval the industry would have take-back responsibilities for sales packaging and mandatory deposits for washing powder, cleansing agents and emulsion points would come into force.

Financial crisis

From the very beginning of its operations, the DSD ran into a number of serious problems which resulted in a financial crisis. The whole scheme was just going short of bankrupt. There were a number of reasons for that. First, the DSD had to co-ordinate its waste collection activities with the activities of the local authorities. The local authorities were allowed to ask an adequate charge for the use of its waste collection infrastructure, for instance, bottle banks. The DSD needed a declaration from the local authority stating that their systems were co-ordinated. This made the DSD dependent on the local authorities. Many of them imposed charges which were according to most observers, far too high (Financial Times 16.3.1997).

The second reason for the financial crisis of the DSD was related to the successful collection of waste. The DSD collected much more waste and at a much earlier point in time than was initially expected. Though the amount of waste collected went beyond the required collection quota, the DSD took responsibility for all sales packaging collected. It was reasoned that the public would not understand it if the packaging waste it separated from the household waste would be dumped or burned. There was, however, a lack of recycling capacity, and as consequence packaging waste had to be stored and exported. In particular plastic waste was a problem. The DSD had organized a capacity for recycling 100,000 tonnes, but the DSD received 400,000 tonnes in 1993 (IW 2.12.1993). Since, due to an intervention of the *Bundesrat,* thermal recycling, i.e. incineration with energy recovery, was not allowed, all plastic packaging waste had to be treated by material recycling. There were simply not enough sorting and recycling capacities. In the Summer of 1993 the contract partner of the DSD, who should have guaranteed the sorting and recycling of plastic waste, stopped its activity after a number of irregularities were discovered including the illegal dumping of plastic waste in France. The DSD was forced to create a new organization in which it took not only the financial responsibilities for the separate collection as in the case of the other materials, but also for storage, and recycling. This further increased the bill for the DSD (DSD 1994: 9).

Moreover, consumers did not put only packaging waste in the DSD dustbin. They also disposed of other products from the same material or other garbage. This fraction amounted to some 20 per cent, and had also to be disposed of by the DSD (IW 2.12.1993). For these and other reasons the running costs of the DSD extended according to its own figures the initially expected 2 billion German marks. The running costs were 3.2 billion German marks in 1993 and stabilized at a level of 4 billion German marks from 1994 on (DSD 1994-1996).

The unexpected high level of expenditure was accompanied by an unexpected low level of income. The main reason was the free-rider attitude of many companies. After the *Länder* had approved the scheme, the companies' payments to the DSD decreased. There were companies who neither joined the DSD, nor established their own take-back and recycling systems. Another group used the "green dot" symbol but did not pay the license or paid only for a part of the packaging they attached to their products (Interview DSD 5.12.1996). Officially, some 14,600 companies had the license to put a green dot on their products in October 1993. The DSD estimated that 70 per cent of all packaged products carried a green dot in that year, while the charge was paid for only 55 per cent to 60 per cent of the packaging the charge was paid (DSD 1994: 9-11). The lack of income was estimated at 600 - 800 million German marks annually because of free-riding behaviour (European Packaging & Waste Law April 1997: 23).

In the autumn of 1993, the debts of the DSD to the private and public waste management companies amounted to 800 million German marks. The very existence of the DSD was endangered. In September 1993, the Minister of the Environment brokered a compromise between the DSD, the BDI, the DIHT, the peak associations of the local authorities and representatives of trade associations and companies. An important point of the compromise was that the private waste companies should temporarily abandon their financial claims. In exchange for this concession three waste management companies had taken chairs in the board of the DSD as long as the DSD had not paid its bill (DSD 1994: 11). The idea that those who benefited from the DSD were also represented on the board was not likely to increase the legitimacy of the scheme. Moreover these companies were also part of the new plastic material guarantor DKR. The German weekly *Die Zeit* perceived this as the culmination of greed. It wrote: *'Schließlich, der Gipfel, plazierten sich die Großen der Entsorgungsbranche im Netz der wuchernden Abfallbürokratie so, daß sie ihren eigenen Tochterfirmen bequem mit Aufträgen versorgen konnten: eine Lizenz zum Gelddrucken, die dem Kartellamt schon lange ein Dorn im Auge war'* (Die Zeit 16.9.1994). The Federal Cartel Office (*Bundeskartellamt*) had criticized the fact that the customers of the DSD were presented on the board of DSD. It was able to prevent the *permanent* representation of waste management companies on the board (Die Zeit 22.10.1993) and also succeeded in making them withdraw from the DKR, i.e. the organization that had to guarantee the plastic waste recycling DKR (Die Zeit 16.9.1994).

Competition problems

The membership of the waste management companies on the board was not the only competition problem associated with the DSD. The requirement of the Packaging Ordinance to establish a nation-wide scheme, allowed the DSD to establish a monopoly for organising the collection, sorting and recycling of sales packaging. This monopoly was criticized by the Federal Cartel Office (*Bundeskartellamt*, BKA) as well as the Economic Ministries at the federal and the *Länder* level. The BKA felt a dilemma which an official described as follows '*Als die Verpackungsverordnung in Kraft gesetzt worden war, stand die Kartellbehörde vor zwei maßgeblichen Fragen: Soll sie eine im Grundsatz begrüßenswerte Privatisierung durch ein Veto blockieren? Und soll sie die Umsetzung umweltpolitischer Ziele aus wettbewerblichen Gründen schon im Ansatz behindern? Wir meinten, daß wir den flexiblen Weg der Tolerierung, daß heißt der behördlichen Duldung, gehen sollten.*' (Quoted in Die Zeit, 22.10.93)

When contract partners of the DSD started to enter the market for transport packaging, however, the BKA endorsed an administrative prodecedure against the DSD. The director of the BKA reasoned that '*Es liegt doch auf der Hand, daß der Entsorger der Verkaufsverpackungen seinen gewerblichen Kunden dazu drängen würde, ihm auch die Entsorgung der Transportverpackungen zu überlassen. Dann ist der Mittelständler, der das bisher gemacht hat, weg vom Fenster, und der Grüne Punkt reißt sich den ganzen Sektor unter den Nagel - mit der Konsequenz, daß die Monopolrenditen letztlich der Verbraucher zahlen muß*' (quoted in Die Zeit 15.1.93). The Federal Cartel Office succeeded in preventing the extension of the activities of the DSD partners (Monopolkommission: 1996: 33).

Another competition problem associated with the DSD concerned the concentration process in the waste management and recycling market. The Packaging Ordinance required ambitious collecting, sorting and recycling quota within a short period of time. These requirements created a boost in the waste management market. Annual growth figures up to 40 per cent were reported. The necessary investments to meet the targets were estimated at 7 billion German marks. To meet the recycling standards, advanced technologies and capital were necessary. This was too much to be expected from small and medium-sized companies who traditionally held a large market share in this sector. Such investments could only be made by companies who had sufficient capital such as large utilities like RWE and VEW or energy groups. The RWE alone bought some 70 small and medium sized companies within four years and became the leading company in the market (FAZ 15.1.1993, see also Financial Times 16.3.1996).

Legitimization crisis

The DSD underwent not only a financial crisis, but a serious legitimacy crisis as well, which was, of course partly related to the financial troubles. Many consumers felt betrayed. They thought that the 'green dot' would be a label for environmentally sound packaging. They thought that the 'green dot' meant that the packaging was easily recyclable and/or environmentally superior to other packaging. This was not the case, however. Refillable bottles, for instance, did not carry a green dot because they did not end in the DSD waste bin. Still, it is rather unlikely that a refillable beer bottle distributed in a regional market is less environmentally friendly than a beer can made from aluminium, though the latter carries a green dot. The 'green dot' was only the symbol for the participation in the DSD. Every packaging item, even those which were completely unnecessary and very harmful to the environment, could get a licence (Die Zeit 3.4.1992). The DSD, however, did not do very much to clarify this misinterpretation by the consumers, since it needed the cooperation of the consumer for the separate collection of packaging waste.

The bad image of the DSD was further fuelled by media reports about the strange fate of some of the packaging, which the consumers had so carefully separated from the rest of their waste. The scandals almost always had to do with plastic packaging waste. This material had been heavily subsidized by the DSD. Contract partners of the DSD got up to 1,000 German marks per tonne to recycle sales packaging (Ökologische Briefe 14.9.1997: 17). This was a high economic incentive to take up the packaging but there were not sufficient domestic recycling capacities. In the first years, a large part of this plastic was transported to Eastern Europe. The largest client was, however, China. Some 100,000 tonnes, or one third of all collected used plastic packaging had this destination (Die Zeit 21.1.1994, Ökologische Briefe 14.9.1997: 17). When China decided to stop the import, North Korea became the main client of the DKR (Communication DKR August 1996). German yoghurt containers were transformed into Asian plastic sandals (Der Spiegel 29/1995: 17).[27] In 1995 still half of DSD's plastic packaging waste was exported for recycling. Though the German Environment Minister Töpfer promised his colleagues in 1993 that German plastic exports to other EU countries would stop in the next year, official figures of the DKR revealed that still 67,000 tonnes (or 14 per cent) of this material were recycled in those countries in 1995 (DKR 1996: 21). Other material was illegally dumped, for instance, in France or Hongkong (SZ 11.11.1996). Some plastic material was stored rather than recycled. In a number of plastic depots fire erupted. Rumours emerged that the owners of the depots were responsible for this cheap way of 'thermal' recycling (communication with Environment Ministry NRW August 1996). The DSD contract partner for plastic packaging (DKR) found eleven examples of fraud through own investigations. Companies paid to recycle DSD sales packaging waste, recycled

[27] The Chinese government was not very happy with the German 'secondary raw material' what they considered as 'waste'. At a conference, a representative of the DKR told the ordinance that the German Environment Ministry asked the DKR not to transport DSD packaging to China during the period when the German Chancellor Kohl visited the country, in order to avoid '*unschöne Bilder und unnötige Krise*'. The DKR stopped the export. But its Chinese contract partners were so angry about that, that they complained to the German government and the German chancellor still got this topic on his agenda (cited in Umwelt [the Magazine for Environmental Engineers, not the BMU Journal with the same title] 3/1993: 78).

instead (cheaper) industrial transport packaging (communication with DSD and DKR August 1996, SZ 28.6.1996).

These examples should not suggest that all accusations are empirically justified. The whole process was very controversial. Exaggerations and wrong accusations can therefore not be precluded. The important thing is that, whether right or wrong, they all added to the negative public image of the DSD. It would go beyond the scope of this study, however, to investigate the claims in detail. This would be very difficult, because all these processes suffer from a lack of transparency. Accordingly, the Environment Ministry of NRW and other actors criticized the fact that the monitoring figures were incomprehensible. They also maintained that they had evidence that the DSD calculated with wrong figures and quota (communication with Environment Ministry NRW August 1996). In the German Land Lower Saxonia, for instance, half of the separately collected sales packaging waste apparently disappeared during the transport from collection to sorting in 1993 (Ökologische Briefe 14.9.1994: 17).

One has to bear in mind, however, that the DSD is a new organization with no historical precedent or counterpart in any other part of the world. Certain problems, in particular in regard to intra-industrial co-ordination, monitoring and enforcement can be related to teething troubles. Still, these implementation problems were reported almost monthly in the press, and contributed to a bad image of the DSD, which ultimately threatened the cooperation of the consumer, and therewith endangered the achievement of high collection quota on which the whole system was based.

The DSD tried to increase its legitimacy by establishing an advisory committee with well respected figures. The committee was made up of members from business, trade unions and political parties. It was initially chaired by the environmentalist and respected environmental economist E.U. von Weizsäcker. In 1994, however, he stepped back, criticizing the dual system as a '*dangerous manipulation*' that '*serves the throw-away society*' (cited in Financial Times 16.3.1994). This criticism was directed at the lack of transparency of the DSD and the emerging concentration of the recycling industry in the hands of a few utilities and energy groups.

Meeting targets

Despite the serious financial problems, the bad image and the irregularities, the DSD was able to achieve the recycling targets set by the Packaging Ordinance, at least according to the official figures. 1996 was the first year in which those recycling quotas had to be fully met, which were set by the Packaging Ordinance for the second phase of operation. In this year, the DSD collected 86 per cent of all German sales packaging from households and small businesses. This is some 5.5 million tonnes, corresponding to a per capita collection of 71.2 kilograms. This result meant that the Green Dot company was able to fulfil the collection quotas set by the Packaging Ordinance and to forward the requisite quantities for recycling. The figures are provided by the material mass flow verification prepared by the DSD. This document is a record of the total mass flow from collection to recycling of the packaging materials. The individual recycling figures are as follows: 2.69 million tonnes for glass (1995: 2.57 million tonnes), 1.32 million tonnes for paper and cardboard (1995: 1.26 million tonnes),

535 thousand tonnes for plastic (1995: 504 thousand tonnes), 302 thousand tonnes for tinplate (1995: 260 thousand tonnes), 36 thousand tonnes for aluminium (1995: 32 thousand tonnes) and 445 thousand tonnes for beverage cartons and other composites (1995: 297 thousand tonnes) (DSD 1997).

Table 29 Targets of the 1991 Packaging Ordinance and Performance of the DSD

	levels 1988	targets 1993+1994	result 1993	result 1994	targets 1995	result 1995	targets 1996	result 1996	result 1997
glass	38	42	61,6	71,7	57	82	72	85	89
paper/cardboard	6	18	54,8	70,6	41	90	64	94	93
plastics	2	9	29	51,8	36,5	60	64	68	69
tin plate	-	26	35	56,3	49	64	72	81	84
composites	0	6	25,8	38,8	35	51	64	79	78
aluminium	3	18	6,8	31,5	45	70	72	81	86

Source: estimations of the 1988 level are from Klepper and Michaelis referred to in The Economist 3.7.1993. The other figures are from the DSD (1995-1997, and press release 7.5.1998).

The overall amount of sales packaging consumed in Germany decreased from 6.4 million tonnes (1995) to 6.3 million tonnes (1996). The sharp increase in the recycling of composites is the most significant development of the years 1996 and 1997. According to a spokeswoman for the DSD, there were two main reasons for that: first, a successful campaign by Tetra Pak, including a game for consumers to increase the collection of its drink packaging; second, a change in the definition of composites. Up to 1994 tinplate-based packaging (i.e. cans with an aluminium lid) fell under the category 'tinplate'. Since 1995 this waste stream has been counted as 'composites'.

The running costs were about 4 billion German marks. An average four-person household pays approximately 200 marks for the packaging scheme each year. Consumers are not compensated by a decrease in the waste management costs of the residual household waste. In fact households see their waste charges rising every year, due to high fixed costs of disposal, higher environmental standards and the use of waste charges to subsidize other local services (SZ 15/16.7.1995).

The total amount of packaging entering the market has been reduced as an intended side effect of the Packaging Ordinance. The figures of the independent *Gesellschaft für Verpackungsmarktforschung* (GVM) revealed a reduction in the amount of packaging of 8 per cent or 1 million tonnes, from 1991 to 1993 and of 10 per cent from 1991 to 1995, most of which concerned sales packaging. Since then, the overall amount of packaging stabilized. Note that, as in the case of the Netherlands, factors other than the Packaging Ordinance also influenced the packaging choices of companies (see chapter 10).

Table 30 Development of the use of packaging in Germany

	1993	**1995**
glass	91,3	86,1
tin plate	93,2	83,3
aluminium	93,1	74,6
plastic	92	96,2
paper/cardboard	92	94,3
composites	99,6	94,2
total	92	89,6

Sources: BMU 1994, BMU 1996, includes transport packaging (40% of total amount), 1991=100

The reduction in the amount of packaging entering the market was achieved at least partly, because of the charging system of the DSD. Initially these charges were calculated according to the volume of the packaging. From late 1993 on, however, the charges were calculated according to the collection, sorting and recycling costs of different packaging waste materials. Packers, fillers and importers of packaging goods who were member of the DSD had to pay for each kilo of plastic they added to their products 3 German marks. The respective figures for the other materials were: glass 0.16 mark, paper and cardboard 0.33 mark, tin plate 0.56 mark, aluminium 1.00 mark, composites 1.66 marks (IW 2.12.1993). Note, however, that plastic packaging and composites are usually much lighter than glass or metal packaging. Still, the costs of plastic packaging were not trivial. A plastic container of one litre of conditioner was initially charged 0.02 mark. According to the new scheme the charge was 0,16 mark. Or to give another example, a chocolate box made from a paper/plastic composite packaging was initially charged 0.02 mark. This price increased to more than 0.08 mark. The huge cardboard box for a television was comparatively cheap. The charge decreased from the initial 0.20 mark to less than 0.18 mark (Der Spiegel 25/1993). Accordingly, industry argued that '*die neue material und gewichtsbezogene Preisstaffelung bieten einen ökonomischen Anreiz zur Reduzierung von Verpackungsmaterial und zur Vermeidung entsorgungsaufwendiger Packmittel - bei Hersteller und beim Verbraucher*'(IW 2.12.93).

Companies were very inventive to ensure that the recycling quota were met. When Tetra Pak, the large packaging manufacturer of drink containers made from composites, was afraid that the targets for composites would not be met, it organized a nation-wide game which cost 20 million marks. Consumers had to write down their name and address on the used Tetra Pak packaging, before they threw it in the DSD bin. In the course of the sorting of the used packaging the winners were picked out of the material stream. The prices were enormously high. Weekly winners got 100,000 marks and the final winner 1 million marks (FAZ 12.6.1996, SZ 16.4.1996).

11.2 The first draft of the Amended Packaging Ordinance: saving the DSD

The Minister of the Environment presented the first draft for an Amended Packaging Ordinance in December 1993, six months after the last provisions of the 1991 Ordinance came into force. This draft was not a reaction to the EC Commission's initiative to adopt a European Packaging Directive. In official statements there were not even references to European developments. The objective was the stabilization of the DSD. The main elements were measures against free riding and a weakening of the recycling standards for sales packaging (BMU 1994: 67-68).

In order to avoid free- riding behaviour, the draft clarified the responsibilities of those packers and fillers and importers of packaged goods who did not participate in the DSD. The draft stipulated that they had to prove that they took back at least 30 per cent of all sales packaging they brought onto the market. From 1996 on this quota was to be 80 per cent. In order to identify free-riders more easily, the draft stipulated that participants had not only to prove to the DSD but also to the relevant public authority that they are participants in a dual system. The new provisions are especially aimed at small shops (e.g. bakers and butchers) and large retailers (e.g. Norma and Peek & Cloppenburg) who use last minute/service packaging without paying DSD or establishing an own system (EU Packaging Report Feb. 1996: 12, Interview DSD 5.12.1994).

The draft also clarified that the collection of packaging by the DSD is restricted to households and small enterprises. This was done to counter the criticism of the Federal Cartel Office, stating that the DSD could get a monopoly of all packaging including transport packaging. The draft also relieved the DSD of recycling collected material which exceded the quota and of material collected which is not packaging. Collected material exceeding the quota could be incinerated with energy recovery. Non-recoverable packaging material may go - as collected material which is not packaging - to the municipal waste stream.

The changes of the recycling standards were especially directed towards plastic packaging. This problem was addressed by saying that material recycling would also mean 'raw material recycling' which includes chemical recycling (hydration) and the use in steel production (gasification), which basically comes down to incineration, but the draft carefully avoided of calling it that (see below). Moreover, the recycling quota were substantially reduced at least for the period until 1998 (see table below). In order to prevent all plastic packaging ending as oil products or as fuel, the draft stated that half of the new quota had to be achieved by material recycling rather than 'raw material' recycling.

Not only the targets for plastic packaging but also the targets were changed, mostly lowered in fact, and/or the deadline was postponed. The following table compares the old and the new recycling targets. Note, that in order to reduce the complexity of the monitoring of requirements, the system of collection, sorting and recycling targets was replaced by a system focusing solely on recycling targets.

Table 31 Proposed targets in the (German) 1993 draft Amended Packaging Ordinance

	from 1993 on (old)	**from 1993 on (new)**	**from 1.7.1995 on (old)**	**from 1996 on (new)**	**from1998 on (new)**
Glass	42	40	72	70	70
Tinplate	26	30	72	70	70
Aluminium	18	20	72	70	70
Paper/Cardboard	18	20	72	50	60
Plastic	9	10	64	50 (25 mat.rec.)	60 (30 mat. rec.)
Composites	6	10	64	50	60

The new draft was strongly criticized. Environmental and consumer groups, as well as the Foundation for Re-use (Stiftung Mehrweg) which represent the interests of companies benefiting from increase in re-use criticized the basic idea of stabilizing the Dual System. They argued that the system was too costly and the environmental benefits doubtful. These groups proposed alternatives to the Dual System such as tax on packaging, or a system based almost entirely on re-usable packaging (Politische Ökologie 35/1993: 12-13).

Another group of critics did not radically objected to the idea of Dual System but were critical about the provisions of the draft especially the allowance of raw material recycling for plastic packaging. Among this group were a number of SPD-led *Länder* (for instance Baden-Württemberg, Müllmagazin 3/1995: 5).

They criticized for instance, that the steel company Bremer Stahlwerke was allowed to use DSD plastic packaging to replace heavy oil as a reduction agent in iron production. The waste is not burnt in the blast furnace, but gasified. At 2100 degrees Celsius the granules split into carbon and hydrogen, and the mixture reacts with the iron ore, converting it to elementary iron. Rather than paying 130 marks for the heavy oil, the company received 200 marks for using the DSD waste. This was seen by critics as unintended subsidy by the German consumer for the steel industry. They also said, that it was totally contrary to the ideas of waste avoidance and re-use (EU Packaging Report Nov. 1995: 11). Moreover, it was argued that the environmental advantages of 'raw material recycling' were rather doubtful. It was argued for instance, that the energy needed for the recycling clearly outweighed the energy saved. Despite the critiques gasification was also tested by Eko-Stahl and Krupp-Hoesch. Another example for 'raw material recycling' was chemical recycling done by VEBA and RWE. VEBA produced through hydration one tonne of synthetic oil from plastic waste for 900 marks, while the raw material would cost some 170 marks. The difference was paid by the DSD. Adding the costs of collection and recycling, the transformation of used plastic packaging to synthetic oil costs some 2300 marks per tonne. This is a lot more expensive than incineration or landfill, even when prices may decrease due to better economies of scale. Moreover, as in the first example ecological advantages of this method are contented (Die Zeit 21.1.1994, 16.9.1994, Ökologische Briefe 14.9.1997: 17).

The draft went through the official hearing on 4 February 1994, its adoption was delayed however, due to the programme development of the EU Packaging Directive which was expected to be near completion. Though the government did not plan to make any EU-induced substantial changes to the draft, it did not want to adopt the Amended Ordinance before the final shape of the Directive was clear. From the year 1994 on another element of the Packaging Ordinance gained importance, the German refillable quota.

11.3 Government vs. EU Commission: Germany's bottle battle

The German refillable quota stipulated that the aggregated market share of beer, water, soft drinks, juice and wine sold in refillables could not fall below 72 per cent. Otherwise mandatory deposits for one-way drink packaging would be introduced. The share of refillables had fallen continuously since the 1970s. Since the second half of the 1980s the level stabilized. In the course of the 1990s, however, the share of one-way packaging started to rise again (Bundesregierung 1996b: 20). Beer cans in particular increased in popularity. Especially in Eastern Germany companies tried to establish themselves on the new markets by using cans. They were easier to market than refillables since no take-back logistic had to be established. This facilitated sales especially at gas stations and kiosks. The market share of cans rose to almost 40 per cent in some East German regions (Die Zeit 17.2.1995, see also World Drinks Report 9.11.1995: 11).

Due to the popularity of cans the German 72 per cent quota came under serious pressure. The quota decreased from 73,55 per cent (1993) to 72,65 per cent (1994) (Bundesregierung 1996b: 20). In order to avoid the mandatory deposit on drink containers which would be the consequence of falling below the quota, the German drinks producers and retailers agreed on a fee of 0.10 marks on a can to steer consumers away from it. This agreement had the support of the German Minister of the Environment. The fee would be charged from 1996 on, if the Federal Cartel Office approved the scheme (SZ 29.11.1995). The Cartel Office criticized the agreement, however, saying that this is '*crystal clear price-fixing cartel*'. It also said that the deal would not be agreed by the Brussels competition authorities and that there '*was no chance of an exemption*' (quoted in World Drinks Report 21.12.1997: 15).The Federal Cartel Office threatened with fines up to 1 million German marks (Verpakken 1/1996: 66).

After the Federal Cartel Office made clear that it would not approve a fee on one-way packaging, companies started an action against litter '*Halte Dein Land Sauber - Mitdenken statt Wegwerfen*'. The initiative was launched to make the problem of one-way packaging less visible. The initiative also implied, however, that one-way packaging is not good. The campaign therefore indirectly supported multiple-use packaging. The chairman of the initiative confirmed that the threat of mandatory deposits on one-way packaging was one reason for the initiative (EU Packaging April 1996: 11).

It took until 12 December 1995 for the European Commission to react to the many complaints by companies that the German refillable quota erected a barrier to trade. The Commissioner responsible for the internal market sent - a long awaited - official letter of complaint (based on Article 169 of the Treaty) to the German government, arguing that the quota is not justified for environmental reasons and an exemption to Article 30 of the Treaty (free movement of goods) would therefore not be possible. The Commission also said that the quota would protect the current market situation. This would be indicated by the fact that the quota for milk is only 17 per cent while the quota for the other drinks is 72 per cent. It asked the government to prove within two months by a LCA for each individual product that the provision is suitable and necessary for the protection of the environment. Otherwise the Commission would launch an infringement procedure (FAZ 15.12.1995, World Drink Report 11.1.1996).

As a reaction to the letter, the German Ministry commissioned the UBA to calculate LCA's for mineral water and carbonated soft drinks. The UBA had already carried out LCA's for beer and milk in 1994. The - heavily contested - official result of these LCA's was that refillables clearly have environmental advantages over one-way packaging as far as beer is concerned, while for milk not only refillables but also one-way plastic sachets are ecologically sensible (FAZ 29.1.1996).

Though the official deadline for the reaction to the Commission's letter of complaint was 12 February 1996, the Commission granted the German government extension of the deadline (EU Packaging Report March 1996). The government reacted as late as 29 April 1996 (Bundesregierung 1996b). It claimed that there was no substantial barrier to trade for foreign companies. These companies could use third parties within Germany for all their obligations. They were therefore not discriminated in favour of German companies. Moreover the 1985 Directive explicitly mentioned deposit systems as instrument to ensure the take-back of packaging. Even if barriers could not be entirely precluded, this would not violate Article 30 of the Treaty because the measures were suitable, necessary and proportionate and therefore justified. In regard to the refillable quota the government argued that a total replacement of refillables by one-way packaging would increase the annual use of packaging by 4,3 million tonnes. This would mean more than a doubling of packaging waste for the overall drink sector. A reduction of the shares of refillables by only 20 percentage points would result in an additional 1,2 million tonnes of waste annually and would therefore counterbalance the waste prevention efforts in other areas. Even if one assumes that 60 per cent of the one-way packaging would be recycled, an additional 500,000 tonnes of waste had to be disposed of (21-22). The government also referred to the above mentioned LCA's for beer and milk and translated these results to other types of drinks. It argued that these LCA's prove that a 72 per cent refillable quota makes ecological sense. It also admitted that more sophisticated LCA's for drinks other than beer and milk would be necessary. They therefore commissioned new LCA's which were to be finished in early 1997 (27-37).

The Commission met later that year with the German government to discuss the German position. The Commission did not undertake any more official steps, however. Observers reasoned that the German government made quite a strong case and that the Commission was impressed by the arguments put forward by government. An official at the Commission also said that the issue was so politically sensitive that political considerations could be taken into account. '*The Commission as a political body will reach an agreement about what measures to take. It depends on the political mood - a climatic assessment if you wish*' (quoted in EU Packaging Report March 1996: 5).

It is quite likely that the Commission did not very much like the idea of becoming the scapegoat for million of tonnes of additional packaging waste. It was probably also important that the Commission did not want to put too much effort in a politically very sensitive infringement procedure against the 1991 Packaging Ordinance, because in the meantime the German government had developed a draft of an Amended Packaging Ordinance in which the government also addressed the refillable quota. It is likely that the Commission therefore wanted to use the notification procedure of the new regulation to scrutinize the trade effects of the German packaging waste policy. For this reason the issue of refillables will be now described in the context of the general development of the German packaging waste policy.

11.4 Government vs. Government: Inter-ministerial bargaining

The decision-making process of the Environment Ministry's draft showed a considerable delay. The forthcoming European Packaging Directive was not the only reason for that. The general elections in autumn 1994 also slowed the process down. Moreover the 1986 Waste Act on which the 1991 Ordinance was based, was to be replaced by the Materials Recirculation Act (*Kreislaufwirtschaftgesetz,* BGBl. I 1994, p. 2705). The discussion of this Act in 1994 overshadowed the packaging waste debate to some extent. The new Act aimed at reinforcing the philosophy of manufacturers' product responsibility, as already realized in the case of used packaging. Industrial designers were expected to design goods in such a manner that waste would be minimized; products could be recycled and disposed of in an environmentally friendly manner. Goods should be capable of re-use, have as long a lifetime as possible, be easy to repair and low in hazardous substances. Industry was made responsible for the disposal of its waste and had to bear the costs in this respect. Instead of municipalities contracting with private waste management companies, industry could contract out specific waste streams direct to third parties. The law was the result of a considerable debate between the German parliament and the *Bundesrat*. Passage of the legislation was accompanied by intense lobbying from interested parties of German industry (EU Packaging Report Nov. 1996). The most important consequence of the 1994 Waste Act on the packaging waste policy process was that the government not only needed the assent of the *Bundesrat* but also the assent of the German parliament to adopt an Ordinance.

The packaging debate also slowed down because the government was internally divided about the packaging waste policy. The standards set out in the BMU draft were criticized by the Ministry of Economic Affairs as being to high. The Ministry especially opposed the refillable quota, and was more generally not in favour of any quota at all (see EU Packaging Report March 1996: 12, Interview BMWi 6.12.1996). A number of German *Länder* wanted an increase of the refillable quota and were also very critical about the DSD. The year 1995, the year following the adoption of the Packaging Ordinance, was spent looking for a compromise within government and between the government and the *Länder*.

It was against this background that it took two years before the Ministry of the Environment presented a new draft. The new draft did not substantially deviate from the old one despite all the developments in the last two years, most notably the adoption of the European Packaging Waste Directive. What is more, the Minister of the Environment did not consider an amendment of the 1991 Packaging Ordinance necessary to implement the Directive. She said '*Ich strebe keine Novelle um jeden Preis an. Meine umweltpolitischen Ziele kann ich grundsätzlich auch nach wie vor mit der geltenden Fassung erreichen*' (BMU press release 28.2.1996). Accordingly, only small changes were made. The modified recycling targets were exactly the same as before and the provisions in regard to free-riding were also maintained. The refillable quota also remained intact, despite the official complaint of the EU Commission (BMU press release 28.2.1996). The only change was that the linkage between the underachievement of the quota and the introduction of the mandatory deposits was slightly relaxed. Deposits would not be required automatically any more but were left to the discretion of the government of the respective Land in which the refillables decreased (EU Packaging Report Feb. 1996: 11, FAZ 29.1.1996). The Environment Ministry announced, however, that parallel to the amendment process, it would examine with the *Länder* whether ecological advantageous drinks packaging could be adequately supported in a more flexible way (BMU press release 28.2.1996). The Ministry therefore commissioned the Federal Environment Agency to assess the effectiveness of different instruments, including mandatory deposits, taxes, licenses, and voluntary agreements (BMU press release 3.7.1997).

Another change concerned the internal working of the DSD: a new provision aimed at increasing competition. Collective schemes had to invite public tenders for the collection, sorting and recycling services. The costs of collection, sorting, and recycling/ disposal of each packaging material had to be clarified. Collected packaging also had to be traded under conditions of competition. This change was induced by pressure from DG III of the EU Commission, who argued that the DSD system comprises a price-fixing cartel, which would restrict both the establishment of a market for raw materials and competition according to Treaty Article 85.1. The DSD system prohibited the collector of separate sales packaging waste to market this secondary raw material. They had to give it away for nothing to the recycling guarantor, the so-called 'interface zero'. The Commission reasoned that therefore no market for secondary raw material could be established at which the material is traded between the DSD, the waste collectors and the recycling guarantors. Since in theory and practice this material can be traded between Member States, the DSD provisions are likely to restrict EU competition (CEC 1994).

In the aftermath of the draft the pressure on the refillable quota increased. Stimulated by the Commission's letter of complaint critical companies increased their lobbying, such as Coca Cola Germany (EU Packaging Report Feb. 1996: 11) and trade associations of beverage can producers (FAZ 12.6.1996). Most of the industry was also against the quota for plastic material recycling though the quota was already substantially lower than in the 1991 Ordinance (EU Packaging Report March 1996: 12), whereas the BUND supported the BMU by giving the Minister a refillable beer bottle marked with her name (BMU press release 3.7.1996).

The situation grew very tense, there was a lot of speculation and false information. The FAZ reported, for instance, that foreign companies would be exempted from mandatory deposits and the refill quota (FAZ 9.8.1996). This speculation was countered by the Ministry of the Environment (BMU Press release 16.8.1996). The Ministry also had to counter the rumour that the Commission had already decided to forbid the quota (BMU press release 31.7.1996).

In September 1996, the well-known German economic research institute *Ifo-Institut* presented its research finding about the most efficient instruments to restrict the sale of one-way drink packaging. They proposed a licensing system where the government fixed the absolute amount of drinks to be filled in one-way packaging. For this amount licenses will be disposed of by auctions and are then freely tradeable (FAZ 17.9.1996). It comes as no surprise that the licensing approach was very much criticized by the producers of one-way drink packaging. The chairman of Schmalbach Lubeca, Germanies large producer of cans and PET packaging, strongly criticized both the refillable quota and the proposed licensing system. He said '*Mit Mehrwegquote und Lizenzmodell machen wir uns zum Gespött in Europa. Das sind planwirtschaftliche Instrumente, die keiner braucht. Die Abfallprobleme haben wir in Deutschland längst im Griff*' (Lebensmittelzeitung 7.3.1997).

It is important to note that the draft was written by the BMU and that it did not yet have the agreement of the Ministry of Economic Affairs (BMWi). As already said earlier, the BMWi opposed, similar to most parts of the German industry, the refillable quota. Moreover, the BMWi wanted to allow incineration with energy recovery for plastic material. These were the two main sticking points in the preparation of the Cabinet draft between the two ministries (EU Packaging Report July 1996: 11).

After tough inter-ministerial negotiations both Ministries reached a compromise on both points and the draft passed the Cabinet stage without any problems on 6 November 1996. The Cabinet draft (Bundesregierung 1996a) modified the quota for refillable containers and relaxed its conjunction with the deposit obligations. Under the new packaging law, mandatory deposit obligations come into force if the *aggregated* market share of beer, soft drinks, mineral water, juice and wine sold in refillable containers falls below 72 per cent. In future those types of drinks where the percentage of refillable containers does not fall below the 1991 level can be excepted from this deposit obligation. Moreover, the deposit obligations come into force only if the quota is not achieved for two consecutive years. The quota for milk was increased from 17 to 20 per cent. Not only refillables but also one-way plastic sachets fell under the quota. The German Ministry of the Environment said according to the FAZ '*Der Sachets aus PET sei eine ökologisch gleichwertige Verpackung. Es sei auch ein Signal and die Europäische Kommission; wenn es neue wissenschaftliche Erkenntnisse gebe, würden diese auch berücksichtigt*' (FAZ 16.9.1996). Initially a quota of 25 per cent was planned by the BMU (FAZ 15.9.1996). The Ministry of Agriculture, representing interests from the dairy industry, wanted to maintain the 17 per cent quota. They compromised on 20 per cent (Interview BMU 12.11.1996).

As far as plastic is concerned the quota remained unchanged as compared to the BMU draft (1.1.1996 50 per cent; 1.1.1998 50 per cent). There were, however, two modifications. First, rather than a minimum of 50 per cent at least 40 per cent of each year's recycling quota have to be recovered by material recycling. This amounts to 20 per cent for the first two years and 24 per cent thereafter. Secondly, the rest cannot be recovered only by 'raw material recycling', but also by incineration with energy recovery, so called 'thermal recycling'. Besides these two elements which point to a weakening of standards, the new draft increased the standards for paper/cardboard and glass. The respective targets for paper and cardboard changed from (50 and 60 per cent, to 60 and 70 per cent) and the 1998 target for glass increased from 70 percent to 75 per cent. The new draft also stated the general maximum recovery (65 per cent) and recycling (45 per cent). These overall targets are, however, not very important, given the mandatory material specific targets.

After the Cabinet stage was passed, parliament was the next hurdle. As already said, the 1994 Waste Act required, unlike the 1986 Waste Act, the assent of parliament for the adoption of a new ordinance. On the parliamentary agenda, however, was not only the Packaging Ordinance, but also a contented federal nature conservation act, an ordinance for the recycling of cars and global environmental problems. Given the full agenda, the SPD and the Green Party did not use the parliament for a specific and intensive debate about the Packaging Ordinance, but generally criticized the environmental policy. Also, the SPD-led *Länder* had a majority in the *Bundesrat,* and the SPD therefore wanted to concentrate its effort on this veto point. Parliament passed the draft Ordinance with the votes of the CDU/CSU/FDP majority (Das Parlament 20/27 Dezember 1996: 13-15). Then, the draft was sent to the *Bundesrat* and notified to the EU Commission.

11.5 The double decision trap

That the *Bundesrat* would not approve the government's draft became likely in February when its Environment Committee categorically rejected the draft without any detailed discussion. The Economic Affairs Committee was a bit more positive and advised the *Bundesrat* to give its assent after some amendments (communication with Ministry of the Environment NRW 21.5.5.1997).

Six weeks before the *Bundesrat* had its plenary session, on 17 March 1997, the Commissioner for the Single Market (Bangemann) sent a detailed opinion to the German government. With this step a three month standstill period started. The government had to postpone the enactment of the draft until 17 June 1997. Bangemann reasoned that the draft would not be in compliance with the European Packaging Ordinance. The proposed modification of the refillable quota and the mandatory deposit would not be sufficient to clear up the complaints. All concerned General Directorates, with the exception of the Environment Directorate (DG XI), were critical about the refillable quota. In addition to the Commission's complaint, the Brussels institution had also received an 'absoluten Rekord' of detailed opinions of other Member States. Seven Member States used the notification procedure to voice their complaints, including France and the United Kingdom. The major point of critique of all detailed opinions, was the refillable quota and the threat of mandatory deposits ((FAZ 27.3.1997, Lebensmittelzeitung 4.4.1997: 28, Long and Baily 1997: 219).

The federal government was not entirely unhappy with the Commission's letter. An official at the BMU said '*Die Stellungnahme der Kommission läßt für zusätzliche Forderungen von Bundesländern, auf die Mehrwegquote noch aufzusatteln, keinen Raum*' (quoted in LZ.4.4.1997: 28). He also said that the government wanted to maintain the refillable quota, that the government and the Commission had close contact, and that he was positive that the Commission would ultimately accept the quota (LZ.4.4.1997: 28). This was to become a nice example of a Member State government safeguarding autonomy *vis-à-vis* domestic actors by referring to EU institutions. But things developed differently.

The *Bundesrat* did not follow the Commission's complaints. It rejected the draft of the Amended Packaging Ordinance at the plenary session on 25 April 1997, though extremely narrowly. The draft achieved 33 of the 35 votes necessary, despite heavy lobbying by the German Minister of the Environment, Angela Merkel. According to an official at the Ministry of the Environment of the state Northrine-Westfalia the amended Packaging Ordinance had been rejected because of the lacking of emphasis on environmental protection. She said that the draft was too closely based on the existing system, (Dual System), which was too costly, not environmentally effective and lacked clarity. The majority of the *Länder* wanted more far-reaching measures including an ordinance that stimulated re-use, and charges on packaging which were environmentally harmful and not recyclable (communication with Ministry of the Environment NRW 21.5..1997).

A month after the rejection in the *Bundesrat,* the government adopted a new draft (21.5.1997), in which it took up a number of remarks of the EU Commission and the *Bundesrat* (Bunderegierung 1997a). There was no modification of the refillable quota. As in the case of the 1991 packaging debate, the government promised further measures to encourage ecologically advantageous packaging. The decision on a separate ordinance for multiple-use packaging, was postponed until the findings of new LCA's had been examined which were expected to be available in the spring 1998 (BMU press release 21.5.1997).

The next session of the *Bundesrat* was expected in the autumn of 1997. In the meantime the debate on the refillable quota gained in importance because figures for 1996 revealed that the quota had decreased to 72,03 per cent, and was therefore only slightly above the minimum. In a press release the Minister emphasized that a mandatory deposit would be introduced if the marker share of one-way packaging increased (BMU press release 30.9.1997). Due to negotiations between the *Länder* and between different committees of the *Bundesrat,* the decision was scheduled as late as 28 March 1998. In the forefront of the plenary session both the Environment Committee and the Economic Affairs Committee advised giving assent to the draft if a number of provisions were be changed (*Bundesrat* 1997). The advice of the committees were, however, contradictory on a number of points. The Environment Committee advised, for example, fixing a separate refillable quota for beer of 78 per cent. The Economic Affairs Committee countered that new scientific studies questioned the ecological superiority of re-use systems for beer. The Environment Committee also wanted not 24 per cent but 36 per cent of material recycling of plastic packaging waste. The other Committee said that quota for any specific recycling technologys are problematic because they restrict the search for the economically and ecologically optimal recycling treatment. These and other contradictory statements reveal that there was not only a line of conflict between SPD-led countries and CDU/CSU led countries, but also between the Ministries of the Environment and of Economic Affairs within the various governments. This conflict was exacerbated by the fact that a number of Environment Ministers were from the Green Party. In at least one of these cases there were also substantial conflicts between the Minister and her own civil servants. This was a *Land* where steel producers would benefit from 'thermal recycling' of plastic waste, as well as from the production of more drink cans, and where a number of large breweries used drink cans to penetrate other markets. The packaging waste debate came in the wake of the federal elections of October 1998. Due to these conflicts, the decision of the Amended Packaging Ordinance was twice withdrawn from the *Bundesrat* agenda. On 29 May 1998, the *Bundesrat* finally decided to agree on the government's draft conditional on the inclusion of a number of rather minor amendments. The *Bundesrat* increased the quota for material plastic recycling from 24 per cent to 36 per cent as wanted by its Environment Committee. It also relieved the providers of service packaging (bakers, butchers etc.) of verifying the achievement of the recycling quotas. Rather than these small shops, the producers of service packaging were now financially responsible for the recycling. Most important for this, however, is that amendments did not introduce stricter refillable quotas. The majority of the *Länder* did not follow the request of Bavaria and the Environment Committee to introduce a 78 per cent refillable quota for beer. Moreover, the refillable quota would not be counted for each Land separately. Under the present regime, the take-back obligations and the mandatory deposits come into force nationally when the national 72 per cent quota is not achieved, or regionally, when the refillable share falls below the 1991 level in the respective region. The latter provision was skipped. This means an additional relaxation of the refillable quota (BMU press release 29.5.1998, *Bundesrat* press release 29.5. 1998, communication with BMU 5.6.1998). Since these amendments implied only minor changes the assent of the Cabinet (12.6.1998) and parliament (24.6.1998) was a matter of formality. The regulation was then notified to the Commission.

Given the considerable delay, the ordinance will be enacted more than two years after the official implementation deadline set by the European Packaging Directive. Whether the European Commission would accept the relaxation of the refillable quota as sufficient was not clear in July 1998. In January 1998, a high level meeting took place in which the Commissioners discussed the German refillable quota. DG XI supported by its Legal Service proposed a withdrawal of the infringement procedure against the old German refillable quota. The majority was against this proposal. It was agreed that the Commission would send a new letter of complaint, restarting the infringement procedure.

11.6 Continuities and changes in German packaging waste policy

After six years of intensive debate about the unintended consequences of the German Packaging Ordinance, and two years after the adoption of the European Directive which aimed to harmonize national packaging waste policies, the basic German approach adopted in 1991 remained unchanged. Germany still has one of the most ambitious and strictest packaging waste regimes in the world. Take-back and recycling responsibilities for the packaging chain remain intact as well as a quota for refillables and threats of mandatory deposists. The DSD has survived, though it has been criticized for a) ending up almost bankrupt; b) being too costly; c) being not environmentally sound and d) being monopolistic. Even though environmentalists, domestic and foreign companies, sub-national actors and political parties objected to the German approach, the Amended Packaging Ordinance will only bring incremental changes.

Table 32 Continuities and changes in German Packaging Waste Policy

Dimension	**Germany Packaging Ordinance 1991**	**Amended Packaging Ordinance (draft) 1998**
FORMALIZATION	**high**	still high
degree of legal codification	Ordinance: generally binding under public law	Ordinance: generally binding under public law
STRICTNESS OF STANDARDS	**strict**	**weaker**
packaging reduction targets	no	no
re-use target	maintenance of share of drinks sold in refillable (72%)	maintenance of share of drinks sold in refillable (72%)
material recycling targets for one-way packaging (levels already achieved)	sales packaging: 64/72% depending on material (1989= 1%-63%)	overall target 45 per cent sales packaging 36-75% depending on material
recovery	no target	65 per cent
incineration and landfill	no target	no target
time horizon	short (1.7. 1995)	by 1.1.1998
prescription of implementation	detailed	detailed
flexibility	possibility for self regulation within tide legal framework	possibility for self-regulation within stricter legal framework
ALLOCATION OF RESPONSIBILITIES	**Private recycling scheme Public (Länder) control**	**still private recycling scheme still public (Länder) control**
Dispute settlement/ enforcement	administrative courts	administrative courts
MODES OF INTEREST INTEGRATION	**corporatist**	**more pluralized**
intensity of integration	medium	high
mode of interest integration	corporatist, peak associations	pluralized predeliction for aggegrated interests, associations
ORIENTATION TOWARDS TARGET GROUP	**impositional**	**still impositional**
general orientation	impositional	impositional
character of incentives	repressive: obligations, mandatory deposits	repressive: obligations, mandatory deposits

Formalization

Germany did not have to change the legal basis of its packaging waste policy. The German legalistic approach perfectly matched the European requirements for the implementation of Directives. This is the biggest difference with the Netherlands and the United Kingdom. The German regulatory culture matched the European command and control approach. It is also very unlikely that Germany would have withdrawn its legalistic approach in the absence of a European Directive. On the contrary, the legalistic approach to waste prevention by producer responsibilities has been strengthened by the 1994 Waste Act and has already been concretized for other products. Ordinances ruling the take-back and recycling of cars and batteries came into force on 1 April 1998 and 3 April 1998 respectively (see BMU press releases from 31.3.1998 and 2.4.1998 respectively). Hence, it is very likely, that there would still be an ordinance for packaging waste even without a European Directive.

The formalization of the German packaging waste has even increased. The 1991 Packaging Ordinance concentrated very much on the collective system to be run by the DSD. The Amended Ordinance added provisions dealing with those companies who are not joining a collective scheme. Moreover, a number of provisions dealt with the functioning of the DSD and were aimed at the increasing competition. These changes were the result of three factors. First, companies who joined the DSD put pressure in government to prevent free-riding behaviour of other companies (Interview DSD 5.12.1996). Second, the government drew lessons from the practical problems with the DSD. Initially the government thought that German companies would obey German law (Interview BDI 6.12.1996). The government then learned that additional provisions are necessary to ensure that all companies comply with the Packaging Ordinance. The government also learned that competition problems arose to an extent which was not anticipated and which needed additional measures. Finally, the increase in competition was demanded by the German and European competition law. This point was increasingly made by the German Office of Fair Trade as well as the EU Commission.

The formalization also increased in regard to the monitoring system. The German *Länder* criticized the lack of transparency of the packaging material flow between consumer, waste collector, waste sorter and waste recycler. The monitoring system became increasingly detailed. The Packaging Waste had to be weighted at all interfaces. The DSD said that 3 million forms have to be filled in each years certifying the amount of packaging material at a certain stage. These documents form the basis of the annual material flow analysis, through which the achievement of the recycling quota for each individual German Land had to be proved (DSD 1998).

Strictness of standards

The strictness of standards slightly decreased. This transformation was, however, only partly induced by European integration. The European influence was most important for the modification of the refillable quota. Under the old packaging regime, mandatory deposit obligations will automatically come into force when the aggregated market share of beer, soft drinks, mineral water, juice and wine sold in refillable containers falls below 72 per cent. When the draft become law, those types of drinks where the percentage of refillable containers does not fall below the 1991 level cent can be exempted from this deposit obligation. The underachievement has to last for at least two years consecutively. The quota for pasteurized milk increased from 17 to 20 per cent. Not only refillables fell under this quota but also PET plastic sachets packaging.

One reason for the modification of the refillable quota, was the pressure from the Commission. The infringement procedure (Article 169 of the Treaty) in combination with Article 30 of the Treaty provided the institutional means to threaten the German refillable quota of the 1991 Packaging Ordinance. The notification procedure for technical norms, as required by Directive 83/189, enabled the Commission to challenge the new proposed quota. Companies and other Member States also put pressure on Germany via these institutional channels. The refillable quota was heavily criticized by producers of one-way drink packaging, large drink producers and even the Peak Association of German Industry. They found their allies in the Commission, with the exception of DG XI (Environment), other Member States, in particular France and the UK, and the federal Economic Ministry and its counterparts on the *Länder* level.

Given these forces, it is surprising, that the refillable quota was only slightly modified. It is important to note that industry interests were not united in this case. Most parts of German industry were against the quota. The Bavarian, not the German, Brewer Association, the association of beverage discounters (*'Getränkegrosshandel'*) and the German mineral water producers were, however, in favour of the refillable quota because it protected their markets. Although their resources were rather restricted as compared to the other economic interests, their relative power was quite strong, because they could channel their interests not only via the BMU into the government but also through the *Bundesrat.* They had not only Green Party executives and most of the SPD as an ally, but also the Bavarian government. The environmental movement was in favour of refillables, though the environmental groups did not concentrate as much on the packaging issue as they did during the late 1980s.

That the Commission hesitated to bring Germany before the Court had two related reasons. First, this would be a politically sensitive step which would have its effects on other areas of European integration. Second, the German government made a rather strong case in arguing that millions of tonnes of additional packaging waste would be generated if the quota were abolished. It would be politically negative for the legitimacy of the Commission and the process of European integration in general, if the public associated the free market with millions of tonnes of additional waste from one-way drink packaging. This argument suggests that it is politically problematic to reduce environmental standards which are established and which are directly related to waste prevention. The German pro-environment interests wanted the refillable quotas because they strongly believed that the environmental benefits outweigh the economic disadvantages and the German government had the capacity, i.e. a well resourced Environmental Agency, to provide enough evidence to impress the EU Commission.

The recycling targets were slightly modified and more time was given to achieve them. The new Ordinance stipulates that by 30 June 2001, 65 per cent of the overall packaging waste should be recovered and 45 per cent should be recycled. These targets are not of practical importance, however, since the draft Ordinance entails material specific mandatory recycling targets. These targets were changed to allow for a sufficient increase in recycling capacity. The targets and deadlines read as follows (the respective targets and the deadline of the 1991 Packaging Ordinance are set in brackets).

Table 33 Old and new German recycling targets compared

	level 1993	**old (by 1.7.1995)**	**new (by 1.1.1998)**
glass	61,6	72	75
tinplate	35	72	70
aluminium	6,8	72	60
paper/cardboard	54,8	64	70
plastic	29	64	60 (36 mat. rec.)
composites	25,8	64	60

As far as plastic is concerned, at least 60 per cent of the quota has to be recovered by material recycling. This is 36 per cent of the overall amount. The rest can be recovered by other technologys such as hydrolysis and pyrolysis (chemical recycling) or incineration with energy recovery (thermal recycling). It was stated in the amended Ordinance that the government would reassess the plastics targets after 1 January 1999 in the light of new experience and insights.

The modifications of targets were already proposed by the government before the final shape of the Packaging Ordinance was clear. The reason for this change was domestic as well as international. As said earlier, other Member States complained that insufficient capacity in Germany led to subsidized exports of waste to other Member States endangering their recycling emerging recycling infrastructure. Their complaints were potentially powerful because they had, via Article 30, the institutional possibility to challenge Germany before the European Court of Justice, though the Danish bottle case has shown that the Court went very far to protect national environmental measures.

The Dual System is still based on high recycling targets. This was legally possible, according to the European Directive, if sufficient capacity can be provided. There was as strong pressure by interest groups from all parts of the packaging chain to get rid of quantified targets, or at least to lower them. That the targets are still as high as they used to be cannot be related to the problem pressure any more. On the contrary, the amount of municipal waste decreased from 50 million tonnes 1990 to 30 million tonnes in 1996. Incinerators currently only use 70/80 per cent of their capacity, also because waste is exported (Der Spiegel 48/97).

As in the case of the refillable issue, demands by various sectors of industry were balanced by opposite demands of the sectors which benefit from high targets and strict regulations, like the waste collection, sorting and recycling industry. The technology-oriented end-of-pipe approach to environmental policy had created new economic interests which supported the pro-environment advocates in government on this issue. Earlier government policies had induced a path-dependent process. Modifications are difficult due to sunk costs and interests that benefit from the policies. This point will be discussed in the next section.

There was one substantial modification in targets and this concerned plastic packaging waste. The target for *material* recycling decreased from 60 per cent to 36 per cent. That the material recycling quota was not reduced to 24 per cent as proposed by the government, was due to the intervention of the majority of *Bundesrat,* which was in favour of material recycling. Since the export of plastic packaging waste was criticized by other EU Member States, the modifications which aimed at self-sufficiency can be read as a reaction to complaints of other Member States and thus as a concession to European integration. But there were also domestic factors involved, namely a) pressure from domestic industry; b) technological developments and (c) policy-oriented learning. The 64 per cent target of the 1991 Ordinance was very ambitious, given the actual recycling rate of some 1 per cent in those days. Most of industry strongly opposed the recycling targets, arguing that establishing a sorting and recycling infrastructure would be very expensive and that there were no markets for the recycled material. In addition, life-cycle assessment revealed that packaging waste recycling was not always more environmentally friendly than other options, such as incineration with energy recovery and new technologies, such as hydrolysis and pyrolysis. European integration did indirectly contribute to the learning processes, because the Commission and the ECJ demanded that environmental measures are suitable and necessary to meet the targets. LCA's are means of meeting these demands.

Allocation of responsibilities

The Amended Packaging Ordinance is based on the same producer responsibility approach as developed in the 1991 Packaging Ordinance. All producers and distributors of packaging are obliged to take back used packaging and are responsible for recycling. As far as sale packaging is concerned, individual companies can be exempted from these individual obligations if they join a collective scheme. So far only one scheme has been established, the Duales System Deutschland (DSD). The first factor which explains the persistence in the policy approach, is the persistence of the belief system and the action capacity of the pro-environment advocacy coalition. The basic idea of a comprehensive producer responsibility is based on deeply-rooted principles such as the polluter pays principle and the notion of the internalization of costs. Another factor is path dependency. Some 420 billion consumer packs are marked with a green dot annually (Verpakkingsmanagement 3/1997: 33). Industry invested approximately 7 billion German marks in establishing the appropriate collection, sorting and recycling infrastructure, with running costs amounting to 4 billion marks annually. These sunk costs made it difficult to change the regulation radically since this would impose capital losses. A specialized journalist got the point when he said *'Handel, Industrie und Entsorgungswirtschaft haben nun einmal notgedrungen in den vergangenen Jahren Milliarden in den Aufbau des Dualen Systems investiert. Das war politisch so gewollt und ist nicht mehr rückgängig zu machen'* (Lebensmittelzeitung 2.5.1997). In addition vested interests emerged. Most notably the DSD, the '*administrative beast*' (Interview British industrialist 1996) itself and its more than 300 employees, who had the institutional interest to survive. Also, the waste management industry benefits and is even represented on the board of the DSD. In total some 17,000 jobs were created, according to the Ministry of the Environment (BMU press release

29.4.1998). Pilot projects in the Netherlands and the United Kingdom have shown that a dual system is inefficient and not necessary to achieve high collection targets. These two countries learned from the negative experience in Germany and drew lessons from their own projects. They left the responsibility for collection of waste with the local authorities. Germany had to stick to its path, however. Besides vested interests and sunk material costs, two other factors contributed to this path-dependent development. First, the government feared losing credibility if it acknowledged that the DSD was a failure. Secondly, consumers were used to separating their waste. According to DSD figures, roughly 90 per cent of the consumers separate their packaging waste into the three fractions required, glass, paper and cardboard, and lightweight-packaging, i.e. plastic, metal and composites (mainly drink packaging, Tetra Pak) (DSD 1998:3). If the DSD were to be abolished, a change in consumer behaviour would take place which was not seen as environmentally sound. Packaging waste would probably not be separated except for paper and glass. Some of the consumers would bring the packaging back to the retailers, especially those with deposits, but much packaging would be thrown in the domestic dustbin and would end in incinerators or dumping sites.

Modus of interest integration

The first Packaging Ordinance and the DSD were primarily developed in close contact between the peak associations of industry and the German government. The set of relevant actors slightly diversified when the *Bundesrat* became involved. As in the Dutch case, the government chose a close network to strike a bargain with highly integrated interests. The new mode of interest integration is very much shaped by the fact that the government did not want to change the basic system which was established. It did not actively seek to include societal interests or company interests to develop a new policy, rather it reacted to industry demands. The mode of interest integration changed from corporatist bargaining to pluralistic lobbying, in which trade associations still played a more important role than individual companies. The importance of the peak association BDI as a focal point of the government decreased. It was more difficult for them to aggregated the highly diverse interests. This was one of the reasons why the number of actors increased. The main reason for the pluralization of interests was that more actors became aware of the intended and unintended consequences of the 1991 Packaging Ordinance. All packers and fillers were confronted with the DSD, and more than 16 thousand signed a contract with them. Packaging material and packaging producers were confronted with changing demands by their clients and/or participated in the DSD recycling system. The latter was also the case for many private waste collection, sorting and recycling companies. Virtually the whole German industry had to do with the German DSD in one way or another. The effects did not stop at the German border, however. The inter-linkage of environmental and trade aspects resulted in many activities of importers of packaged goods. They felt supported by the harmonization objective of the European Packaging Directive. The Treaty of the European Union (Art 170: infringement procedure) and secondary regulation (Notification Directive) provided them with the institutional channel to challenge the German approach. The Europeanization resulted in a new strategic context for all actors. German companies could use the EU Commission or the governments of other Member States to put pressure on the German government. An interesting case in point is provided by the detailed opinion which the German government received from the Netherlands. The Dutch government criticized in this detailed opinion the German refillable quota. The Dutch government itself is divided about policies concerning re-use. In the Netherlands the Ministry of the Environment is the leading department for packaging waste policies. The Ministry of Economic Affairs (EZ), is however, solely responsible for the adoption of detailed opinions. EZ can therefore express its free trade interests more easily on the EU level, than within the government, where it has to make compromises with the Ministry of the Environment. It goes without saying, that the Dutch Ministry of the Environment was rather unhappy that EZ wrote a detailed opinion (communication VROM 7.5.1998).

Orientation towards the target group

The government became tougher towards the target group. In contrast to the 1991 Ordinance, the stricter provisions that are adopted prevent free-riding behaviour. The new Ordinance states that manufacturers and distributors who do not want to participate in a collective scheme (e.g. DSD) have to prove that they fulfil their take-back and recycling obligations. However, the new Ordinance provides an interim regulation for these companies. This change matches the EU requirements to make directives enforceable, but is primarily a reaction to strong criticism from (German) companies who participate in the scheme. The annual loss caused by free-riders is estimated at 600-800 million German marks which is approximately 20 per cent of the annual running costs of the scheme. The government stuck to its threat to introduce mandatory deposits if the refillable quota is not achieved. The imposition of these deposits would result in economic losses of retailers, many filling companies, in particular beer and soft drinks, and the producers of one-way packaging.

Government also reduced the burden on industry, however. The companies who neither joined the DSD nor fulfilled their responsibilities individually have to meet only 50 per cent of each of the recycling quotas in the transitional phase, from 1997 to 1998. Hence, these companies are actually rewarded for non-compliance. The burden on industry was further reduced by allowing for 'raw material' recycling of plastic waste.

12. Convergence and persistent diversity

12.1 Convergence of packaging waste policies?

12.1.1 Introduction

In the first empirical part, this study focused on the research questions as to whether and why the German, the Dutch, and the British packaging waste policies differed in the early 1990s. In the second part, the Europeanization of national packaging waste policy has been described from two perspectives, the European and the domestic one. Having made a cross-national comparison for the early 1990s and longitudinal comparisons for each country, we are now equipped to answer the research question as to whether European integration will bring about a convergence of the German, Dutch and British packaging waste policies. What was the impact of European integration on national packaging waste policies? Did national packaging waste policies converge? Which were the stimulating and the hindering forces of change? (10.1) These questions will be answered first for each of the policy elements separately. Then, the perspective will turn to the independent variables: the various factors will be examined one by one (10.2). In the final sections of this study the findings of the case study will be placed in the more general discussion about regulatory policymaking in Europe (10.3).

Table 34 Convergence or persistent diversity: An overview

	Germany	**The Netherlands**	**United Kingdom**	**Convergence or persistent diversity**
	from Ordinance (1991) to Draft Amended Ordinance (1998)	from Covenant (1991) to Packaging Regulation (1997) and Covenant II (1997)	from White Paper (1990) and Codes of Practice to Packaging Regulation 1997)	
Formalization:	still high (public law)	from medium to high (public law)	from low to high (public law)	*→ convergence towards strong formalization*
Strictness of standards	slightly weaker	slightly weaker	slightly stricter	*→ still diversity, more narrow range*
Allocation recycling responsibilities in waste management	still private	still public and private	still public and private mix, more private responsibilities	*→ persistent diversity*
Allocation of enforcement responsibilities	still public (*Länder*)	Regulation: new: Public (National Ministry), Covenant: still public private (Committee)	new: public (Environment. Agency)	*→ convergence towards public enforcement, but diversity of public actors responsible.*
Mode of interest integration	more pluralized	more actors but still corporatist	pluralist	*→ still diversity, trend towards pluralistic mode*
Orientation towards target group	still impositional	still mediating	slightly more impositional	*→ still diversity trend towards more imposition*

12.1.2 Convergence towards formalized policies

The general impression in popular writing that we live in period of deregulation is not confirmed in the case of packaging waste policies. This field provides yet another case where EU Member States introduced rather more generally binding provisions than less. There has been a convergence towards the German legalistic approach. Both the Netherlands and the UK followed the example of Germany and introduced regulations based on public law. The voluntary agreement between Dutch industry and the Ministry of the Environment has been replaced by a ministerial regulation based on the 1994 Dutch Environment Management Act. The same happened in the United Kingdom where lukewarm commitments by industry were replaced by a statutory instrument placing producer responsibilities for packaging waste on industry. The legal base for this was provided in this case by the 1995 Environment Act.

The convergence towards more legal formalizm of national policies mainly took place for institutional reasons: the adaptation to positive European integration. The Netherlands and the United Kingdom had to meet the legal requirements concerning the transposition of directives into national law. These requirements have been concretized by the European Court of Justice.

As pointed out earlier, even though Member States are free to choose their own legal means and the form in which to implement a directive, they have to consider basic legal requirements such as clarity, legal security, and legal protection of third parties (Jans 1994: 133) when implementing directives. Member States are also held liable if the desired result is not achieved (Sewandono 1993: 83). According to ECJ case law, directives have to come to full effect. This implies certain obligations for Member States to make the directive enforceable. The ECJ has shown that it accepts very few arguments if a directive is not properly implemented. The Member State concerned cannot evade its responsibility by blaming private actors or decentralized governmental institutions. Under these circumstances state agencies are very cautious in choosing other implementation instruments than command and control measures based on public law (Steyger 1993). The institutional rules narrow down the choice of Member States and clearly privilege command and control measures over voluntary agreements or voluntary commitments by industry.

These legal requirements are less important when directives explicitly offer the possibility of voluntary measures for implementing them. In the case of packaging waste there was the 1985 Directive on Containers of Liquids for Human Consumption, which explicitly allowed for voluntary measures. Why was a more legalistic approach taken in regard to the European Directive on Packaging and Packaging Waste? The first reason is that the problem of waste in general and packaging in particular was high on the political agenda in the late 1980s and early 1990s. A number of Member States introduced legalistic and interventionist measures. Examples are the Danish ban on drink cans and the German plastic bottle ordinance. ECJ decisions did not hinder this development and the need for harmonization was felt by many Member States. The General Directorate for Environment of the Commission saw a chance to revitalize its packaging waste policies and to increase its legal codification, in order to make it more effective than the 1985 Directive. Pro-environment countries favoured a legalistic approach towards the problem of packaging waste. Germany and Denmark wanted to transform their national legalistic approach into European legislation. The Netherlands, though employing a voluntary approach on the national level, wanted regulation on the EU level, in order to avoid economic disadvantages for those national companies who joined the Covenant. The Dutch government committed itself in the 1991 Packaging Covenant to strive for regulation at the EU level. Other countries led by the United Kingdom emphasized harmonization. They wanted to ensure by effective legal means, that the free movement of goods would not be hindered by national regulations. The British government wanted to combine this with rather weak voluntary guidelines and codes of practice in regard to packaging prevention and recycling. During the negotiations, however, the kind of targets - mandatory or voluntary - was not an important topic. For the "green" advocates the mandatory character of targets was out of the question, and the British government and its allies concentrated much more on preventing a high level of targets, than preventing the legal codification of those targets. This suggests that it was more important for the government to win in the short term, i.e. getting the targets reduced, than to look at the long term consequences of the kind of targets, namely legally codified quantitative ones, which were in contrast to the traditional British environmental policy style. The negotiating position of the

British government in Brussels was also weakened by the fact that its domestic industry was not able to agree on a voluntary packaging waste policy which would be effective. The outcry of British industries: 'regulate us, please!' was not likely to strengthen the British argument that packaging waste policies could be effective by voluntary means alone.

In order to answer the question of whether positive integration was the decisive factor for the convergence towards formalization, one has to try to answer the question of what would have happened without the Directive? There is no reason to believe that Germany would have stepped back from its legalistic approach. The legalistic approach fitted in the overall regulatory culture and was also the result of a learning process. The government drew lessons from the failure of voluntary agreements and successes of statutory measures in regard to drink packaging. Furthermore, in the aftermath, legally codified producer responsibility was introduced for other products such as cars or batteries. Moreover, an Amended Packaging Ordinance was already planned in 1993 for domestic reasons. This draft would have made the German approach more legalistic, by including measures to prevent free- riding and to increase competition.

In the British case it is quite unlikely that regulation would have been introduced. Environmental issues decreased in importance in the mid 1990s and the British government was against regulation and intervention into market processes. In areas other than packaging, such as paper waste from newspapers, the government was satisfied with voluntary commitments by industry (DoE 1996b: 96).

Counterfactual reasoning is probably the most difficult for the Dutch case. There are some factors that point to stronger formalization of packaging waste policies also without a EU directive. It is very likely that the packaging prevention target of the 1991 Covenant would not have been achieved. A modification of the covenant would have been necessary. One route would have been stronger formalization, by making the Covenant generally binding as it was the case with the Covenant for old cars. This would broaden the base and facilitate the achievement of the target. In my opinion, however, it is more likely that the government would have re-negotiated the Covenant because it would have been politically too costly to include laggards, and to make it binding for 300,000 rather than 300 companies. This would probably have exceeded the capacity of the consensus-oriented approach. Moreover, the government's will to do so would have been weak, because environmental issues decreased in importance on the political and the societal agenda. Anyway, the European Packaging Directive was an elegant opportunity for industry to be relieved of the ambitious packaging prevention target. Moreover, for those companies who participated in the first Covenant and for the government it was an opportunity to deal more effectively with the problem of free riders.

In short, the Dutch and the British system developed towards the German approach. Adaptation to positive integration was the most important factor for the convergence of packaging waste policies towards a formalized generally binding policy.

12.1.3 Persistent diversity of packaging waste standards

In contrast to the degree of formalization, there is no convergence in regard to the targets of packaging waste policies. The diversity in the strictness of standards continues to persist. Germany still has quantified quotas for refillables and very ambitious recycling targets for sales packaging, though the levels of most targets have been lowered slightly. The Netherlands still has a packaging prevention target and maintained the same ceiling on the amount of packaging waste for landfill and incineration as stated in the 1991 Covenant. The recycling targets have even been slightly increased. On balance, the standards are, however, slightly less strict because the packaging prevention target is related to economic growth. Still both countries have targets which exceed the maximum quota set in the European Packaging Directive. The United Kingdom adopted the least strict targets allowed for by the Directive. These targets are, however, more ambitious than the packaging targets implied in the government's recycling target for municipal waste. Hence, there is a diversity of targets but the range of targets is narrowed down a bit. The British are doing a bit more, and the Dutch and the Germans are doing a bit less. It was neither a race-to-the-top, nor a race-to-the-bottom. There were only incremental changes.

Given the objective of the European Packaging Directive to harmonize national regulations, this persistent diversity is surprising. In order to explain the persistent variation of standards several questions have to be answered. Why does the Packaging Directive only set a range of targets - recovery 50-65 per cent, recycling 25-45 per cent, and why does it provide the possibility to go beyond these targets? There is also the question of why Germany and the Netherlands wanted these exemptions, and why they used this leeway to maintain high environmental standards. A majority of the Member States wanted a high degree of harmonization on a relatively low level. They accepted however, the call for an exemption by the 'green' Member States. This behaviour of the majority can be related to the institutional logic of EU policymaking. In Brussels there is an informal procedural consensus between the delegates of Member States not to overrule minorities when strong interests are at stake. The Member State officials participate in many parallel decision-making processes. It is therefore rational to take into account the consequences of the strategy in one decision-making process for other neighbouring processes. The institutional structure of the EU makes a cooperative policy style necessary. Member States depend on each other to reach decisions or to conduct policies (Kerremans 1996: 233). Therefore, exemptions for minorities are granted even under conditions of majority voting.

The strong preferences of high standards by the 'green countries' was driven by their wish to Europeanize the rules which their companies had to obey, in order to prevent economic disadvantages for them or to give them even a competitive advantage above companies who have to get used to new legal requirements. The ambitious national standards were related to the high public support for strict environmental policy making and to other factors, which have been described in detail in the first empirical part of this study.

Germany and the Netherlands could have used the European Packaging Directive, however, to weaken their standards. Even though the standards were slightly weakened, this was only indirectly related to the European Directive. Persisting to maintain high standards in both countries cannot be explained by the persistent problem pressure as such. In both countries environmental and waste issues decreased in importance. More important were policy inertia, in particular in the German case. New interests emerged which benefited from the policies adopted, especially the DSD and waste management companies who had an economic interest in high recycling standards. The Packaging Ordinance has created resources and incentives that provide a strong motivation for benificaries to mobilize in favour of maintenance (see Pierson 1995: 40, chapter 1). In the German case, the maintenance of the refillable quota can be explained by a coalition of certain protectionist economic interests with environmentalist interests which gained crucial influence via the *Bundesrat* The *Bundesrat* provided an effective veto point against a substantial weakening of the refillable quota. The German case resembles therefore, the findings of D. Vogel (1993) for the US, that in a federal system environmental policy making is a ratchet-like phenomena: advances are difficult to reverse.

More generally speaking, it is quite difficult to lower standards in such a visible area as packaging waste. It is quite easy to attract public attention by calculating how much more waste is produced when standards are lowered. This was the case with the German refillable quota. An important reason why the Commission hesitated to take Germany before the Court was the German government's argument that the abolishment of the refillable quota would result in millions of tons of additional waste. The same argument would have been made by domestic actors, if the government had considered weakening or abolishing the refillable quota. In Germany as well as in the Netherlands, a radical lowering of standards would have meant a loss in credibility for the government, even in times when environmental issues were not so high on the political agenda any more.

It was important in the German case, that the existing system fitted the general idea of ecological modernization. The German Environment Ministry emphasized that the Packaging Ordinance created 17,000 new jobs and that the waste industry in general was the fastest growing sector of the economy. Hence, environmental protection and the pursuance of economic policy objectives were reconcilable at least in the reading of the Ministry. The economic arguments were of particular importance in a political discourse dominated by issues of international competitiveness and mass unemployment.

These factors explain why the Dutch and the German government stuck to high packaging waste standards. But why were there at least some modifications of the standards? In Germany, the reduction of the material recycling quota was the most substantial modification. The export of plastic packaging waste was criticized by other EU Member States. Therefore, the reduction of this quota can be read as a reaction to complaints of other Member States and thus as a concession to European integration. But there were also domestic factors involved, namely a) pressure from the domestic industry; b) technological developments, and (c) policy-oriented learning. The 64 per cent target of the 1991 Ordinance was very ambitious, given the actual recycling rate of only 1 per cent in those days. Most of industry strongly opposed the recycling targets, arguing that establishing a sorting and recycling infrastructure would be very expensive and that there were no markets for the recycled material. In addition, life-cycle assessment revealed that packaging waste recycling was not always more environmentally friendly than other options, such as incineration with energy recovery and new technologies, such as hydrolysis and pyrolysis.

In the Netherlands, the most substantial modification was that the new packaging prevention target was no longer an absolute reduction target but was related to economic growth. The new target was a compromise between VROM and Dutch industry. A compromise was necessary because both parties disagreed on the prevention target, and an exit from the negotiations was not perceived as a feasible alternative by either party. Business as well as the Ministry of the Environment wanted to avoid the bureaucracy associated with a Ministerial Regulation. Compared to the negotiations resulting in the 1991 Covenant, industry forces gained in influence. Not only the SVM but also MKB and the VNO-NCW sat around the table. The latter organizations represented not only the pro-active part of industry but more or less all of industry. Business attitudes were shaped by the experience that a reduction of the total amount of packaging against the background of steady economic growth was difficult to achieve. Against these strong countervailing forces the Ministry of VROM had to give in and step back from its quantitative and absolute packaging prevention target. Note, however, that the Dutch Ministry succeeded in getting an absolute ceiling accepted on the amount of packaging to be landfilled or incinerated. Hence any increase in packaging had to be compensated by increased recycling efforts, in order to avoid a rise in the amount of packaging waste.

The strong domestic forces would have also triggered the weakening of the targets in Germany and the Netherlands without a European Packaging Directive. In fact, in Germany, weaker standards were already included in the first draft of the Amended Packaging Ordinance, which not only preceded the European Packaging Ordinance but was also unrelated to it. In the Dutch case, the European Packaging Directive did not directly induce a weakening of the targets. Nevertheless, the Directive was an opportunity for industry to escape from the targets agreed on in the 1991 Covenant, in particular the prevention target for the year 2000. It is very likely that industry would not have achieved the prevention target of the 1991 Covenant.

The high standards in Germany and the Netherlands contrast with the low ones in the UK. The UK Regulation entails standards which are slightly above the minimum standards set in the Directive. In fact, it was rather for cosmetic reasons that the Regulation stipulated a 52 per cent recovery target and a 26 per cent recycling target, rather than the 50 per cent and 25 per cent respectively stated as minimum in the EU Packaging Directive. These targets were the direct consequence of the European Packaging Directive. Without the need to adapt to this piece of positive integration, there would have been no quantified standards for packaging on this level. These targets are still relatively ambitious for the UK, given the rather low recycling level in this country. It is important to note in this context that the UK had to undertake more recycling than the recycling target suggests. Because of a lack of incinerators with energy recovery facilities, most of the recovery target had to be achieved by material recycling. Hence, rather than 26 per cent, more than 40 per cent of the packaging waste had to be recycled, still significantly less than in Germany and the Netherlands, but ambitious in the domestic context, given a recycling rate of some 17 per cent in the early 1990s (EC Packaging Report March 1994: 12). Given the low level of recycling in the UK, the unimportance of environmental issues on the political and the societal agenda, and the pro-industry stance of government, it is very unlikely that anything like this would have been adopted in the absence of the EU Packaging Directive.

12.1.4 Allocation of Responsibilities: Diverse public-private mix

In all three countries industry plays a substantial role in packaging waste management. This increasing role of business reflects a change in the policy paradigm from end-of-pipe solutions to preventive measures. This trend is the expression of a learning process in all three countries over the last two decades that pollution and resource depletion is most effectively and efficiently dealt with at the source. The paradigmatic shift shaped the allocation of responsibilities between private and public actors. The trend from end-of-pipe policies focusing on waste disposal to prevention and recycling-oriented packaging waste management has increased the importance of industry. While waste collection and disposal are manageable by public organizations, for packaging prevention and packaging waste recycling the government needs the private sector, because it is business that makes choices concerning packaging and that controls crucial information and technologies needed for recycling. Against this background the idea of producer responsibility gained importance in the three countries, though to a different extent. The principle of producer responsibility also fitted with the general belief of governments that its role in society should be minimized and public tasks should be privatized as much as possible.

Diversity continues to exist, however, concerning the extent to which waste management functions are allocated to the private sector. In Germany, a private company, the Duales System Deutschland (DSD), is responsible for the collection, sorting and recycling of sales packaging waste, while in the Netherlands and the UK local authorities remain responsible for the collection of household packaging waste. The European Packaging Directive had no direct influence on the allocation of waste management responsibilities. In line with the principle of subsidiarity, Brussels did not impose certain waste management arrangements.

Although Germany was a forerunner with its industry-led system, the Netherlands and the UK did not follow Germany. In this case, being a laggard was an advantage. Neither countries as restricted by the legacy of the past. These countries drew lessons from their own pilot projects and the perceived failings of the German DSD, which was seen as too expensive and too bureaucratic. So the Netherlands and the United Kingdom decided that local authorities should still play an important role in the separate collection of packaging waste. Germany, however, continued its privately organized collection system, despite heavy criticism that it is a) becoming almost bankrupt; b) too costly; c) not environmentally sound and d) monopolistic. The main reason for the German continuity is - as in the case of the maintenance of high standards, path dependency. Industry invested approximately 7 billion German marks in the appropriate collection, sorting and recycling infrastructure, with running costs amounting to 4 billion marks annually. New interests emerged. Most notably the DSD, the '*administrative beast*' (Interview British industrialist 1996) itself, has an institutional interest to survive. Path dependency was induced not only by sunk costs and newly vested interests but also by the new routines of tens of millions of German consumers who learned to separate their packaging waste. A turn-back would result in a severe loss of legitimacy. These factors greatly increased the cost of adopting once-possible alternatives and inhibited exits from a current policy path (see Pierson 1995: 42).

The codification of the national packaging waste policies in public law implied that public actors became responsible for licensing and enforcement. The allocation of these responsibilities varied, however, between the three countries. In Germany these tasks are situated at the sub-national level, that of the *Länder*. In the Netherlands, these tasks are performed by the Environmental Inspection of the Ministry of VROM and in the UK they are allocated to the newly created national Environment Agency. The European Packaging Directive did not pre-structure these choices. In the German case it reflected a regulatory tradition which is fixed in the German Constitution which grants the *Länder* substantial implementation tasks. In this case, however, initially the federal government wanted to have the responsibility for licensing. It was part of the compromise with the *Bundesrat* that the DSD was put under the control of the *Länder*. The Dutch choice for the Environmental Inspection is a standard operating procedure in this country. The allocation of the licensing and enforcement function to the new Environment Agency in Britain reflected a domestic learning process which was independent of the packaging waste issue, namely, that the decentralized and fragmented environmental implementation infrastructure had to be replaced by a more central and integrated one. This centralization and integration was, however, also triggered by the need to adapt to the requirements of a number of other EU environmental directives (see Knill 1995).

12.1.5 Modus of interest integration: Towards EU induced pluralism?

There is still persistent diversity in the intensity of government-industry interaction, the type of actors involved and the function of societal actors in decision-making.

The UK employed a minimalist approach to the problem of packaging waste in the early 1990s. The intensity of interests integration was therefore comparatively low. Anticipating the European Packaging Directive, the government increased its effort to come to industry-led solutions. The government still did not engage in close, regular and institutionalized contacts. There was no co-production of policy as was the case with the German 1991 Ordinance or the old and the new Dutch Covenants. One of the problems was that the traditional British mode of interest integration had always been between individual firms and government officials. The properties of the policy problem, especially its 'chain' character required, however, cross-sectoral interest aggregation. Given the weakness of British trade associations as compared to German and Dutch, even sectoral interest aggregation became a problem, let alone cross-sectoral aggregation of interests. This resulted in a pluralist pattern of interest integration, an *ad hoc* network made up of individual firms and government officials, which either had too large a number of participants to make decisions, or did not represent all interests affected, or a combination of both. Hence the need for cross-sectoral interest associated with the European

Packaging Directive challenged the traditional mode of British interest integration that was based on the interaction of individual companies with government.

The European Packaging Directive also threatened the well-balanced system of interest integration developed in the Netherlands. The legal requirement that provisions of directives have to be enforceable and come to full effect, meant in principle that regulations must be generally binding. The voluntary agreement, however, has been signed only by parts of industry, represented by the Foundation of Packaging and the Environment. The need for a generally binding rule for all packers and fillers and importers of packaged goods implied a broadening of the interests affected as compared to the 1991 Covenant. The function of the network remained the same: the co-production of policy - in this case a new packaging covenant. But now government had to negotiate with a larger part of industry. The policy network became wider, also because the Ministry of Economic Affairs gained a stronger position. More interests had to be considered. The policy process became more complicated and a bit more contentious. Rather than bilateral negotiations between VROM and the SVM, industry now organized itself in the packaging platform, where representatives of the peak associations of Dutch industry, VNO-NCW and MKB, also participated. Still the network remained corporatist. A small number of key figures from trade associations negotiated with the government. The trade associations were, however, able to accommodate intra-industry conflicts. The SVM remained a central actor in the network since it could build on the already established relationships with VROM.

In contrast to the Dutch and British cases, the German government did not need the business sector for the co-production of new policies, it had rather to defend its existing policies. Therefore it did not actively engage in close and selective networks any more. However, it was increasingly exposed to a plurality of interests because business became aware of the intended and unintended consequences of the German Packaging Ordinance (Interview BMU 12.11.96). Unlike in the UK and the Netherlands, domestic developments changed the mode of interest integration. Nevertheless, negative integration also provided through the infringement procedure a channel for companies to express their concerns about the 1991 Ordinance, the refillable quota in particular. In addition, other Member States and the economic interests they presented, used the notification procedure extensively to challenge the new draft. Hence European integration increased the number of interests to be integrated in the German packaging waste policy process. But even without the EU, the policy network would have been pluralized because of the large influence of the German packaging ordinance on industry. In contrast to the UK and in line with the Netherlands, however, the government maintained its preference for interacting with trade associations rather than individual companies (Interview BMWi 6.12.1996).

12.1.6 Orientation towards the target group: persistent diversity

The orientation of government to the target group has not substantially changed in Germany and the Netherlands. The German government is still impositional, while the Dutch government maintains its mediating style. The British government maintained for some time a consensus-oriented approach. But when industry was not able to firm-up a proposal, it exerted power on industry to develop a packaging waste management system which would be in compliance with both the degree of legal codification required by the EU Directive and the minimum recycling and recovery targets. It also hammered out a compromise of financing principles which did not reflect the positions of all parts of industry. In the UK case therefore, the more impositional style was induced by positive European integration.

The German government had imposed the 1991 Packaging Ordinance on business. Though this programme met with a lot of resistance, the government did not change its approach substantially in the last years. Moreover, the government repeatedly made clear that it would impose mandatory deposits if the refillable quota were no longer achieved. The degree of imposition might even increase, because the government included provisions in the Amended Packaging Ordinance aiming at a more effective control of free-riders. It should be noted, however, that the government also decreased the burden on industry by allowing for 'raw material' recycling of plastic packaging up to 60 per cent of the packaging recycling target.

The Dutch government maintained its mediating and consensus-oriented approach. After some irritation about the Packaging Waste Regulation by parts of industry, industry returned to the traditionally active and consensus seeking attitude towards the government. Though the network increased, VROM and the SVM remained the central actors on behalf of the government and industry. Using an academic as intermediary, these two organizations co-produced the new Packaging Covenant, which is the basic means for implementing the Packaging Waste Regulation. Hence, even though the European Packaging Directive induced the number of participants in the network, the government did not change its orientation towards the target group.

12.2 Factors of convergence and persistence revisited

12.2.1 European integration

This study has shown that European integration played an important role in the development of national packaging waste policies. The Europeanization of the packaging waste issue provided access for new actors in the policy process and also new mechanisms by which ideas and interests could find their way in the political arena even though it was not the only force that shaped these policies.

Positive Integration

The need to adapt to the legal requirements of the European Packaging Directive was the decisive factor for the convergence of national policies towards a formalized approach as outlined in section 10.1.2. The British government would not have introduced packaging waste regulations without the European Packaging Directive. It is also unlikely that the Dutch government would have replaced its 1991 Packaging Covenant based on private law by generally binding regulation. The European Packaging Directive did not shape the packaging waste standards in the Netherlands and Germany. For reasons outlined earlier, the EU Packaging Directive allowed front- runner countries to maintain or introduce higher standards than stipulated in the Directive. The introduction of the packaging waste standards in the UK, however, was induced by the Directive. Without the Directive, the British government would not have introduced anything as ambitious as a 52 per cent recovery target for packaging waste. The European Packaging Directive also had no influence on the allocation of recycling responsibilities between public and private actors. The required legal codification of packaging waste policies in public law implied, however, that public agencies assumed licensing and enforcement tasks. The Directive also influenced the mode of interest integration in the Netherlands and Britain. The need for generally binding regulation increased the number and types of actors potentially affected. The close relationship between VROM and the SVM was temporarily threatened. New organizations, in particular the MKB and the VNO-NCW, were included in the policy process. The SVM remained, however, the main negotiation partner of VROM. The British government was induced by the Directive to increase its national recovery and recycling performance. The government therefore intensified its contact with industry, which was organized in a number of *ad hoc* networks. Not only the mode of interest integration but also the orientation towards the target group changed in Britain. The government increasingly put pressure on industry to develop feasible solutions to meet the targets set out in the Packaging Directive. Neither the impositional approach of the German government nor the mediating approach of the Dutch government were seriously affected by the EU Directive.

Negative integration

Negative integration was less important in this case, than positive integration. In fact, the attempt to discipline interventionist national packaging waste policies by the mechanisms of negative integration failed. The ECJ judgement in the Danish Bottle Case effectively sheltered national environmental protection policies against Court driven deregulation. Hence, positive integration was necessary to harmonize national regulations in order to break down barriers to trade. But the provisions aiming at negative integration provided through the infringement procedure, in particular Article 169 of the Treaty, an institutional channel for foreign interests to national policy making processes and partly shaped the political agenda. Companies complained in particular about the negative trade effects of the too ambitious German packaging recycling quotas and its refillable quota. These objections provided even in the absence of an actual infringement procedure, a continuous threat to national governments. Though the ECJ judgement protected national measures, it was not clear whether the Court would judge in the same way after the European Packaging Directive had been adopted. There was a certain degree of uncertainty which was one of the factors which forced the German government to relax the refillable quota and to weaken the recycling standards, in particular for plastic packaging, in its draft of the Amended Ordinance. Another institutional channel for foreign interests was provided in the form of the notification procedures required by the

Directive for all measures designed to transpose it into national legislation. Member States used this procedure for detailed opinions criticizing the German refillable quota. This was another way in which European integration contributed to the relaxation of the refillable provision in the draft Amended Ordinance. Such institutional channels not only affected the content of the German standards, but also resulted in an increasing number of interests entering the German system of interest integration. The German policy network became wider.

There was another, more indirect effect of negative integration. Although the provisions ensuring the free movement of goods did not keep the Government from introducing a refillable quota aiming at the *maintenance* of the current share of refillables in the early 1990s, it dampened the enthusiasm of the Ministry of the Environment, to introduce measures aiming at an *increase* in refillables in the early 1990s. It is difficult to establish, however, whether the Ministry of Environment really wanted those provisions, or whether it promised the *Bundesrat* the development of such an ordinance to get the assent to the 1991 Packaging Ordinance while being aware of the fact that Brussels would never go along with it.

12.2.2 Policy-oriented learning vs. path dependency

'You try and think from scratch over a blank piece of paper. And you look at what other countries have done' (Interview DTI 15.5.1996)

Policy-oriented learning

Packaging waste policy making is a technically complex exercise which takes place in specialized policy subsystems comprising government officials, scientists, specialists from industry, and sometimes representatives from environment and consumer groups. In these subsystems ideas about packaging problems and solutions are generated and diffused. Experience from adopted policies or pilot projects are assessed and evaluated. Since packaging waste regulation is a comparatively new area of public policy making, there is much to learn. The announcement of the European Packaging Ordinance increased the attention for packaging waste problems and the interaction within the policy subsystems. New fora were created. This Europeanization of the issue of packaging therefore also accelerated the generation and cross-national diffusion of information about packaging waste problems and solutions.

Governments, trade associations, and environmental and consumer groups commissioned surveys of the packaging waste policies of the several Member States. These actors were reading specialized journals, some of them solely devoted to the issue of packaging waste, such as the newsletter European Packaging & Waste Law. Moreover, organizations such as the DSD and the SVM were regularly visited by delegations from governments and companies who wanted information on the system and its strengths and weaknesses. In addition, there have been many specialized conferences. At least four times a year representatives from the Commission, national ministries, business, and environmental and consumer groups are invited to conferences in Brussels to speak about recent developments in

European and national packaging waste policymaking. There are also national conferences; at least in the three countries covered in this study.

Against this background it is not surprising that policy-oriented learning was one of the factors that shaped the European Directive and national policies. This is not to say that certain policy provisions can be entirely traced back to policy-oriented learning. In most cases, information about the issue of packaging and the environment was embodied in the broader policy paradigms and was used to strengthen one's arguments or to weaken the arguments of others.

The most important source of learning was the German Packaging Ordinance, in particular the institutional arrangement for the recycling of sales packaging organized by the DSD. The German system was the first of its kind and its intended and unintended consequences were tremendous. The German DSD also made clear that learning from failure was at least as important as learning from success.

The failure of the German system to create sufficient recycling capacities to meet its ambitious plastic recycling target contributed to the introduction of a maximum recycling quota in the European Packaging Directive and the provision that Member States could exceed this quota only when sufficient recycling capacities existed. In the Netherlands and the United Kingdom, the perceived failure of the German system to create cost effective waste management systems together with the outcome of domestic pilot projects in these countries resulted in a distribution of responsibilities that differed from that in Germany. The Dutch and British local authorities kept responsibility for the separate collection of packaging waste materials. No packaging waste dustbin has been put in British and Dutch households by private companies.

All Member States under review learned from life cycle assessments. The new Dutch Packaging Ordinance included the provision that the packaging prevention quota is not conditional solely on economic growth. If an increase in the amount of packaging is due to a replacement of one-way packaging by heavier refillable packaging, the resulting surplus in weight will be corrected for. The weakening of the German target for the material recycling of plastic was partly shaped by the lesson that material recycling is not always more environmentally advantageous than other methods of waste treatment such as incineration with energy recovery. A more general lesson, which the German government drew from LCA's, was that the Amended draft of the Ordinance spoke of the protection of 'environmentally advantageous packaging' rather than the protection of refillables as it did in the 1991 Packaging Ordinance. This indicated that the Government was prepared to learn from LCA's that under certain circumstances types of packaging other than refillables could be environmentally advantageous. This lesson had an impact on the range of packaging which is allowed under the quota for milk packaging. Not only refillables but also one-way plastic sachets were covered by the quota, because they scored well in life cycle analyses commissioned by the German Federal Environment Agency. The contradictory results from LCA's reassured the British government that there is now sound scientific evidence for the environmental superiority of refillable packaging. The LCA's for refillable packaging illustrate that policy learning is very difficult when core conflicts (Sabatier 1993) are involved. The ideas about how the economy should be organized clashed in this case. Even if the process explicitly aims at achieving consensus, as in the Dutch case (see 5.3.6), there are diverging opinions about the lessons to be drawn.

These instances of policy learning concerned rather instrumental aspects of the policies. The learning processes are, however, embedded in a larger learning process over the last two decades, namely the paradigmatic change from end-of-pipe approaches to source-oriented approaches in environmental policymaking. This learning process resulted in a larger role of private actors in waste management (see 10.1.4) and the integration of business interests in the policy making process (see 10.1.5).

Path dependency

The case of packaging waste also indicates that regulatory change necessitated by policy-oriented learning could be impeded by path dependency. The German government was constrained by the legacy of the past. Billions of marks have been put into the German Dual System. Waste management companies and the Dual System itself benefit from the system, and consumers learned to separate packaging waste. Thus sunk costs, vested interests and habits prevented Germany from converging to the Dutch and British waste management system that was based on one bin per household.

But path dependency also mattered in a more general way. It is quite difficult to lower adopted standards in the packaging waste area, which is more visible than, for instance, air pollution or global warming. It quickly leads to pointing to the additional amount of packaging or packaging waste created by lower standards. At least when the modifications are not formulated in such complicated terms as in the case of the Dutch packaging prevention target, where the members of the parliament's Environment Committee, for instance, did not understand the practical meaning (Verpakkingsmanagement 10/1997: 1,5).[28]

12.3 REGULATORY POLICYMAKING IN EUROPE: What is this a case for?

When this study started at the end of 1993, research of European regulatory policymaking that also took account of the national dimension was in an embryonic phase. Meanwhile, however, more studies have been carried out. This section will show how the findings of the packaging case fit into the broader picture of EU regulatory policymaking. I will relate the findings to the idea of a competition of regulatory approaches as the logic of multi-level policymaking, examine whether a single type of governance is emerging, discuss whether European integration has strengthened or weakened the state, and address the tensions between convergence and national traditions.

---

[28] Even an official at the Ministry itself was not able to explain this target on request. The person said *'figures are not my strength'*(communication VROM).

12.3.1 Multilevel policy making and the competition of policy approaches

The European Union created a new institutional context for public and private actors, which provides them with new opportunities and new constraints. The overall picture which emerges is that of multilevel policy-making in which political pressure goes from the national level to the EU level and vice versa. This vertical interaction is complemented by a horizontal dimension, the interaction between Member States (Grande 1996, Héritier *et al.* 1994, 1996). The choice to include the implementation perspective in this study sheds light on the role of sub-national actors with power of veto, in this case the majority of the German *Bundesrat* Hence, the supra-national and the national level is supplemented by the sub-national level. The choice for the issue of packaging waste further increased the complexity of the multi-level process. It has both environmental aspects and trade aspects which mobilize two different sets of actors. On a national level and the German sub-national level, Ministries for the Environment have to negotiate with Ministries of Trade or Economic Affairs. This pattern of conflict is reflected in the Commission itself between DG III (Industry), DG IV (Competition) and DG XV (Internal Market) on the one side, and DG XI (Environment) on the other. Neither the State nor the Commission are monolithic actors. National and Commission positions were often the result of fragile compromises. This complexity created constraints and opportunities. The most remarkable case of constraints is Germany, where the Government was - at the time of writing - still unable to transpose the EU Packaging Directive into national law. The Government was trapped between the Commission on the one hand, demanding a relaxation of the 72 per cent refillable quota, and the majority of the *Bundesrat* on the other hand, asking for the maintenance or even strengthening of this quota.

As Tsebelis (1990: 8) emphasized, the Europeanization of policymaking also provides more opportunities. They were, for instance, employed by the Dutch Ministry of Economic Affairs who used the infringement procedure to criticize the German refillable quota on behalf of the Dutch government. The institutional provision that this Ministry has sole responsibility within the Dutch government for infringement procedures allowed them to voice a position via Brussels against Germany, which would not be an official Dutch position in intergovernmental bargaining. In that case the Ministry would need a consensus opinion in its own Cabinet, which it would not get given the pro-refillable stance of the Dutch Ministry of the Environment. Another and probably more important example is that of large companies. They used many channels of access to pursue their preferences. Multinationals are usually part of numerous national and European trade associations or more loosely coupled networks. Affiliates of the SVM for instance, bound by the Dutch Covenant to reach high packaging prevention and recycling targets in the Netherlands, were also members of European initiatives such as EUROPEN or the PCF which lobbied in the course of the formulation of the European Packaging Directive against targets for prevention and in favour of low targets for recycling. Hence these companies used the European arena to escape from national constraints (see also Eichener 1993: 74). Environmental and consumer organizations, though also present at both levels, did not have the resources to equal the efforts of large companies.

This research finding suggest that the Europeanization of policymaking changes the balance of interests towards business interests.

Héritier *et al.* (1994, 1996) have suggested that the logic of this multilevel policymaking can be described as a competition of national policy approaches in which those Member States who are pace setters in a certain area and who are able to form a coalition with the EU Commission are more successful than other countries and are able to force other countries, the laggards, to adapt to their policy approaches (see chapter 1.2.4). This image of European policy making is partly confirmed by the case study. Member States tried to Europeanize their policy approaches. The German government was in favour of high, quantified and binding targets. When this was deemed impossible it fought hard to get exemptions in order to be able to maintain its national approach. The British government was against this legalistic and rigid approach and had a preference for framework provisions and flexible general guidelines for waste prevention and recycling measures. The Dutch position did not entirely mirror the national policy approach, however. The Dutch government wanted high and generally binding standards, despite the fact that the national standards were binding only on those companies who signed the 1991 Covenant. The rationale for the Dutch position was as follows: The government did not believe that the Dutch Covenant approach could be transferred to other countries. The government doubted that most of the packaging chains in other countries would enter into private agreements and would be committed to reach the targets. The government wanted binding regulations in the EU, in order to avoid competitive disadvantages for those companies who had joined the Dutch Covenant. Note that the Dutch government had committed itself in the Covenant to strive for European packaging legislation. Another reason for the Dutch position was possibly that binding targets would strengthen the domestic position of the government against those companies who did not (yet) participate in the Covenant.

The first drafts of DG XI of the Commission were indeed informed by the pace setters Germany and the Netherlands. They were hybrids in the terminology of Rose (1993), combinations of two programmes (see 1.4.2.). The collection and recycling targets were virtually identical to the Dutch Covenant and the take-back requirement for packaging imitated from the German Packaging Ordinance. It is possible that the development would have taken the path described by Héritier for the case of air pollution, namely that a directive would have been adopted that reflected the policy approach of the pace setters and DG XI of the Commission. But things turned out differently, because the environmental issue of packaging waste was linked to the issue of fair competition and the free movement of goods, more so than the issue of air pollution from stationary sources analysed by Héritier. Subsidized German packaging waste flooded the market of other Member States and endangered the embryonic domestic waste collection and recycling infrastructure, and the German refillable quotas were seen as barriers to trade. Hence, the issue increasingly became a single market issue, involving new directorates of the Commission and strengthening the position of countries emphasizing the need for harmonization rather than environmental protection. Because of competing problem definitions and the decision-making logic of the EU, a compromise emerged which was neither fish nor fowl. There were neither strict environmental protection standards, reflecting the approach of the pace setters, nor provisions effectively ensuring the harmonization of national approaches. This case study suggests therefore that the outcome of the competition of policy approaches is more difficult to predict for policy problems which have more than one core aspect.

12.3.2 Towards transparent regulatory states in Europe?

The dis-aggregation of policies into different dimensions, as done in this study, has the analytical advantage of providing a refined concept of the dependent variable but may not blur the general picture. What has emerged in the case of packaging waste is a convergence towards a uniform mode of public-private co-ordination, or governance. The Netherlands and the UK still allow for self-regulation, but now within a hierarchic setting. The new governance system is in fact quite similar to the German one: self-co-ordination of private actors within a tight legal framework. This development confirms the argument of Majone, that this type of regulation might become the dominant mode of governance, both on the level of Member States and on the EU level (Majone 1994, 1996a, 1996b, see also McGowan and Wallace 1996).

Since all three Member States have developed this kind of governance one might argue that a system of steering has evolved which is considered more effective than either pure command and control policy or pure self-regulation. The emergence of this new mode of governance is, however, not the result of a master plan, but emerged over time and was shaped by a number of factors which are only loosely coupled.

The tight legal framework is the result of the European Packaging Directive. The reasons underlying this formalization of packaging waste policies have been outlined earlier. They match the overall picture of European integration as dominated by legal integration. But why is legal integration dominant? It has been argued that Member States accept the binding force of legal commitments to secure compliance to both other Member States and domestic actors out of distrust of each other. In the case of packaging waste the problem of free riding and the resulting distortion of competitive positions was already important on the domestic level. The strong economic interests at stake, such as the potential financial burden on industry, make a certain hierarchical and transparent setting much more necessary on the supranational level, '*because problems of compliance and verification loom large in international agreements*' (Weale 1992a: 176). Another reason for the predilection for a legalistic approach may be that Member States want to make sure that future changes will not evolve in a way that might be detrimental to them, compared to the status quo.

Hence it is a Europeanization of the German regulatory culture, emphasizing values such as predictability and reliability (Dyson 1992: 10), that drives the EU into a transparent and legalistic setting. The legalistic approach of the EU is not surprising, given that the codified/roman law tradition was prevalent in the six founding countries France, Italy, Germany, Belgium, the Netherlands, and Luxembourg (Héritier *et al.* 1996: 151).

These arguments explain why hierarchy may be needed. But why was only a legal framework set and not detailed legal norms? All governments acknowledged that some form of self-co-ordination is necessary, given the properties of the policy problem at stake. As many other modern problems, scientific, political and economic uncertainty and dynamic developments demand a certain degree of flexibility at least concerning the means by which targets have to be achieved. An active role of industry is also necessary because industry controls crucial resources, especially information about the composition and the proper treatment of packaging.

12.3.3 Strengthening or weakening the state?

In the process of political and legal integration nation states transfer formal power to intergovernmental (Council) and supranational (EP, Commission, ECJ) institutions. In the literature on European integration there is a lively debate about the question of how this process impinges on the distribution of material power (autonomy) between national and supra-national institutions and between the nation states and private interests. A number of authors argue that European integration strengthens the nation state (Moravcsik 1994) or even means the rescue of it (Milward 1992). Milward claims that the EC helped the nation state to offer a degree of security and prosperity which has justified the survival of the nation state after the Second World War. In his reading, the EC has played an indispensable part in the nation states post-war construction. States would further surrender their sovereignty only if this seemed necessary in an attempt to survive (Milward 1992: 3). Moravcsik says that EU integration is driven by bargains among Member State governments, no Member State has to integrate more than it wants to, because decisions are always taken on the lowest common denominator. He acknowledged that cooperation may limit the *external* flexibility of executives, but he emphasized that this cooperation simultaneously confers greater *domestic* influence. *'In this sense, the EC strengthens the state'* (Moravcsik 1994, quoted in Marks, Hooghe and Blank 1996: 346). According to Moravcsik, the unique institutional structure of the EC is acceptable to national governments only insofar as it strengthens rather than weakens their control over domestic affairs, permitting them to attain goals otherwise unachievable (Moravcsik, 1993: 518).

Moravcsik's empirical work is concentrated on the great bargains like the Single European Act and the Treaty of Maastricht. He makes a strong case when he points to the control of the nation states at these events. His arguments seem less plausible in regard to day-to-day policymaking. It might be exaggerating to argue that integration results in an accumulation of supra-national powers in Brussels, Strasbourg and Luxembourg, but a state-centred view seems equally unrealistic. As has been said above, the case of packaging waste provides more substance to the claim of adherents to a multi-level governance perspective, namely that sovereignty is diluted in the EU by collective decision-making among national governments and by the autonomous role of the European Parliament, the European Commission, and the European Court of Justice (see Marks, Hooghe and Blank 1996 342-343). It is certainly true that Member States have ensured a great deal of autonomy, that the Council played an important role and that the Commission was rather weak. All these factors give evidence to the 'statist' claims. One should not forget, however, a) that the Netherlands and the United Kingdom were forced to take a course of action they would not have taken without the

Directive; b) that the EP accept the Council's position on the Packaging Directive with only a small majority; and c) that the ECJ case law helped to construct environmental competencies on the EU level. Moreover, ECJ case law has given people standing to assert EU rights before national courts (the earliest case was 26/62 [Van Gend en Loos], ECR 1963: 1, see also Daintith 1995: 14, Maher 1996: 85). National courts, guided by ECJ decisions, can compel their governments to comply with EU rules they have opposed (Stone Sweet and Sandholtz 1997: 312).

Moreover, one should not confuse collective Council authority with the preservation of national sovereignty and national autonomy (Golub 1996: 331). Qualified majority voting implies that some become losers. *'Having lost the fights, these unfortunate states are faced with binding European legislation which curtails their own choice of policies thus eroding national autonomy and national sovereignty'* (332). In the case of packaging waste, even Member States which belonged to the qualified majority, for instance, the UK, had to make choices in their national policies which they would not have made without the Directive. Even decisions on the lowest common denominator limit options, preventing back-sliding, setting compliance deadlines and targets which otherwise might have been repealed or amended by the national executives: hence, leaving a restricted range of options for implementers (Golub 1996: 333). The loss of national autonomy becomes more evident if one considers that new governments are bound by EU obligations entered into by previous administrations: Given the 27,000 EC regulations and directives which were in force by the end of 1995 (Golub 1996: 334), which make up the lion's share of national legislation, it seems very unlikely that the EU does not restrict the policy options of national executives.

Moravcsik also argues that European integration can be described as a sequential two level game, in which first interests are aggregated domestically and a national interest is defined, and then national executives bargain according to the resulting preferences in the European arena (Moravcsik 1993, 481-483). The packaging case, however, strengthens the argument of other authors that Member States are loosing their grip on domestic interest representation in international relations (see Marks, Hooghe and Blank 1996: 341). Large internationally-oriented companies in particular have used multiple points of access to the European decision-making process, directly and/or through more or less formalized networks and associations. Unilever, Coca Cola, and Procter & Gamble for example are driving forces behind most of the following groups: the German Working Group Packaging and the Environment (AGVU), the British INCPEN, the Dutch SVM, and the European AIM, EUROPEN, PCF, and ERRA. Not only governments but also companies instrumentalize the integration process to escape from domestic constraints.

There is not only evidence supporting the claim of a weaking of the state, however. There is also evidence for Moravcsik's claim that national governments can use supranational institutions and policies to weaken domestic pressure. This study gives evidence that the legalistic approach is bound to strengthen the state *vis-à-vis* economic interest groups. Although the legal requirement to impose enforceable obligations on companies clashed with the consent-based mode of governance in the Netherlands and the UK, it strengthen the state. In the case of the Netherlands, the Ministry of the Environment welcomed a more legalistic approach to tackle the increasing problem of free-riding, while continuing with the covenant approach on a higher regulatory level. After having accepted the need for regulation, the UK government welcomed the need to set uniform quantified targets (in percentages) for each company. It relieved the Environment Agency of the administrative burden of negotiating individual obligations with several hundreds of thousands of companies. Hence, Member States can gain autonomy *vis-à-vis* domestic interests by arguing that they are tied down to EU obligations. Grande (1996: 383-392) has made similar observations in the case of research and development policies and called this phenomenon the '*paradox of weakness*'.

The legalistic approach may also strengthen business interests, however. As Gehring (1997) points out, a producer or importer, believing himself to be adversely affected by the domestic packaging waste policies of Member States, may test whether the Directive has modified the legal situation as compared to the status based upon the Danish bottle case jurisdiction (351-352). This might be particularly the case with the German refillable quota. This system of indirect enforcement is expensive and the remedies are not always adequate (see Maher 1996: 585). But given the significant commercial interests at play, and the number of packaged goods traded in the EU, such challenges are not unlikely.

12.3.4 Convergence and national traditions

The new type of governance very much reflected the German approach but challenged national policy traditions, especially in the UK. Legally binding uniform and quantified targets are foreign to the British tradition, where national law provided the framework for decentralized negotiations between the implementation agency and the company concerned. The new mode of governance matched, however, developments in other areas of environmental policy, such as air pollution, and even beyond environmental protection, namely the change from an informal negotiation state to a transparent regulation state (Knill 1995: 314, see also Dunleavy 1989: 287, S. Vogel 1996). All these go in the direction of a more centralized approach, away from secret and informal negotiations between companies and inspectors towards transparent, detailed and clear obligations for industry. The 1996 government paper '*A Waste Strategy for England and Wales*' shows that Britain is prepared to announce national quantitative targets, also in areas not subjected to European integration (DoE 1996b). There is a case of lesson-drawing from the issue of packaging waste. As a civil servant of the DoE said: '*The British approach has always been the best practical environmental option for each individual case. And, in practise, for most of industry the best practical environmental option was landfill. The hierarchy is meant to set out what the priorities are... Some targets are needed to change the attitude of a slow changing industry. Three years ago, we had another attitude*' (Interview DoE 1996).

In the Dutch case, the European Packaging Directive challenged the Covenant approach. It is important to note, however, that the Covenant approach reflects the general consensus-oriented Dutch political culture, but was rather new in Dutch environmental policy making. Throughout the 1970s and the early 1980s the Ministry of the Environment employed a command and control approach. It was the lack of control and the idea of moral internalization '*verinnerlijking*' that led to the emergence of voluntary agreements in the form of covenants. These agreements were, however, subject to domestic critics, not least lawyers who objected to the lack of clarity and legal protection involved in these agreements. In addition, the problem of free-rider behaviour and unfair competition ranked high among the points of criticism. A stricter legal framework was therefore not strongly opposed by the Dutch Environment Ministry. Furthermore, companies who participated in the Covenant wanted to prevent free-riding behaviour and were therefore not totally against some form of generally binding regulation. The introduction of regulation based on public law was therefore less problematic than in the UK. It also matched a more general development in the Dutch environmental policy debate emphasizing the need for stricter enforcement of existing regulations. Moreover, the development of the Dutch Covenant has its parallels in Dutch history. There are many examples where self-regulation by industry, partly forced by government, resulted in free-rider problems, which made additional Government regulation necessary. Examples are dairy quality regulation or, more generally, the development towards statutory trade associations and generally binding collective wage agreements (CAO's) (see Fernhout 1980, van Waarden 1985).

The British case makes clear that policy traditions can change. The British government resisted the industry' proposal for bilateral negotiations between individual companies and implementation agencies on individual recovery and recycling targets. It included quantified national targets into legislation and made them legally binding on individual companies. Hence, institutional tradition might not be as sticky as is assumed by many institutionalist authors.

The suggestion of convergence towards a uniform mode of governance in Europe made in this section needs two crucial qualifications, however. This study paid more attention to the legal implementation of the European Directive on Packaging and Packaging Waste. It may well be that Member States evade pressure on their national traditions by lax practical implementation and enforcement. This argument is made by Eichener (1996) for the case of occupational health and safety as well as environmental protection, and by Golub (1997) for the case of environmental protection. Moreover, it is possible that certain properties of this policy issue imply a bias towards the convergence hypothesis. The most important point is that environmental protection in general, and packaging waste policy in particular, are relatively new policy fields (less than 30 years). It can be argued that actor relations and routines are much more stabilized and embedded in national institutions and traditions in older policy fields and therefore show more resistance towards change (see Lehmbruch *et al.* 1988:255, Nassmacher 1991:205).

13. SAMENVATTING

Inleiding
Internationaal vergelijkend onderzoek naar overheidsbeleid in de jaren zeventig en tachtig heeft aangetoond dat landen een specifieke, nationale manier hebben om te reageren op problemen zoals werkloosheid, kostengroei in de gezondheidszorg of milieuvervuiling. Hierbij rijst de vraag of economische, politieke en sociaal-culturele globalisering zal leiden tot een vermindering van deze nationale verschillen. Dit algemene thema wordt in dit onderzoek toegespitst op de vraag of Europese integratie tot een convergentie van nationaal beleid leidt. Het onderzoek richt zich in concreto op het verpakkingsafvalbeleid van Nederland, Engeland en Duitsland. Vraagstukken omtrent verpakkingsreductie, statiegeld en recycling-plichten behoren tot de meest controversiële gebieden van regulering van bedrijven. In het verpakkingsafvalbeleid komt bovendien de spanning tussen Europese eenheid en nationale verscheidenheid tot uitdrukking, alsmede de spanning tussen milieubescherming en de vrije marktwerking in Europa, in het bijzondere het vrije verkeer van goederen.

Om een gedifferentieerd beeld van beleid en beleidsveranderingen te verkrijgen werd op basis van internationaal vergelijkend literatuuronderzoek een vijftal dimensies onderscheiden: de graad van formalisering van beleid, de strengheid van normen, de verdeling van taken tussen publieke en private actoren, de wijze waarop maatschappelijke en economische belangen bij het ontwikkelen van beleid worden betrokken, en de houding van de overheid tegenover de doelgroep.

Het doel van het onderzoek is echter niet alleen om verschillen en ontwikkelingen in beleid te beschrijven, maar ook om deze te verklaren. Het verklaringsmodel is op recente ontwikkelingen in de vergelijkende beleidsanalyse gebaseerd, in het bijzonder het neo-institutionalisme. Deze invalshoek benadrukt, in tegenstelling tot het pluralisme en het marxisme, dat de overheid niet enkel een speelbal is van maatschappelijke belangen is maar een zekere handelingsvrijheid heeft.

Volgens een centrale aanname van het neo-institutionalisme wordt de macht van maatschappelijke belangen niet alleen door factoren als organisatiegraad en hulpbronnen bepaald maar vooral ook door de mogelijkheden die het nationale institutionele landschap biedt om belangen in het beleidsproces te effectueren. Instituties zijn de regels van het politieke spel en bepalen als intermediaire variabelen de relatieve sterkte van belangen en ideeën. Recente ontwikkelingen in het neo-institutionalisme wijzen erop dat niet alleen politieke instituties maar ook ideeën of paradigma's een intermediaire functie bezitten. Net als instituties structureren zij de perceptie van de omgeving en beïnvloeden op deze manier de preferenties van politieke actoren. In dit verband zijn naast ideeën die op specifieke beleidsterreinen spelen ook rechts- en staatstradities en de politieke cultuur van belang.

Met de nadruk op de rol van de overheid is het neo-institutionalisme verenigbaar met twee andere recente theoretische concepten: 'beleidsgericht overheidsleren' als een van de mogelijke factoren voor beleidsverandering, en 'padafhankelijkheid', dat optreedt wanneer bestaand beleid beleidsverandering belemmert.

Om de validiteit van de causale uitspraken te verhogen zijn drie methodische strategieën toegepast. Ten eerste werden de drie landen gekozen in het licht van de '*most similar systems design*'. Duitsland, Nederland en Engeland hebben een vergelijkbaar economisch en technologisch niveau en zijn alledrie liberale democratieën. Eventuele nationale verschillen laten zich dus niet tot de bovengenoemde algemene factoren herleiden. Ten tweede werden de beleidsprocessen in detail bestudeerd om de wisselwerking tussen causale factoren zoals ideeën, belangen en instituties in kaart te brengen (*process tracing*). Ten derde werd met zogenaamde contrafaktische argumentatie gewerkt. Bij deze techniek wordt gepoogd aannemelijk te maken dat een bepaalde uitkomst niet tot stand zou zijn gekomen zijn zonder een bepaalde factor. Voor de dataverzameling werd gebruikt gemaakt van uiteenlopend schriftelijk materiaal (wetsteksten, rapporten, persberichten, publicaties van belangengroepen, vak- en krantenartikelen, enzovoort) en bovendien van 20 half-gestructureerde diepte-interviews met sleutelpersonen in alledrie de landen.

Verpakkingen en Milieu

Al de bovenstaande zaken worden in het inleidende hoofdstuk aan de orde gesteld. In het *tweede hoofdstuk* worden de verschillende dimensies van het vraagstuk van verpakkingen en milieu geïntroduceerd. Het laat zien dat de groei van het verpakkingsafval nauw verbonden is met het proces van modernisering. Sinds productie en consumptie niet op dezelfde plaats en tijd plaats hoeven te vinden, zijn transport en distributie een noodzaak geworden en daarmee conservering en bescherming van goederen. De hoeveelheid verpakkingsafval is met name sinds de jaren vijftig sterk toegenomen. In de jaren negentig 'produceert' elke EU burger gemiddeld zo'n 150 kg verpakkingsafval per jaar.

Zoals vele andere politieke onderwerpen kenmerkt ook het vraagstuk van verpakkingsafval zich door dynamiek, wetenschappelijke onzekerheid en pluriforme probleempercepties en belangen. Tot de dynamiek dragen onder ander bij: a) wisselende wereldmarktprijzen voor verpakkingsgrondstoffen die de economische calculatie van het bedrijfsleven voordurend doen veranderen, b) beleidsveranderingen op gerelateerde terreinen in binnen- en buitenland en c) innovaties in de verpakkingstechnologie.

Naast dynamiek kenmerkt het vraagstuk zich ook door wetenschappelijke onzekerheid omtrent de meest milieuvriendelijke alternatieven. Uitkomsten van zogenoemde levenscyclusanalyses zijn bijna altijd controversieel. Zij worden mede bepaald door aannames, afbakeningen en waardeoordelen waarvoor geen eenduidige wetenschappelijke normen bestaan.

Een oorzaak voor de politisering van het onderwerp is dat het vraagstuk van verpakkingen en milieu gekenmerkt wordt door uiteenlopende probleempercepties en een grote pluriformiteit van belangen. De industrie benadrukt meestal de belangrijke functie van verpakkingen voor de samenleving, wijst op het kleine aandeel ervan in de totale afvalstroom (ongeveer drie procent) en is strikt tegen iedere vorm van overheidsingrijpen. Voor veel kritische consumenten en de milieubeweging zijn verpakkingen echter een symbool van onze wegwerpmentaliteit. Zij spreken van onnodige verpakkingen, verspilling van hulpbronnen en energie, beslag op milieuruimte en schadelijke emissies bij storting en verbranding, en eisen verpakkingsverboden en uitbreiding van statiegeld. Het onderwerp wint aan complexiteit door het feit dat er ook binnen de industrie verschillende belangen zijn, zoals bijvoorbeeld tussen de verpakkingsketen (van grondstofproducent tot detailhandel) en de afvalverwerkingsindustrie. Laatstgenoemde heeft meestal baat bij strenge milieunormen. Daarnaast zijn er ook binnen de verpakkingsketen diverse en vaak tegenstrijdige belangen.

NATIONAAL VERPAKKINGSAFVALBELEID

Dynamiek, wetenschappelijke onzekerheid en pluriforme belangen maken het aannemelijk dat er geen eenvoudig, functioneel verband bestaat tussen probleem en oplossingen. Het nationale beleid wordt dan ook mede bepaald door ideeën, belangen en instituties. In de hoofdstukken 3 tot en met 7 wordt nagegaan hoe Duitsland, Nederland en Engeland met het verpakkingsvraagstuk omgingen voordat Europese instituties nadrukkelijk op het terrein actief werden. Dit deel begint met een vergelijking van het nationale verpakkingsafvalbeleid van de drie betreffende landen aan het begin van de jaren negentig (*hoofdstuk 3*). Dit hoofdstuk vormt tevens de basis voor de vraag of het beleid sindsdien sterker op elkaar is gaan lijken (convergentie). Uit het onderzoek blijkt dat in het begin van de jaren negentig nog op uiteenlopende wijzen op het verpakkingsafvalprobleem gereageerd werd. Een legalistisch en streng Duits stelsel (Duale System, Grüne Punkt) contrasteerde met een pragmatisch, iets minder ambitieus Nederlands systeem (Convenant Verpakkingen) en een minimalistische aanpak in Engeland.

Begin jaren negentig had Duitsland een verordening aangenomen waarin het bedrijfleven gedwongen werd een verzameling- en recyclingsysteem voor verpakkingen op te zetten. Dit systeem moest binnen vier jaar aan eisen voldoen die internationaal gezien zeer streng waren. Het hield in dat 72 procent van glas-, papier-, aluminium- en metaalverpakkingen moest worden hergebruikt alsmede 64 procent van kunststof- en drankverpakkingen. Bovendien moest tenminste 72 procent van alle drankverpakkingen hervulbaar zijn. Dit beleid was het product van onderhandelingen tussen overheid en het bedrijfsleven, waarbij brancheorganisaties een belangrijke rol vervulden. De onderhandelingen werden door de overheid in feite afgedwongen.

Het Nederlandse regime was pragmatischer, minder ambitieus en werd op een meer consensus- georiënteerde manier bereikt. In plaats van een algemeen bindende regeling, werd met delen van het bedrijfsleven een convenant afgesloten, waarbij de aangesloten bedrijven zich verplichtten om de hoeveelheid verpakkingen die op de Nederlandse markt kwam in het jaar 2000 tot het niveau van 1986 terug te brengen en 60 procent ervan te recyclen. Anders dan in het Duitse systeem kon het Nederlandse bedrijfleven zich daarbij op verpakkingen concentreren, die vrij makkelijk te recyclen zijn, zoals transportverpakkingen van karton en verkoopverpakkingen van papier, glas en metaal. Ook was het bedrijfsleven niet verantwoordelijk voor het inzamelen van verpakkingen. Ambitieus was echter wel de doelstelling om in een situatie van economische groei de hoeveelheid verpakkingen terug te brengen tot het niveau van 1986. De Nederlandse overheid had een groot aantal actoren uitgebreid betrokken bij het voorbereiden van het beleid. Toch moesten de meest omstreden punten bilateraal tussen het ministerie van milieu en de verpakkingsorganisatie SVM (Stichting Verpakking en Milieu) opgelost worden.

Het Engelse bedrijfsleven had te maken met het minst ambitieuze en meest vrijblijvende verpakkingsafvalbeleid. Er bestond noch een dwingende regeling noch een bindende afspraak tussen overheid en het bedrijfsleven. Er was een vage doelstelling die in eerste instantie voor de gemeenten was bedoeld. Het kwam erop neer dat in het jaar 2000 rond een kwart van het huishoudelijke afval hergebruikt moest worden. Daarvoor had de overheid een financiële prikkel ingesteld, maar van beperkt omvang. Op de vraag van de overheid aan het bedrijfleven om het verpakkingsafval terug te dringen werd met een business plan en twee gedragcodes gereageerd die echter vaag en vrijblijvend waren. De Engelse overheid had geen gestructureerd contact met de industrie en brancheorganisaties speelden geen rol van betekenis bij de totstandkoming van het beleid.

Duitsland

In de *hoofdstukken 4 tot met 7* worden de bovengenoemde verschillen vanuit neo-institutionalistisch perspectief verklaard op de basis van een reconstructie van de totstandkoming van het beleid.

Het Duitse milieubeleid is een goede bevestiging van de neo-institutionalistische bewering dat de overheid een zekere autonomie tegenover de samenleving heeft. Geïnspireerd door ontwikkelingen in de VS en zonder maatschappelijke druk werd begin van de jaren zeventig de basis van de milieu- en afvalregulering vanuit het politieke systeem zelf ontwikkeld. Ambtenaren formuleerden algemene principes voor dit beleidsterrein, die richting gaven aan specifieke wetgeving. In het bijzonder het voorzorgsbeginsel diende als legitimatie voor strenge en gedetailleerde eisen aan de industrie. Deze eisen waren gebaseerd op de best beschikbare technologie. In overeenstemming met het voorzorgsbeginsel werden maatregelen ook genomen als een concreet milieugevaar niet eenduidig was bewezen. Het voorzorgsbeleid kwam tot stand onder invloed van een drietal factoren: a) het sectorspecifieke beginsel was ingebed in de algemene, het beleidsterrein overstijgende conceptie van de overheid als enige vertegenwoordiger van het algemeen belang, b) het voorzorgsbeginsel was juridisch vastgelegd en kon voor de rechter worden ingeroepen en c) het principe werd toegepast binnen de Duitse legalistische traditie. Volgens deze traditie is wetgeving een deductief proces; algemene principes worden - in een bijna apolitiek proces - door wetgevingsjuristen vertaald in concreet beleid. Het Duitse milieubeleid was dus al streng voordat de maatschappelijke druk toenam, onder andere via goed georganiseerde nationale bewegingen. Dat het milieubeleid in de jaren tachtig ondanks een economische recessie op de agenda bleef, is onder andere te danken aan de opkomst van een groene partij. Daarnaast had de continuïteit in het strenge milieubeleid ook te maken met de opkomst van de idee van ecologische modernisering, die benadrukt dat milieubescherming en economisch groei goed samen kunnen gaan. Deze idee werd tevens gesteund door een snel groeiende en sterk exportgeoriënteerde milieu- en afvalverwerkingsindustrie, die van de strenge eisen in Duitsland profiteerde. Meer dan twintig jaar curatief milieubeleid dat internationaal gezien voldeed aan strenge eisen, leidde echter tot een tekort aan sowieso kostbare capaciteit voor afvalverwijdering en -verbranding in de jaren tachtig. Dit was een van de redenen waarom de overheid zich sterker ging concentreren op recycling en het voorkomen van afval. Een tweede reden was dat al sinds de oliecrisis van 1973 afval in toenemende mate als waardevolle hulpbron beschouwd werd. Dat het verpakkingsafvalbeleid de speerpunt van een afvalverminderings- en recyclingstrategie werd, heeft ook met de symbolische betekenis van verpakkingen te maken. Interventionistisch overheidsoptreden op het terrein van verpakkingen deed het goed bij de kiezers en werd door vele kritische consumenten en de milieubeweging toegejuicht. Om de industrie tot een effectieve oplossing te dwingen dreigde de Duitse Minister voor Milieu dat de detailhandel gebruikte verpakkingen terug moest nemen. Daarop oefende de detailhandel, gesteund door de koepelorganisatie van de industrie, druk uit op de leveranciers om tot een alternatieve oplossing te komen. Op deze manier werd het bedrijfsleven gedwongen een privaat systeem voor verpakkingsafvalverzameling en -recycling op te bouwen dat parallel aan het publieke huisvuilsysteem zou functioneren, het zogenaamde *Duale System*. De gedetailleerde en ingrijpende doelstellingen voor recycling waaraan het *Duale System* moest voldoen waren niet gebaseerd op kosten-batenafwegingen, maar werden min of meer arbitrair vastgelegd, en gelegitimeerd met het voorzorgsbeginsel. Onder invloed van de Bondsraad werden de toch al strenge eisen van de regering nog verscherpt: verbranding met energieterugwinning mocht niet als recyclingoptie gelden.

Nederland

Het Nederlandse milieubeleid werd in eerste instantie ook gevormd door het politiek systeem zelf, met name door het Ministerie van Volksgezondheid en Milieu. De rol van het parlement was echter sterker dan in Duitsland. Groene ideeën die in de maatschappij speelden, werden al vroeg opgepikt door partijen als D'66 en de PPR, die door het ontbreken van een kiesdrempel gemakkelijk toegang kregen tot het parlement.

Het Nederlandse milieubeleid was aanvankelijk, net als het Duitse, op interventionistische en rigide wet- en regelgeving gebaseerd. Deze autoritaire stijl paste echter niet in de Nederlandse consensuscultuur, met de nadruk op coöperatie en overleg en pragmatische rechtstradities. Niet alleen de overheid maar ook private actoren werden als legitieme vertegenwoordigers van het algemeen belang beschouwd, waarbij, anders dan in Engeland, wel een actieve en plannende rol voor de overheid was weggelegd. Het duidelijkste teken dat de autoritaire stijl van het Ministerie van Milieu niet aansloot bij de consensuscultuur was een implementatietekort, dat veel groter was dan in Duitsland. Het implementatietekort was onder andere mogelijk door een zwakke bestuursrechtelijke controle van ambtenaren, een pragmatische rechtscultuur (gedogen) en ambtenaren die veel begrip toonden voor de individuele omstandigheden van bedrijven. De gebrekkige handhaving van de strenge eisen leidde ertoe dat de kosten achterbleven bij het Duitse niveau. De top-down reguleringsstijl van het Ministerie van Milieu kwam al in de vroege jaren tachtig mede door het grote implementatietekort, onder druk te staan. Deze druk werd nog versterkt door de komst van het op deregulering georiënteerde kabinet Lubbers in 1982, en de nieuwe liberale minister van milieu Winsemius. De oriëntatie op de doelgroep veranderde in een meer op consensus gerichte aanpak.

De nieuwe aanpak werd voor het afvalbeleid in de Notitie Preventie en Hergebruik (1988) geconcretiseerd. Deze notitie gaf indicatieve preventie- en recyclingdoelstellingen. De afvalproducenten werden uitgenodigd om in overleg tot concrete doelstellingen en maatregelen te komen. De aanpak kreeg vorm in zogenaamde strategische afvaldiscussies, waarin in een breed samengesteld gezelschap van overheid, industrie, wetenschap en samenleving, over de aard en omvang van het verpakkingsafvalprobleem en mogelijke maatregelen werd gediscussieerd. Ondanks intensieve discussies kon er geen overeenstemming worden bereikt over verpakkingsreductie en uitbreiding van de statiegeldregeling. Om toch nog tot een zichtbaar resultaat te komen, werden de strategische discussies afgebroken en het Ministerie van Milieu onderhandelde bilateraal verder met de Stichting Verpakking en Milieu (SVM), die veel, vooral grote, bedrijven in alle schakels van de verpakkingsketen vertegenwoordigde. Zowel het Ministerie als de SVM verkozen een convenant boven wetgeving. Het Ministerie benadrukte, geheel in de algemene Nederlandse traditie van de consensuele beleidsstijl, dat een vrijwillige overeenkomst legitimatie en draagvlak bij het bedrijfsleven zou creëren. Voor de industrie was het belangrijk dat in ieder geval de eigen inbreng gewaarborgd was en dat een convenant in meerdere opzichten flexibeler was dan formele wetgeving. De industrie kon echter niet vermijden dat er een preventiedoelstelling kwam. Een uitbreiding van statiegeldregelingen wist ze echter te voorkomen.

Engeland

Engeland is het land met de langste traditie van nationaal milieubeleid. Vanaf de jaren zeventig was de regulering van bedrijven echter minder ambitieus dan in de meeste andere OECD landen. Dat lag om te beginnen aan een andere conceptie van het milieubeleid. Terwijl in Duitsland milieubeleid gezien werd als technische reductie van emissies en afval, was milieubeleid in Engeland in eerste instantie het beschermen van natuurgebieden. Alleen om deze reden al werden minder middelen uitgetrokken om het bedrijfsleven te reguleren. Belangrijker was echter een andere opvatting over de rol van de overheid. In Engeland is de staat niet een boven de maatschappij zwevende vertegenwoordiger van het algemeen belang en wordt zij niet geacht als enige het algemeen belang te dienen. Sterker nog dan in Nederland, wordt de overheid gezien als een bemiddelaar tussen verschillende legitieme belangen, die samen het algemeen belang vormen. De Engelse overheid heeft daardoor ook grotere moeite om een beroep op abstracte principes te doen, bijvoorbeeld om onder verwijzing naar het voorzorgsbeginsel maatregelen te nemen die het bedrijfsleven met kosten belasten hoewel de noodzaak van de maatregel niet eenduidig wetenschappelijk is aangetoond. In tegenstelling tot Duitsland ontbrak in Engeland ook de juridische inbedding van het voorzorgsbeginsel. De Engelse *case law* traditie kent een inductieve logica: algemene principes komen voort uit concrete gevallen. Algemene principes staan niet aan de wieg van een nieuw beleidsveld. Het was mede daarom dat het Britse milieubeleid een minder interventionistisch karakter had. Milieueisen waren niet gebaseerd op de best beschikbare technologie, maar op de meest praktische optie. Daarbij werd in het bijzonder met de kosten voor het bedrijfsleven rekening gehouden. Normen waren ook minder gedetailleerd en werden minder rigide gehandhaafd dan bijvoorbeeld in Duitsland. Er was meer ruimte voor inspecteurs om over concrete maatregelen te onderhandelen. Dit werd mede mogelijk gemaakt door een relatief zwak ontwikkeld systeem van administratief beroep en de grote reputatie van civil (sic) servants.

Naast verschillen in staats- en rechtstradities waren er ook verschillen in belangen en instituties die tot een minder interventionistisch milieubeleid leidden. De Engelse milieuorganisaties richtten zich, overeenkomstig de Engelse conceptie van milieubeleid, meer op natuurbescherming dan op regulering van de industrie. Bovendien viel het recyclingbeleid onder het Ministerie van Handel en Industrie, een departement waartoe de milieubeweging minder toegang had dan tot het Ministerie van Milieu. Ook was de institutionele drempel die het Engelse districtenstelsel opwierp voor een groene partij praktisch prohibitief. Daarmee was er ook geen dwang voor de grote Engelse politieke partijen om met allerlei groene onderwerpen de kiezers bij de kleine partijen weg te halen. Omdat de Engelse milieu-industrie niet sterk ontwikkeld was, ontbrak ook de vraag vanuit het bedrijfsleven naar strenge eisen. Een andere belangrijke factor was de dominantie van het paradigma dat milieubeleid en economisch groei niet te verenigen zijn. Anders dan in Duitsland en Nederland was de idee van economische modernisering veel te zwak in Engeland vertegenwoordigd om een streng milieubeleid ook in tijden van economische recessies te legitimeren. De institutioneel gewaarborgde sterke rol van de centrale regering zorgde er bovendien voor dat de op economische groei gestoelde ideeën zonder effectieve veto's in beleid kon worden omgezet. De bovengenoemde factoren leidden tot lage normen voor afvalverwijdering en gebrekkige handhaving. Een gevolg hiervan was dat de kosten voor afvalverwijdering slechts tien procent waren van die in Duitsland in de jaren negentig. Engelse regeringen beschouwden mede daarom verpakkingsafval niet als een serieus politiek probleem. Zij waren bovendien doorgaans van mening dat de werking van de markt voldoende was om materiaal te besparen en afval te reduceren. Dit was echter niet geval, mede als gevolg van de lage stortkosten. Eind jaren tachtig steeg ook in Engeland het onbehagen over 'onnodige' verpakkingen; milieugroepen voerden steeds vaker actie tegen eenmalige verpakkingen, en de groene partij kreeg 15 procent van de kiezers achter zich bij de Europese verkiezingen van 1989.

Het Ministerie van Handel en Industrie, dat verantwoordelijk was voor recycling, wilde het bedrijfsleven echter niet met regelgeving belasten. Door deze competentieverdeling moest de Minister van Milieu vooral proberen via gemeenten recycling te stimuleren. De regering deed dit via een krediet-programma. Dit programma was echter vrij klein omdat de regering publieke uitgaven tot een minimum wilde beperken.

Tegenover de industrie stelde de overheid zich terughoudend op. Bedrijven werd weliswaar gevraagd het gebruik van onnodige verpakkingen te verminderen, maar de dreiging om de industrie te reguleren was niet erg overtuigend. Tegen de achtergrond van deregulering en de beperkte politisering van het probleem verpakkkingsafval, geloofde de industrie niet in overheidsoptreden. Het antwoord van het bedrijfsleven bestond dan ook slechts uit een vrijblijvend plan en een aantal gedragscodes. Naast het ongeloof in het overheidsoptreden was de matige reactie van het bedrijfsleven ook te wijten aan het zwakke zelfregulerende vermogen van het Engelse bedrijfsleven. Zelfregulering was bij verpakkingsafval bijzonder moeilijk omdat adequate oplossingen de medewerking van de hele verpakkingsketen vereisten. Engelse branche-organisaties waren echter niet goed in staat om belangen van meerdere sectoren te aggregeren en hun leden te binden aan afspraken met de overheid. Dit had ook de maken met de individualistische industriële cultuur in Engeland.

EUROPEANESERING VAN HET VERPAKKINGSBELEID

Na het bestuderen van het nationale beleid richt het onderzoek zich in *hoofdstuk 8 tot en met 11* op de interactie tussen het nationale en het Europese niveau. Het blijkt dat nationale verschillen in het verpakkingsafvalbeleid negatieve gevolgen hadden voor het vrije verkeer van goederen binnen de EG. Problemen werden in het bijzonder veroorzaakt door het Duitse beleid. Door een gebrek aan binnenlandse capaciteit voor het bereiken van de hoge recyclingdoelstellingen werden andere landen overspoeld met gesubsidieerd Duits verpakkingsafval, hetgeen de opbouw van hun eigen recyclinginfrastructuren belemmerde. Bovendien vonden vele bedrijven en regeringen dat het Duitse beleid buitenlandse bedrijven discrimineerde.

De Europese Verpakkingsrichtlijn

Het was onzeker of het Duitse beleid juridisch aan banden kon worden gelegd. Dit was een reden waarom de Europese Commissie met het voorstel voor een Europese Richtlijn kwam. De tweede reden was, dat het Directoraat voor Milieubescherming van de Commissie de mogelijkheid zag haar eigen verpakkingsafvalbeleid te intensiveren. De Europese Richtlijn werd een van de meest controversiële Europese wetsteksten ooit. De discussie draaide daarbij vooral om de strengheid van de milieunormen. In het eerste voorstel van de Commissie stond de bescherming van het milieu centraal en de milieueisen waren gebaseerd op het Duitse en het Nederlandse beleid. In het verloop van het beleidsproces werden onder druk van een coalitie van multinationale bedrijven, een aantal regeringen - onder andere Engeland - en een aantal afdelingen van de Europese Commissie - in het bijzonder die voor industriebeleid en voor de interne markt - de milieueisen sterk verlaagd. De drie koplopers van het Europese milieubeleid, Duitsland, Nederland en Denemarken, lukte het echter een uitzonderingspositie te bedingen om hun hoge eisen te kunnen handhaven. De grenzen die hun daarbij gesteld werden, waren echter niet duidelijk omschreven. De vage formulering diende ertoe om een compromis in de Europese Ministerraad te bereiken. Door deze en andere uitzonderingsclausules werd naast het doel van milieubescherming ook het harmonisatiedoel niet bereikt. Het grote verschil in nationale belangen en de informele regel tussen lidstaten elkaars essentiële belangen te respecteren, verhinderden deze harmonisatie. Ondanks het sterke compromiskarakter van de Richtlijn waren alle lidstaten aan een zekere druk onderhevig zich aan te passen, een druk waarvan de aard en omvang echter per lidstaat varieerden.

Engeland

Het grootst was de druk voor Engeland. Dat is nogal verrassend, omdat dit land, anders dan Duitsland en Nederland, voor de Richtlijn gestemd had. De druk had twee oorzaken. Ten eerste waren de - internationaal vergeleken vrij lage - minimumeisen van de Richtlijn (50 procent materiaal- en energierecycling, waarvan tenminste de helft materiaalrecycling) voor Engeland vrij ambitieus. Ten tweede, en belangrijker, stond de meest adequate manier om de Richtlijn te implementeren - algemeen dwingende, gedetailleerde en uniforme normen - haaks op de Engelse traditie van milieubeleid. Daarom had Engeland al voordat de Richtlijn was aangenomen geprobeerd alsnog een vrijwillig systeem van de grond te krijgen. Dit lukte niet. Sterker nog, bedrijven vroegen zelf om wetgeving om zwartrijden tegen te gaan.

De Engelse regering werd door de Europese Richtlijn gedwongen actiever te worden. Door onderhandeling met een kleine groep managers en met hulp van een onafhankelijk persoon van hoog aanzien lukte het na vele jaren van conflicten tot een wettelijk vastgelegde verdeelsleutel te komen voor de kosten die het bereiken van de minimumeisen van de Richtlijn met zich mee zouden brengen. Deze verdeling van de kosten werd echter niet door het gehele bedrijfsleven gesteund. Voor de conflictueuze gang van zaken en het niet verder gaan dan de minimumdoelstellingen van de Richtlijn zijn enkele oorzaken aan te wijzen. Ten eerste ontbrak een duidelijke houding van de regering. Zij had geen animo voor de milieudoelstelling van de Richtlijn. Milieu, afval en verpakkingen stonden niet hoog op de binnenlandse politieke agenda. Anders dan in Nederland en Duitsland had de Engelse overheid ook geen duidelijke, uniforme en gekwantificeerde eisen gesteld waaraan een systeem zou moeten voldoen. Dit zou namelijk haaks staan op de Engelse politieke tradities. Dit probleem werd nog versterkt door het feit dat voor dit soort eisen een eenduidige wetenschappelijke basis ontbrak. De verpakkingsketen moest dus zelf proberen zowel eisen vast te leggen als ook de verdeling van de kosten daarvan te regelen.

Bovendien werd, zoals ook al in het begin van de jaren negentig, duidelijk dat, anders dan in Nederland en Duitsland, geen slagvaardige organisaties bestonden die bedrijven uit alle schakels van de verpakkingsketen vertegenwoordigden en in staat waren geïntegreerde oplossingen te vinden en hun leden daaraan te binden. Weliswaar waren de meeste multinationale producenten van consumptiegoederen in een gemeenschappelijke regeling geïnteresseerd maar lieten verpakkingsproducenten en de detailhandel, de machtigste schakel in de keten, het massaal afweten. De individualistische Engelse cultuur was voor de tweede keer geen vruchtbare bodem voor een ketenoplossing.

Nederland

In Nederland was de druk om zich aan te passen geringer dan in Engeland, ondanks het feit dat Nederland tegen de Richtlijn had gestemd. Weliswaar lag Nederland al in 1995 boven de maximumdoelstellingen uit de Richtlijn, maar dit veroorzaakte geen problemen door de uitzonderingspositie die de Europese koplopers hadden bedongen. De aanpassingsdruk werd dan ook niet veroorzaakt door uiteenlopende milieueisen maar door een verschillend niveau van formalisering van beleid. Het Nederlandse Convenant voldeed niet aan de Europeesrechtelijke eisen voor de implementatie van Richtlijnen. Het moest een wettelijke regeling worden en dat veroorzaakte veel commotie in het bedrijfsleven, mede omdat het Convenant naar de mening van de industrie effectief was. In 1995 waren de tussendoelstellingen van het Convenant bereikt, waarbij de kosten veel lager waren dan in het Duitse geval.

Verdere formalisering leverde twee problemen op. Een redelijk effectief instrument, dat bovendien goed in de Nederlandse doelgroepenaanpak paste, moest worden vervangen. Bovendien impliceerde de komst van een algemeen bindende regeling dat het beleidsnetwerk groter moest worden. Niet alleen het bijzonder actieve deel van het bedrijfsleven, vertegenwoordigd in de SVM, maar het gehele bedrijfsleven moest meedoen. Naast de SVM kregen daarom ook de koepelorganisaties VNO-NCW en MKB-Nederland een centrale rol in het beleidsnetwerk, waarbij de SVM echter de primaire partner van de overheid bleef.

Voor de overheid en de bedrijven die bij het Convenant waren aangesloten, had de vervanging van het Convenant door een wettelijke regeling ook een aantal voordelen. Het was namelijk meer dan onzeker of de einddoelstellingen van het Convenant bereikt zouden worden. Het recyclingniveau schommelde al jaren rond de 50 procent (doelstelling voor 2001: 60%) en het was onwaarschijnlijk dat in het jaar 2000 een reductie van de hoeveelheid verpakkingen tot het niveau van 1986 tot stand gebracht zou zijn. De algemeen bindende regeling gaf nu de mogelijkheid om het beleid een bredere basis te geven, dus ook die bedrijven medeverantwoordelijk te maken die zich niet bij het Convenant hadden aangesloten. Bovendien bood de Richtlijn de kans om zonder gezichtsverlies de preventiedoelstelling van het Convenant Verpakkingen te verlagen.

Overheid en het bedrijfsleven waren echter zo gecharmeerd van het Convenant dat - op basis van de algemeen bindende regeling - een nieuw Convenant werd opgesteld. In het nieuwe Convenant werd een recyclingdoelstelling van 65 procent afgesproken. Dat is iets minder dan in Duitsland, maar 20 procentpunten meer dan het maximum van de Europese Richtlijn, en maar liefst 40 procentpunten meer dan in Engeland. Het Nederlandse bedrijfsleven hoeft weliswaar niet de hoeveelheid verpakkingen te reduceren maar wel de hoeveelheid verpakkingsafval voor storten en verbranden. Deze hoeveelheid moet teruggebracht worden tot het niveau van 1986.

Dat het bedrijfsleven met deze ambitieuze doelstellingen instemde heeft een drietal redenen. Ten eerste waren de partijen af van de reductiedoelstelling voor verpakkingen. Ten tweede was het Convenant op een aantal punten flexibeler dan de algemeen bindende regeling en ten derde was het twijfelachtig of de overheid de naleving van het nieuwe Convenant hard zou afdwingen.

Duitsland

Duitsland hoefde zich op papier van de drie landen het minst aan te passen. Dat is nogal opmerkelijk, omdat juist het Duitse systeem begin jaren negentig de nek had uitgestoken, en de gevolgen ervan een reden waren geweest voor het ontwikkelen van de Europese Richtlijn. De lage druk om zich aan te passen had twee redenen. Ten eerste kwam de legalistische aanpak van de EU goed overeen met de Duitse tradities. De op verboden en geboden gefundeerde Duitse aanpak was geschikt voor de implementatie van EU-Richtlijnen. Ten tweede kon Duitsland zich net als Nederland beroepen op een uitzonderingspositie wat de recyclingeisen betrof. Alleen de Duitse regeling dat 72 procent van alle drankverpakkingen op de Duitse markt hervulbaar moest zijn lag onder vuur.

Anders dan in Engeland waren in Duitsland binnenlandse factoren de aanleiding voor veranderingen in het verpakkingsafvalbeleid. Het was tamelijk onomstreden dat het systeem effectief werkte. Alle recyclingdoelstellingen werden gehaald en anders dan in Engeland en Nederland werd de hoeveelheid verpakkingen die jaarlijks op de binnenlandse markt kwam substantieel verminderd: rond de 13 procent tussen 1992 en 1997. De grootste reductie vond echter in de eerste twee jaren plaats.

Het *Duale System* werd echter van het begin af aan bekritiseerd. Zijn succes werd een probleem. Er werden veel meer verpakkingen ingezameld dan van tevoren was ingecalculeerd. De hoge recyclingeisen, en in het bijzonder het verbod op thermische recycling, kostten het bedrijfsleven en burgers vele miljarden marken en leidden bovendien tot - ten dele illegale - afvalexporten naar andere landen, onder andere China, omdat er niet voldoende recyclingcapaciteit in het binnenland bestond. Tegenover de hoge kosten stonden lage inkomsten. Vele bedrijven sloten zich niet aan bij het systeem of betaalden te weinig. Het systeem was na twee jaar bijna failliet.

Het eerste ontwerp voor een herziene verpakkingsverordening dateerde al van 1993. Het was enkel bedoeld om het *Duale System* voor de ondergang te behoeden: de recyclingeisen zouden worden afgezwakt en andere maatregelen zouden zwartrijden voorkomen. Door uiteenlopende belangen en een complexe institutionele situatie liep het beleidsproces vervolgens vast. De meerderheid van de Duitse deelstaten, geregeerd door sociaal-democraten, was tegen een verlaging van de recyclingeisen. Zij wilden ook niets weten van een afzwakking van de regel dat 72 procent van de drankverpakkingen op de Duitse markt hervulbaar moest zijn. Dit quotum werd echter door de Europese Commissie, een groot aantal andere lidstaten en grote delen van het bedrijfsleven bekritiseerd. Omdat de meerderheid van de Duitse deelstaten een veto in de Bondsraad hadden zat de conservatieve Duitse bondsregering in de val. Het op coöperatie gebaseerde Duitse federale systeem werd door het op competitie georiënteerde partijenstelsel ondermijnd.

Een compromis werd pas 22 maanden na de officiële termijn voor de implementatie van de Richtlijn gevonden. Het 72 procentsquotum werd enigszins afgezwakt en de recyclingdoelstellingen werden minder sterk verlaagd dan oorspronkelijk gepland. Zij zijn echter nog steeds hoger dan in Nederland en Engeland. Of de Europese Commissie en de andere lidstaten met dit compromis genoegen nemen was in de zomer van 1998 nog niet duidelijk. Ook twee jaar na de officiële termijn was de Europese Richtlijn in Duitsland nog niet geïmplementeerd.

CONCLUSIE

Hoofdstuk 12 gaat terug naar de centrale vraagstelling: heeft de Europese integratie tot convergentie in het Duitse, Nederlandse en Engelse verpakkingsafvalbeleid geleid? Uit het onderzoek komt naar voren dat de Europese integratie verschillende gevolgen heeft gehad voor de diverse dimensies van beleid. De vergevorderde integratie van Europese markten heeft geleid tot Europeanisering van het verpakkingsbeleid. Er heeft convergentie plaatsgevonden in de vorm van het beleid. Nederland en Engeland werden om Europeesrechtelijke redenen gedwongen hun beleid net als het Duitse systeem op algemeen bindende, uniforme en gedetailleerde geboden en verboden te baseren. Zonder de Europese Richtlijn zou dit convergentieproces niet hebben plaatsgevonden.

De Europese integratie heeft echter niet geleid tot convergentie in de doelstellingen van het verpakkingsbeleid. Duitsland en Nederland benutten hun hard bevochten mogelijkheid om strenge milieunormen te blijven handhaven, terwijl Engeland achteraan blijft lopen. Nationale belangen waren hier sterker dan de Europese harmonisatiekrachten.

De Europese integratie heeft ook niet geleid tot convergentie bij de verdeling van taken tussen overheid en bedrijfsleven. Gelegitimeerd door het subsidiariteitsbeginsel werd ook door niemand serieus overwogen, om vraagstukken omtrent de organisatie van verpakkingsafvalsystemen op Europees niveau vast te leggen. Men had echter kunnen verwachten dat door beleidsgericht leren lidstaten tot een bepaald systeem zouden convergeren. Dit was echter niet het geval. Er is een diversiteit blijven bestaan omdat in Duitsland het oude beleid tot dermate sterk gevestigde belangen en *sunk costs* heeft geleid dat er moeilijk van dit eenmaal ingeslagen pad kan worden afgeweken. Om deze reden blijft de Duitse industrie verantwoordelijk voor het inzamelen van verpakkingsafval, ondanks dat dit systeem hoge kosten met zich meebrengt en een twijfelachtig milieurendement oplevert.

De Europese integratie heeft een indirecte invloed gehad op de manier waarop economische en maatschappelijke belangen bij het beleidsproces worden betrokken. Dit heeft echter niet geleid tot convergentie. Toenemende marktintegratie leidde er in Duitsland toe dat regeringen van andere EU-lidstaten en bedrijven via Europese institutionele kanalen sterker poogden om invloed uit te oefenen op het Duitse beleid. In Nederland leidde de plicht om tot een algemeen bindende regeling te komen tot een uitbreiding van het netwerk, dat echter nog steeds corporatistische trekken heeft. In Engeland werd onder druk van de Richtlijn het beleidsnetwerk verkleind om tenminste op papier tot een compromis te komen. Belangenverenigingen hebben in Duitsland en Nederland nog steeds een centrale rol en individuele bedrijven zijn nog steeds de belangrijkste onderhandelingspartner voor de Engelse regering. Dat over het algemeen milieugroepen nu minder participeren in het beleidsproces heeft minder met de Europeanisering van het beleid te maken dan met nieuwe prioriteiten van de milieubeweging. De aantrekkingskracht van verpakkingen als symbool voor onze wegwerpmaatschappij nam af naarmate de uitkomsten van levenscyclusanalyses niet meer eenduidig het voordeel van hervulbare verpakkingen aantoonden.

Er is ook diversiteit blijven bestaan in de houding van de overheid tegenover het bedrijfsleven. De Europese integratie heeft nationale tradities op dit punt niet kunnen aantasten. Duitsland is nog steeds relatief conflictgeoriënteerd; de dreiging van strengere maatregelen blijft op de agenda. De Nederlandse overheid is nog steeds net als de Engelse regering consensusgeoriënteerd.

De door Europa geïnduceerde convergentie naar een op uniforme en wettelijke normen gebaseerd verpakkingsafvalbeleid wijst in de richting van het ontstaan van een specifiek coördinatie- en reguleringsmechanisme (*governance*) in Europa: zelfregulering binnen strenge en algemeen bindende wettelijke kaders. Deze ontwikkeling is ook voor andere beleidsterreinen aangetoond. Dit impliceert een zeker verandering van de nationale tradities en industriële culturen, vooral in Engeland maar in mindere mate ook in Nederland. Nationale verschillen bij de invulling van het systeem komen echter nog duidelijk naar voren. In Duitsland domineren, afgezien van de groene punt als economisch coördinatiemiddel, contractuele relaties met het *Duale System*, dat praktisch een monopoliepositie heeft (juridische sturing). In Engeland worden de recyclingplichten via certificaten op een kunstmatige markt verhandeld (economische sturing) en in Nederland is weliswaar een wettelijke regeling aangenomen maar in de context van een nieuw Convenant blijft men waarde hechten aan informele samenwerking en communicatieve sturing. De traditionele beleidsstijlen zouden hun kop ook weer kunnen opsteken als het om de handhaving van de regels gaat. Het blijft afwachten of de Nederlandse en de Engelse regeringen dan minder onderhandelen en minder pragmatisch zijn dan in het verleden.

14. References

Aalders, M.V.C. (1993) 'Het milieuconvenant wordt "salonfaehig". Convenanten "nieuwe stijl" in het milieubeleid', in Bressers, J.Th.A., P.Jong, P.-J. Klok and A.F.A. Korsten (eds), *Beleidsinstrumenten bestuurskundig beschouwd*, Assen and Maastricht: Van Gorcum, pp. 75-92.

Afval Overleg Orgaan, (1995) *Tienjarenprogramma Afval*, Utrecht: AOO.

Afval Overleg Orgaan (1996) 'Ministeriële regeling verpakkingen een formaliteit?', *Informatiebulletin Preventie en Hergebruik*, Maart 1996, pp. 8-9.

Arbeitsgemeinschaft Verpackung und Umwelt (1990) *Umweltstrategien der Verpackungswirtschaft. Langfristkonzept der AGVU*, Bonn: AGVU.

Arp, H. (1993) 'Technical regulation and politics: the interplay between economic integration and environmental policy goals in the EC car emission legislation', in Liefferink, J.D., P.D. Lowe and A.P.J. Mol (eds), *European Integration & Environmental Policy*, London and New York: Belhaven, pp. 150-171.

Arthur, W.B. (1989) 'Competing technologies, increasing returns, and lock-in by historical events', *Economic Journal*, vol. 99, pp. 119-131.

Ashby, E.and M. Andersen (1981) *The Politics of Clean Air*, Oxford: Clarendon Press.

Axelrod, R. (1976) 'The cognitive mapping approach to decision making', in Axelrod, R. (ed.), *Structure of Decisions*, Princeton, N.J: Princeton UP.

Badaracco, J.L. (1985) *Loading the dice. A five country study of vinyl chloride regulation*, Boston: Harvard Business School Press.

Baggot, R. (1989) 'Regulatory Reform in Britain: The Changing Face of Self-Regulation', *Public Administration*, vol. 67, pp. 435-454.

Bartlsperger, R. (1995) 'Die Entwicklung des Abfallrechts in den Grundfragen von Abfallbegriff und Abfallregime', *Verwaltungsarchiv*, vol. 86, no. 1, pp. 32-68.

Bell, D. (1973) *The coming of post-industrial society*, New York: Basic Books.

Bekkers, V., J. Bonnes, A. de Moor-van Vught, P. Schoneveld and W. Voermans (1995) 'The case of The Netherlands', in Pappas, S.A. (ed.), *National administrative procedures for the preparation and implementation of Community decisions*, Maastricht: European Institute of Public Administration.

Bennet, C.J. (1991) 'Review Article: what is policy convergence and what causes it', *British Journal of Political Science*, vol. 21, no. 2, pp. 215-234.

Benz, A. (1997) 'Von der Konfrontation zur Differenzierung und Integration - Zur neueren Theorieentwicklung in der Politikwissenschaft', in Benz, A. and W. Seibel (eds), *Theorieentwicklung in der Politikwissenschaft - eine Zwischenbilanz*. Baden-Baden: Nomos, pp. 9-29.

Berger, S. and R. Dore (eds), (1996) *National Diversity and Global Capitalism*, Ithaca: Cornell UP.

Blankenburg, E. (1997) *Patterns of Legal Culture: The Netherlands Compared to Neighboring Germany*, Amsterdam: Duitsland Instituut, UvA. Duitsland Cahier 1/1997

Blankenburg, E. and F. Bruinsma (1994) *Dutch Legal Culture,* Deventer and Boston: Kluwer Law and Taxation Publisher.

Bletz, J. (1996) 'Minimale bewegingsvrijheid voor verpakkingsindustrie', *Intermediair*, vol. 32, no. 25, 21.6.1996, pp. 39-41.

Boehmer-Christiansen, S. and J. Skea (1991) *Acid Politics: Environmental and Energy Policy In Britain and Germany*, London and New York: Belhaven Press.

Bohne, E. (1982) 'Absprachen zwischen Industrie und Regierung in der Umweltpolitik', in Gessner, V. and G. Winter (eds), *Rechtsformen der Verflechtung von Staat und Wirtschaft.* Opladen: Westdeutscher Verlag, pp. 266-281.

Bovens, M. and P. t'Hart (1996) *Understanding Policy Fiascoes,* New Brunswick and London: Transaction.

Bressers T.A. and L.A. Plettenburg (1997) 'The Netherlands', in: Jänicke, M. and H. Weidner (eds), *National Environmental Policies. A Comparative Study of Capacity-Building*. Berlin: Springer, pp. 109-131.

Brettschneider, F., K. Ahlstich, and B. Zügel (1992) 'Materialien zu Gesellschaft, Wirtschaft und Politik in den Mitgliedstaaten der Europäischen Gemeinschaft', in Gabriel, O.W (ed.), *Die EG Staaten im Vergleich. Strukturen, Prozesse, Politikinhalte. Opladen: Westdeutscher Verlag*, pp. 433-637.

Brickman, R., S. Jasanoff, and T. Ilgen (1985) *Controlling Chemicals: The Politics of Regulation in Europe and the United States*, Ithaca and London: Cornell UP.

Bridge, J. (1981) 'National legal tradition and Community law. Legislative drafting and judicial interpretation in England and the European Community', *Journal of Common Market Studies*, vol. 4, pp. 351-376.

British Retail Consortium (1993) *Guidance Notes on Retail Packaging,* Update. London: BRC.

Brückner, C. and G. Wiechers (1985) 'Umweltschutz und Ressourcenschonung durch eine ökologische Abfallwirtschaft', *Zeitschrift für Umweltpolitik*, vol. 8, no. 2, pp. 153-180.

Bünemann, A. and G. Rachut (1993) *Der Grüne Punkt. Eine Versuchung für die Wirtschaft,* Karlsruhe: C.F. Müller.

Bundesministerium für Umwelt (1990) *Umweltbericht 1990*, Bonn: Bundesanzeiger.

Bundesrat (1997) 'Ausschußempfehlungen', Bundesrats-Drucksache 518/97.

Bundesregierung der BRD (1996a) 'Entwurf einer Verordnung ü/ber die Vermeidung und Verwertung von Verpackungsabfällen', unpublished document, 28.10.1996.

Bundesregierung der BRD (1996b) 'Mitteilung der Regierung der Bundesrepublik Deutschland an die Kommission der Europäischen Gemeinschaften vom 29.4.1996', unpublished document.

Bundesregierung der BRD (1997) Veränderungen im Novellierungsentwurf VerpackV vom, 12.5.1997, unpublished document.

Carter, N. and P. Lowe (1995) 'The Establishment of a Cross-Sector Environment Agency', in Gray, T.S. (ed.), *UK Environmental Policy in the 1990s*, Houndsmills and London: MacMillan Press, pp. 38-56.

Cawson, A. (1985) 'Conclusion: some implications for state theory', in Cawson, A. (ed.), *Organized interests and the state. Studies in meso-corporatism*, London: Sage.

Centruum voor Milieurecht (1991) Letter to the SNM, 6.5.1991, Amsterdam: Universiteit van Amsterdam.

Checkel, J.T. (1998) 'Social construction, institutional analysis and the study of European Integration', paper presented at the ECPR Joint Sessions of Workshops, University of Warwick, March 1998.

Chilcote, R.H. (1994) *Theories of comparative politics. The search for a paradigm reconsidered*, Boulder, San Francisco, Oxford: Westview Press.

Cini, M. (1995) 'Administrative Culture in the European Commission: the case of competition and environment', paper presented at the fourth biennal Community Studies Association (ECSA) conference, Charleston, USA, 11-14.7.1995.

Commissie Verpakkingen (1993) *Jaarverslag Commissie Verpakkingen*, Utrecht: Commissie Verpakkingen.

Commissie Verpakkingen (1994) *Jaarverslag Commissie Verpakkingen*, Utrecht: Commissie Verpakkingen.

Commissie Verpakkingen (1994a) *Toetsing Milieu-analyses*, Utrecht: Commissie Verpakkingen.

Commissie Verpakkingen (1995) *Jaarverslag Commissie Verpakkingen*, Utrecht: Commissie Verpakkingen.

Commissie Verpakkingen (1996) *Jaarverslag Commissie Verpakkingen*, Utrecht: Commissie Verpakkingen.

Commissie Verpakkingen (1997) *Jaarverslag Commissie Verpakkingen*, Utrecht.

Commission of the European Communities (1992) *Fifth Environmental Action Programme*, COM (92) 23.

Commission of the European Communities (1994a) 'Letter to the Duales System Deutschland', 3.2.1994, Brussels.

Commission of the European Communities (1996a) *Communication from the Commission to the Council and the European Parliament on Environmental Agreements*, 27.11.1996, COM (96) 561 final, Brussels.

Coopers & Lybrand (1993) *A Survey of English Local Authority Recycling Plans,* London: HMSO.

Cramer, J. (1989) *De groene golf. Geschiedenis en toekomst van de Nederlandse milieubeweging*, Utrecht: Jan van Arkel.

Czada, R. (1991) 'Regierung und Verwaltung als Organisatoren gesellschaftlicher Interessen', in Hartwich, H.-H., and G. Wewer (eds), *Regieren in der Bundesrepublik III. Systemsteuerung und Staatskunst*, Opladen: Leske & Budrich, pp. 151-173.

Daalder, H. (1966) 'The Netherlands: opposition in a segmented society' in Dahl, R.A. (ed.), *Political opposition in Western democracies*, New Haven, Conn.: Yale University Press, pp. 188-236.

Daintith, T. (1995) 'Introduction', in Daintith, T. (ed.), *Implementing EC Law in the United Kingdom. Structures for Indirect Rule*, Chichester: Wiley, pp. 1-23.

Dam, P. en E.van Oevelen (1993) 'De polycarbonaatfles mogelijk goed alternatief? Het pak of de fles', *BNI*, Lente 1993, pp. 5-7.

Department of the Environment (1995) *Producer Responsibility for Packaging Waste-A Consultation Paper*, 18.8.1995, London: HMSO.

Department of the Environment (1996a) *The Producer Responsibility (Packaging Waste) Obligations Regulations. A consultation paper*, Annexe 1. London: HMSO.

Department of the Environment (1996b) *Making Waste Work. A strategy for sustainable waste management in England and Wales*, London: HMSO.

Deth, J.W. van, and J.C.P.M. Vis (1995) *Regeren in Nederland. Het politieke en bestuurlijke bestel in vergelijkend perspectief,* Assen: Van Gorcum.

Deutsche Gesellschaft für Kunststoffrecycling (1996) *Recycling von Verkaufsverpackungen aus Kunststoffen*, Köln: DKR.

Döhler, M. (1990) *Gesundheitspolitik nach der 'Wende': Policy-Netzwerke und ordnungspolitischer Strategiewechsel in Grossbritannien, den USA und der Bundesrepublik Deutschland*, Berlin: Sigma.

Dogan, M. and D. Pelassy (1990) *How to Compare Nations?*, second edition, Chatham, NJ: Chatham House Publishers

Donner, H. and U. Meyerholt (1995) 'Die Entwicklung des Abfallrechts von der Beseitigung zur Kreislaufwirtschaft', *Zeitschrift für Umweltpolitik*, vol. 18, no. 1, pp. 81-99.

Douma, W. T. (1995) 'De EG-Richtlijn Verpakking en Verpakkingsafval', *Milieu & Recht*, no. 6, June 1995, pp. 108-112.

Dreier, V. (1997) *Empirische Politikforschung*, München and Wien: Oldenbourg Verlag.

Duales System Deutschland (1994) *Geschäftsbericht 1993*, Köln: DSD.

Duales System Deutschland (1995) *Geschäftsbericht 1994*, Köln: DSD.

Duales System Deutschland (1996) *Geschäftsbericht 1995*, Köln: DSD.

Duales System Deutschland (1996a) *Recycling Data - Techniques and Trends*, Köln: DSD.

Duales System Deutschland (1997) *Geschäftsbericht 1996*, Köln: DSD.

Duales System Deutschland (1998) *Mengenstromnachweis 1997*, Köln: DSD.

Dunleavy, P. (1989) 'The United Kingdom. Paradoxes of Ungrounded Statism', in Castles, F.G. (ed.), *The Comparative History of Public Policy*, Cambridge: Polity Press, pp. 242-291.

Dunlop, R.E, G.H. Gallup and A.M. Gallup (1992) *Health of the Planet Survey*, New Jersey: Gallup International Institute.

Dyson, K. (1983) 'Cultural, ideological and structural Context', in Dyson, K. and S. Wilks (eds), *Industrial crisis*, Oxford: Martin Robinson.

Dyson, K. (1992) 'Theories of Regulation and the Case of Germany: A Model of Regulatory Change,' in Dyson, K. (ed) *The Politics of German Regulation*, Bodmin, Cornwall: Hartnolls, pp. 1-28.

Eberg, J. (1997) *Waste Policy and Learning. Policy Dynamics of Waste Management and Waste Incineration in the Netherlands and Bavaria,* Delft: Eburon.

Eichener, V. (1993) 'Social Dumping or Innovative Regulation? Processes and Outcomes of European Decision-Making in the Sector of Health and Safety at Work Harmonization', Florence: European University Institute, EUI Working Paper SPS no. 92/28.

Eichener, V. (1996) 'Die Rückwirkung der europäischen Integration auf nationale Politikmuster', in Jachtenfuchs, M. and B.Kohler-Koch (eds), *Europäische Integration*, Opladen: Leske & Budrich, pp.249-280.

Eising, R. (forthcoming) *Die Liberalisierung der Elektrizitätsversorgung in der Europäischen Gemeinschaft.*

Ellwein, T. and J.J. Hesse (1987) *Das Regierungssystem der Bundesrepublik Deutschland*, sixth edition, Opladen: Westdeutscher Verlag.

Epema-Brugman, M. (1996) 'The Packaging Covenant', *Newsletter of the Institute for Applied Economics*, no. 3, p.1.

Evans, P.B., D. Rueschemeyer, and T. Skocpol (1985) 'On the road toward a more adequate understanding of the state', in Evans, P.B, D. Rueschemeyer and T. Skocpol (eds), *Bringing the State Back In*, Cambridge: Cambridge UP, pp 347-366.

Faure, A.M. (1994) 'Some methodological problems in comparative politics', *Journal of Theoretical Politics*, vol. 6, no. 3, pp. 307-322.

Feess-Dörr, E., U. Steiger and P. Weihrauch (1991) 'Strategien zur Reduktion der Umweltbelastungen durch Einwegverpackungen', *Zeitschrift für Umweltpolitik und Umweltrecht*, no. 4, pp. 349-378.

Feick, J. (1992) 'Comparing comparative policy studies - a path towards integration?', *Journal of Public Policy*, vol. 12, pp. 257-285.

Fernhout, R. (1980) 'Incorporatie van belangengroeperingen in de sociale en economische wetgeving', in Verhallen, H.J.G., R. Fernhout, and P.E. Visser (eds), *Corporatisme in Nederland. Belangengroepen en democratie*, Alphen a.d. Rijn and Brussels: Samsom, pp. 119-228.

Financial Times (1996) 'European Results Analysis. A Tour of European Industry in 20 sectors', *Financial Times*, 17.6.1996, pp. 25-27.

Fischer, F., and J. Forrester (eds), (1993) *The argumentative turn in policy analysis and planning*, Durham: Duke University Press.

Fischer, M. (1993) 'Müllberge: Europäische Wegwerf-Gemeinschaft?', *EG-Magazin*, no. 6, pp. 30-32.

Freeman, G.P. (1985) 'National policy styles and policy sectors: explaining structured variation', *Journal of Public Policy*, pp. 467-496.

Friends of the Earth (1992) 'Bring Back the "Bring Back", Briefing Sheet.

Galbraith, J.K. (1967) *The new industrial state*, New York: Mentor.

Gandy, M. (1994) *Recycling and the politics of urban waste*, London: Earthscan.

Geelhoed (1997) 'Economische vooruitzichten stemmen niet tot somberheid', *Staatscourant*, 3 January 1997.

Gehring, T. (1995) 'Die EG-Umweltpolitik im Spannungsfeld von Umweltschutz und Binnenmarktpolitik', paper presented at the conference 'Regulative Politik in Europa' of the Arbeitskreis Europäische Integration of the Deutschen Vereinigung für Politische Wissenschaft, Köln, December 1995.

Gehring, T. (1996) 'Environmental policy in the European Union. Governing in nested institutions and the case of packaging waste', Florence: European University Institute, EUI Working Paper RSC no. 96/63.

Gehring, T. (1997) 'Governing in nested institutions: environmental policy in the European Union and the case of packaging waste', *Journal of European Public Policy*, vol.4, no.3, pp. 337-354.

Genschel, P. and T. Plümper (1997) 'Regulatory competition and international co-operation', *Journal of European Public Policy*, vol. 4, no. 4, pp. 626-642.

Geradin, D. (1993) 'Free trade and environmental protection in an integrated market: A survey of the case law of the United States Supreme Court and the European Court of Justice', *Journal of Transnational Law & Policy*, vol. 2, pp. 144-197.

Giddens, A. (1990) *The Consequences of Modernity*, Cambridge: Polity Press.

Gladdish, K. (1991) *Governing from the Centre. Politics and Policy-Making in the Netherlands*, London: Hurst & Company.

Golding, A. (1992) 'Europa kann einpacken. Die geplante EG-Verpackungsdirektive läßt der Industrie freie Hand', *Müllmagazin*, no. 2, pp. 7-9.

Goldstein, J. and R.O. Keohane (eds), (1993) *Ideas and Foreign Policy. Beliefs, Institutions, and Political Change*, Ithaca and London: Cornell UP.

Goldthorpe, J.H. (1984) 'The End of Convergence: Corporatist and Dualist Tendencies in Modern Western Societies', in Goldthorpe, J,H. (ed.), *Order and Conflict in Contemporary Capitalism*, Oxford: Clarendon Press.

Golub, J. (1996) 'State power and institutional influence in European integration: lessons from the Packaging Waste Directive', *Journal of Common Market Studies*, vol. 34, no. 3, pp. 314-339.

Golub, J. (1997) *The Path to EU Environmental Policy: Domestic Politics, Supranational Institutions, Global Competition*, paper presented at the fifth Biennal International Conference of the European Community Studies Association, 29.5-1.6., Seattle, Washington.

Goverde, H.J.M. (1993) 'Verschuivingen in het milieubeleid: van milieuhygiëne naar omgevingsmanagement', in Godfroij, A.J.A. and N.J.M. Nelissen (eds), *Verschuivingen in de besturing van de samenleving*, Bussum: Coutinho, pp. 49-88.

Grande, E. (1989) *Vom Monopol zum Wettbewerb? Die neokonservative Reform der Telekommunikation in Großbritannien und der Bundesrepublik Deutschland*, Wiesbaden: Deutscher Universitätsverlag.

Grande, E. (1996) 'Das Paradox der Schwäche. Forschungspolitik und die Einflußlogik europäischer Politikverflechtung', in Jachtenfuchs, M. and B. Kohler-Koch (eds), *Europäische Integration*, Opladen: Leske & Budrich, pp. 373-399.

Grant, W. (1993) *Business and politics in Britain,* second edition, Houndsmill and London: MacMillan

Gray, T.S. (1995) 'Introduction', in Gray, T.S. (ed.), *UK environmental policy in the 1990s*, Houndsmill and London: MacMillan, pp. 1-11.

Gronden, J.W. van de (1997) 'Convenanten als verpakking van het EG-milieurecht', in Hessel, B., G.H. Hagelstein, K. Hellingmann, P.C. Ippel and R.J.G.M. Widdershoven (eds), *Het recht over de schutting*, Nijmegen: Ars Aegui Libri, pp. 175-188.

Grove-White, R. (1994) 'Großbritannien', in Hey, C. and U. Brendle (eds), *Umweltverbände in der EG. Strategien, Politische Kulturen und Organisationsformen*, Opladen: Westdeutscher Verlag, pp. 171-214.

Haas, P.M. (1992) 'Introduction: Epistemic Communities and International Policy Coordination', *International Organization*, vol. 46, no. 1, 1-35.

Hague, R., M. Harop, and S. Breslin (1992) *Comparative government and politics,* third edition, Houndsmill and London: MacMillan.

Haigh, N. (1990) *EEC environmental policy and Britain,* second revised edition, Burnt Mill: Longman.

Haigh, N. and S. Mullard (1993) 'UK National Report', in Institute for European Environmental Policy(ed.) *Packaging waste. The proposed EC Directive and the existing schemes in five EC member states*, London: Institute for European Environmental Policy.

Hajer, M.A. (1995) *The Politics of Environmental Discourse: A Study of the Acid Rain Controversy in Great Britain and The Netherlands*, Oxford: Oxford UP.

Hall, P. (1986) *Governing the Economy. The Politics of State Intervention in Britain and France*, New York: Oxford UP.

Hall, P. (1992) 'The movement from Keynesianism to monetarism: institutional analysis and British economic policy in the 1970s', in Steinmo, S., K. Thelen, and F. Longstreth (eds), *Structuring Politics. Historical Institutionalism in Comparative Analysis*, Cambridge: Cambridge UP, pp. 90-113.

Hall, P. (1993) 'Policy paradigms, social learning and the state: The case of economic policymaking in Britain', *Comparative Politics*, pp. 275-296.

Hall, P.A. and C.R. Taylor (1996) 'Political science and the three new institutionalisms', Köln: Max-Planck-Institut für Gesellschaftsforschung, MPFIG Discussion Paper 96/6 (also published in Political Studies, 1996, vol. 44).

Hanf, K. (1989) 'Deregulation as regulatory reform: the case of environmental policy in the Netherlands', *European Journal of Political Research*, vol. 17, pp. 193-207.

Hartkopf, G. and E. Bohne (1983) *Umweltpolitik, Band 1, Grundlagen, Analysen und Perspektiven*, Opladen: Westdeutscher Verlag.

Hartkopf, G. (1986) 'Ein ziemlich wilder Haufen. Für ihre eigenen Ziele finanzierten Bonner Spitzenbeamte grüne Lobbyisten', *Die Zeit*, 14.2.1996, p. 28.

Hayward, J.E.S. (1974) 'National attitudes for planning in Britain, France and Italy', *Government and Opposition*, vol. 9, no. 4, pp. 397-410.

Heclo, H. (1974) *Modern social politics in Britain and Sweden*, New Haven: Yale UP.

Hemerijck, A.C. and M. Verhagen (eds), (1994) 'Themanummer: De institutionele factor: De maatschappelijke imbedding van de economie', *Beleid & Maatschappij*, vol. 21, September/ October.

Hennessy, P. (1990) *Whitehall*, London: FontanaPress.

Henselder-Ludwig, R. (1992) *Verpackungsverordnung. Textausgabe mit einer Einführung, Anmerkungen und ergänzenden Materialien*, Köln: Bundesanzeiger.

Héritier, A., S. Mingers, C. Knill, and M. Becka (1994) *Die Veränderung von Staatlichkeit in Europa. Ein regulativer Wettbewerb: Deutschland, Großbritannien und Frankreich in der Europäischen Union*, Opladen: Leske & Budrich.

Héritier, A., C. Knill, and S. Mingers (1996) *Ringing the changes in Europe. Regulatory competition and the redefinition of the state. Britain, France, Germany*, Berlin and New York: Walter de Gruyter.

Her Majesty's Government (1990) *This Common Inheritance*, London: HMSO, Cm 1200.

Her Majesty's Government (1990a) *This Common Inheritance (Summary)*, London. HMSO.

Herweijer, C. (1996) 'Op weg naar een tweede convenant verpakkingen', *Recycling*, November/December 1996, pp. 32-33.

Heukels, T. (1993) 'Alternatieve implementatietechnieken en art. 189 lid 3 EEG: grondslagen en ontwikkelingen', *Nederlandse Tijdschrift voor Bestuursrecht*, pp. 59-74.

Heuvelhof, E.F ten *et al.* (1992) *Processtandaard voor de uitvoering van de milieu- en markt-economische analyses zoals vastgelegd in het Convenant Verpakkingen*, Rotterdam: Universiteit Rotterdam.

Hey, C. and U. Brendle (1994) *Umweltverbände und EG*, Opladen: Westdeutscher Verlag.

Heijden, A. van (1994) 'Niederlande', in Hey, C. and U. Brendle (eds), *Umweltverbände und EG*, Opladen: Westdeutscher Verlag, pp. 215-276.

Hollingsworth, J.R. and R. Boyer (eds), (1997) *Comparing capitalism: The embeddedness of institutions*, New York: Cambridge UP.

Hoogerwerf, A. (1995) 'Instituties en (ont-)institutionalisering: Kanttekeningen bij het neo-institutionalisme', *Beleidswetenschappen*, no. 2, pp. 93-109.

Houben, T. and P. Leroy (1993) *Vlammende besluiten. Een beleidsevaluatief onderzoek naar de locatie van afvalverbrandingsinstallaties*, Nijmegen: Katholieke Universiteit Nijmegen.

House of Lords (1993) *House of Lords Select Committee on the European Communities. Packaging and Packaging Waste, with evidence*, London: HMSO, HL paper 118.

Hucke, J. (1992) 'Umweltschutzpolitik', in Schmidt, M.G. (ed) *Politisches System und Politikfelder westlicher Industrieländer*, München: Piper, pp. 440-447.

Immergut, E. (1992) *Health Politics. Interests and Institutions in Western Europe*, Cambridge: Cambridge UP.

Immergut, E. (1994) 'Historical Approaches to Public Policy', paper presented at the second workshop 'Politik- und Verwaltungswissenschaftliche Theorieentwicklung' of the Deutsche Vereinigung für politische Wissenschaft, 13-14 January 1994, Konstanz.

Immergut, E. (1997) 'The normative roots of the new institutionalism: Historical-institutionalism and comparative policy studies', in Benz, A. and W. Seibel (eds), *Theorieentwicklung in der Politikwissenschaft - eine Zwischenbilanz*, Baden-Baden: Nomos, pp. 325-355.

Industry Council of Packaging and the Environment (1995) *Doing more with less,* London: INCPEN.

Inglehart, R. (1995) 'Public support for environmental protection: The impact of objective problems and subjective values in 43 societies', *Political Science and Politics*, vol. 28, no. 1, pp. 57-72.

Ingram, V. (1997) 'Case study packaging waste', unpublished manuscript.

Institut der deutschen Wirtschaft (1992) *Der Grüne Punkt*, Köln: Deutscher Institutsverlag.

Institute of Grocery Distribution (1992) *An Integrated Approach to Environmentally and Economically Sustainable Waste Management. The Adur Project,* London: IGD.

Jachtenfuchs, M. (1993) 'Ideen und Interessen. Weltbilder als Kategorien der politischen Analyse', Mannheim: Mannheimer Zentrum für Sozialforschung, Arbeitspapier AB II, Nr. 2.

Jachtenfuchs, M. and B. Kohler-Koch (1995) 'The transformation of governance in the European Union', paper presented at the Fourth Biennal Conference of the European Community Studies Association, Charleston, May 1995.

Jachtenfuchs, M. and B. Kohler-Koch (eds), (1996) *Europäische Integration*. Opladen: Leske & Budrich.

Jachtenfuchs, M., C. Hey, and M. Strübel (1993) 'Umweltpolitik in der Europäischen Gemeinschaft', in Prittwitz, V. von (ed.), *Umweltpolitik als Modernisierungsprozeß*, Opladen: Leske & Budrich, pp. 137-162.

Jänicke, M. (1990) 'Erfolgsbedingungen von Umweltpolitik im internationalen Vergleich', in *Zeitschrift für Umweltpolitik und Umweltrecht*, vol. 13, pp. 213-232.

Jänicke, M. (1993) 'Ökologische und politische Modernisierung in entwickelten Industriegesellschaften', in Prittwitz, V. von (ed.), *Umweltpolitik als Modernisierungsprozeß*, Opladen: Leske & Budrich, pp 15-30.

Jänicke, M. and H. Weidner (1997) 'Germany', in Jänicke, M. and H. Weidner (eds), *National Environmental Policies. A Comparative Study of Capacity-Building,* Berlin *et al.*: Springer, pp. 133-155.

Jans, J.H. (1994) *Europees milieurecht in Nederland,* Groningen: Wolters-groep.

Jasanoff, S. (1991) 'Cross-national differences in policy implementation', *Evaluation Review,* vol. 5, pp. 103-119.

Jordan, A. (1993) 'Integrated Pollution Control and the evolving style and structure of environmental regulation in the UK', in *Environmental Politics*, pp. 405-427.

Judge, D. (1990) *Parliament and industry*, Aldershot: Dartsmouth.

Katzenstein, P.J. (1978) *Between Power and Plenty. Foreign Economic Policies of Advanced Industrial States*, Madison: University of Madison Press.

Katzenstein, P.J. (1985) *Small States in World Markets*, Ithaca, N.Y.: Cornell UP.

Katzenstein, P.J. (1987) *Policy and Politics in West Germany. The Growth of the Semi-Souvereign State*, Philadelphia: Temple UP.

Kavanagh, D. (1996) *British Politics. Continuities and Change*, third edition, Oxford: Oxford UP.

Kelman, S. (1981) *Regulating America, Regulating Sweden: A Comparative Study of Occupational Safety and Health Policy*, Cambridge, MA: MIT Press.

Keman, H. (ed.), (1993a) *Comparative politics. New directions in theory and method*, Amsterdam: VU University Press.

Keman, H. (1993b)'Comparative politics: A distinctive approach to political science?', in Keman, H. (ed.), *Comparative politics. New directions in theory and method*, Amsterdam: VU University Press.

Kempen, A.L van (1992) 'Het convenant verpakkingen', in Huls, N.J.H. and H.D. Stout (eds), *Reflecties op reflexief recht*, Zwolle: Tjeenk Willink, pp. 185-197.

Kern, K. and S. Bratzel (1996) 'Umweltpolitischer Erfolg im internationalen Vergleich', *Zeitschrift für Umweltpolitik und Umweltrecht*, no. 3, pp. 277-312.

Kerremans, B. (1996) 'Do institutions make a difference? Non-institutionalism, neo-institutionalism, and the logic of common decision-making in the European Union', *Governance*, vol. 9, no. 2, pp. 217-240.

Key Note Market Review (1993a) *UK Packaging Industry,* third edition, Hampton Middlesex: ICC Business Publications.

Key Note Market Review (1993b) *UK Waste Management*, Hampton Middlesex: ICC Business Publications.

King, D. (1993) 'Government beyond Whitehall', in Dunleavy, P., A. Gamble, I. Holliday, and G. Peele (eds), *Developments in British Politics 4*, Chatham Kent: Mac Millan.

King, G., R.O. Keohane, and S. Verba (1994) *Designing Social Inquiry. Scientific inference in qualitative research*, Princeton, N.J.: Princeton UP.

Kitschelt, H. (1983) *Politik und Energie. Energie-Technologiepolitiken in den USA, der Bundesrepublik Deutschland, Frankreich und Schweden*, Frankfurt: Campus.

Klages, C. (1988) 'Rechtliche Instrumente zur Abfallvermeidung', *Neue Zeitschrift für Verwaltungsrecht*, vol. 7, no. 6, pp. 481-486.

Klages, C. (1990) 'Kein Pfand in Sicht. Die EG-Kommission ist bisher über Klagen gegen Rücknahmevorschriften nicht herausgekommen', *Müllmagazin*, no.2, pp. 47-50.

Klok, P.-J. (1989) *Convenanten als instrument van milieubeleid,* Enschede: Faculteit der Bestuurskunde, Universiteit Twente.

König, T. (1994) 'Intergouvernmentale versus supranationale Politikfeldmuster', in Eichener, V. and H. Völzkow (eds), *Europäische Integration und verbandliche Interessenvermittlung*, Marburg: Metropolis, pp. 151-180.

Knill, C. (1995) *Staatlichkeit im Wandel. Grossbritannien im Spannungsfeld innenpolitischer Reformen und europäischer Integration*, Wiesbaden: Deutscher Universitätsverlag.

Knoepfel, P. and H. Weidner (1985) *Luftreinhaltepolitik*, Berlin: Edition Sigma, vol. 1-6.

Knoepfel, P., L. Lundqvist, R. Prud'homme, and P. Wagner (1987) 'Comparing environmental policies: different styles, similar content', in Dierkes M., H.N. Weiler, and A.B. Antal (eds), *Comparative Policy Research. Learning from Experience*, Wissenschaftszentrum Berlin, Aldershot: Gower, pp. 171-188.

Kohler-Koch, B. (1996) 'Catching up with change: the transformation of governance in the European Union', *Journal of European Public Policy,* vol. 3, no. 3, pp. 359-380.

Kohler-Koch, B. and R. Eising (eds), (forthcoming) *The transformation of governance in the European Union* (working title), London: Routledge.

Koppen, I.J. (1993) 'The role of the European Court of Justice', in Liefferink, J.D., P.D. Lowe, and A.P.J. Mol (eds), *European Integration & Environmental Policy*, London and New York: Belhaven, pp. 126-149.

Koppen, I.J. (1994) 'Regulatory negotiation in the Netherlands: the case of packaging waste', in Knoepfel, P. (ed.), *Lösung von Umweltkonflikten durch Verhandlung*, Basel: Helbing & Lichtenhahn, pp. 153-169.

Krämer, L. (1997) *Focus on European Environmental Law,* second edition, London: Sweet & Maxwell.

Kriesi, H. (1993) *Political Mobilization and Social Change. The Dutch Case in Comparative Perspective*, Aldershot *et al.*: Avebury.

Lang, T. and H. Raven (1994) 'From Market to Hypermarket. Food Retailing in Britain', *The Ecologist*, vol. 24, no. 4, July/August 1994, pp. 124-129.

Lansink, A. (1996) 'Lansink over Verpakken', *Verpakken*, 2/1996, p. 77.

Lansink, A. (1996a) 'Lansink over Verpakken', *Verpakken*, 3/1996, p. 69.

Le Blansch, K. (1996) *Milieuzorg in Bedrijven. Overheidssturing in het perspectief van de verinnerlijkingsbeleidslijn*, Amsterdam: Thesis.

Lehmbruch, G. (1992) 'The Institutional Framework of German Regulation', in Dyson, K (ed.), *The Politics of German Regulation*, Dartmouth: Aldershot *et al.*, pp. 29-52.

Lehmbruch, G., O. Singer, E. Grande, and M. Döhler (1988) 'Institutionelle Bedingungen ordnungspolitischen Strategiewechsels im internationalen Vergleich', in Schmidt, M.G. (ed.), *Staatstätigkeit. International und historisch vergleichende Analysen*, Opladen: Westdeutscher Verlag, pp 251-283.

Lenschow, A. (1996) 'Transformation in European Environmental Governance', paper prepared for the ECPR Joint Session of Workshops, Oslo, 29.3-3.4.1996.

Leroy, P. (1994) 'De ontwikkeling van het milieubeleid en de milieubeleidstheorie', in Glasbergen, P. (ed.), *Milieubeleid een beleidswetenschappelijke inleiding*, 's-Gravenhage: VUGA, pp. 35-58.

Leroy, P. (1995) *Milieukunde als roeping en beroep*, Nijmegen: Katholieke Universiteit Nijmegen.

Liefferink, D. (1997) 'The Netherlands: a net exporter of environmental policy concepts', in Andersen, M. and D. Liefferink (eds), *European environmental policy. The pioneers,* Manchester: Manchester UP, pp. 210-249.

Lijphart, A. (1975) 'The comparable-cases strategy in comparative research', *Comparative Political Studies*, vol. 8, no. 2, pp. 158-177.

Lijphart, A. (1976) *Verzuiling, pacificatie en kentering in de Nederlandse politiek*, second edition, Amsterdam: J.H. De Bussy.

Lijphart, A. (1984) *Democracies. Patterns of majoritarian and consensus government in twenty one countries*, New Haven and London: Yale University Press.

Long, A. and T. Bailey (1997) 'The single market and the environment: the European Union's dilemma: The example of the Packaging Directive', *European Environmental Law Review*, July 1997, pp. 214-219.

Luhmann, N. (1988) *Ökologische Kommunikation*, second print, Opladen: Westdeutscher Verlag.

Lundqvist, L. (1980) *The Hare and the Tortoise: Clean Air Policies in the United States and Sweden*, Ann Arbor: The University of Michigan Press.

MacMullen, A.L. (1979) 'The Netherlands', in F.F. Ridley (ed) *Government and Administration in Western Europe*, Oxford: Martin Robertson, pp. 227-240.

Maher, I. (1996) 'Limitations on Community regulation in the UK: legal culture and multi-level governance', *Journal of European Public Policy*, vol. 3, no. 4, pp.577-593.

Majone, G. (1989) *Evidence, argument and pursuasion in the policy process*, New Haven, CT: Yale UP.

Majone, G. (1994) 'The rise of the regulatory state in Europe', *West European Politics*, vol. 17. pp. 77-101.

Majone, G. (ed.), (1996) *Regulating Europe,* London and New York: Routledge.

Majone, G. (1996a) 'The future of regulation in Europe', in G. Majone (ed.), *Regulating Europe*, London and New York: Routledge, pp. 265-283.

March, J. G. and J. Olsen (1989) *Rediscovering Institutions*, New York: Free Press.

Margedant, U. (1987) 'Entwicklung des Umweltbewußtseins in der Bundesrepublik Deutschland', *Aus Politik und Zeitgeschichte*, B29/87, 18.7.1987. pp. 15-28.

Marks, G., L. Hooghe, and K. Blank (1996) 'European integration from the 1980s: State-centric vs. multi-level governance', *Journal of Common Market Studies*, vol. 34, no. 3, pp. 341-378.

Mayntz, R. and F.W. Scharpf (1995) 'Der Ansatz des akteurszentrierten Institutionalismus', in Mayntz, R. and F.W. Scharpf (eds), *Gesellschaftliche Selbstregulierung und politische Steuerung*, Frankfurt and New York: Campus, pp.39-71.

Mazey, S. and J. Richardson (1996) 'EU policy-making: a garbage can or an anticipatory and consensual policy style', in Mény, Y, P. Muller, and J.-L. Quermonnee (eds), *Adjusting to Europe. The impact of the European Union on national institutions and policies*, London and New York: Routledge, pp. 41-58.

McGowan, F. and H. Wallace (1997) 'Towards a European regulatory state', *Journal of European Public Policy*, vol. 3, no. 4, pp. 560-576.

Meadows, D, Da Meadows, E. Zahn, and P. Milling (1972) *Die Grenzen des Wachstums,* Stuttgart: Deutsche Verlagsanstalt.

Mény, Y, P. Muller, and J.-L. Quermonnee (eds), (1996) *Adjusting to Europe. The impact of the European Union on national institutions and policies*, London and New York: Routledge.

Michaelis, P. (1995) 'Product Stewardship, Waste Minimization and Economic Efficiency: Lessons from Germany', *Journal of Environmental Planning and Management*, vol. 38, no. 2, 1995, pp. 231-243.

Mierlo, J.G.A. van (1988) *Pressiegroepen in de Nederlandse politiek*, Den Haag: Stichting Maatschappij en Onderneming.

Milieudefensie (1979) 'Verpakkingenbelweken', *Milieudefensie*, no. 1, p. 30.

Milward, A. (1992) *The European Rescue of the Nation-State*, Berkely: Berkely UP.

Mingelen, J.P. (1995) 'Verpakkingen en Milieu in Nederland. Inhoud en Achtergronden', unpublished manuscript.

Monopolkommission (1996) *Hauptgutachten der Monopolkommission*, Bundestags Drucksache 13/5309, 19.7.1996.

Moravcsik, A. (1991) 'Negotiating the Single European Act. National interests and conventional statecraft in the European Community', *International Organization*, vol. 45, pp. 651-688.

Moravcsik, A. (1993) 'Preferences and power in the European Community: A Liberal Intergovernmental Approach', *Journal of Common Market Studies*, vol. 31, pp. 473-524.

Mortelmans, K. (1994) 'De interne markt en het facettenbeleid na het Keck-arrest: nationaal beleid, vrij verkeer of harmonisatie', *Tijdschrift voor Europees en economisch recht*, vol. 42, no. 4, pp. 236-250.

Müller, E. (1989) 'Sozial-liberale Umweltpolitik. Von der Karriere eines neuen Politikbereiches', *Aus Politik und Zeitgeschichte*, no. 47-48, pp. 3-15.

Naßmacher, H. (1991) *Vergleichende Politikforschung. Eine Einführung in Probleme und Methoden*, Opladen: Westdeutscher Verlag.

Noort, W. van (1988) *Bevlogen bewegingen. Een vergelijking van de anti-kernenergie-, kraak- en milieubeweging*, Amsterdam: SUA.

North, D. (1990) *Institutions, institutional change and economic performance,* Cambridge: Cambridge UP.

Nullmeier, F. (1997) 'Interpretative Ansätze in der Politikwissenschaft', in Benz, A, and W. Seibel (eds), *Theorieentwicklung in der Politikwissenschaft eine Zwischenbilanz*, Baden-Baden: Nomos, pp.101-144.

OECD (1994) *OECD Environmental Performance Review. United Kingdom,* Paris: OECD Publication Service.

OECD (1995a) *OECD Environmental Data*, Paris: OECD Publication Service.

OECD (1995b) 'Waste Stream Case Study. Packaging', Documentation prepared for the International Workshop on Waste Minimization, Washington DC 29-31.3. 1995, unpublished manuscript.

Olsen, J.P (1991) 'Political science and organization theory - parallel agendas but mutual disregard', in R. Czada, and A. Windhoff-Héritier (eds), *Political Choice. Institutions, Rules, and the Limits of Rationality*, Frankfurt/M and Boulder, Colo: Campus and Westview, pp. 53-86.

Paqué K.-H. (1995) 'Weltwirtschaftlicher Strukturwandel und die Folgen', *Aus Politik und Zeitgeschichte*, B 49/95, 1.12.1995, pp. 3-9.

Parson, E.A. and Clark, W.C. (1997) 'Learning to manage global environmental change: a review of relevant theory', paper presented at the Summer Symposium on the Innovation of Environmental Policy. Bologna, Italy, 21-25 July 1997.

Pehle, H. (1993) 'Umweltpolitik im internationalen Vergleich', in Prittwitz, V. von (ed.), *Umweltpolitik als Modernisierungsprozeß*, Opladen: Leske + Budrich, pp. 113-136.

Pelkmans (1997) 'Competition among regimes in the EU, between ideas and sober reality', paper presented at the conference Institutions, Markets and (Economic) Performance: deregulation and its consequences, Utrecht, 11-12 December 1997.

Peters, B.G. (1992) 'Buerocratic Politics and the Institutions of the European Community', in Sbragia, A. (ed.), *Euro-Politics: institutions and policymaking in the 'new' European Community*, Washington, DC: The Brookingsinstitute, pp. 75-122.

Peterse, A.H (1992) 'Verpakkingsconvenant:element in een alternatieve sturingsstrategie', in Huls, N.J.H, and H.D. Stout (eds), *Reflecties op reflexief recht,* Zwolle: Tjeenk Willink, pp. 199-209.

Pierson, P. (1995) *Dismantling the Welfare State. Reagan, Thatcher, and the Politics of Retrenchment,* Cambridge: Cambridge UP.

Porter, M. (1994) 'The Packaging and Packaging Waste Directive: Scientific uncertainty, the role of expertise and North-South variations in the EU environmental policy process', paper prepared for presentation at the 'Conference on environmental standards and the politics of expertise in Europe', Bristol, 9-11 December 1994.

Porter, M. (1995) Interest Groups, Advocacy Coalitions and the EC Environmental Process: A Policy Network Analysis of the Packaging and Packaging Waste Directive, PhD dissertation, University of Bath.

Porter, M. (1995b) 'Cross national policy-networks and the EU's Packaging and Packaging Waste Directive', paper prepared for presentation at the workshop 'New Nordic states and the impact on EU environmental policymaking', Sonderborg, Denmark, 6-8 April 1995.

Porter M. and A. Butt Philip (1993) 'The role of interest groups in EU environmental policy formulation: a case study of the draft packaging directive', *European Environment,* vol. 3, no.6, 16-20.

Prechal, S. (1995) *Directives in European Community Law. A Study of Directives and Their Enforcement in National Courts,* Oxford: Clarendon Press.

Producer Responsibility Industry Group (1994) *Real value from packaging waste. A way forward.* London: PRG.

Przeworski, A. and H. Teune (1970) *The Logic of Comparative Social Inquiry,* Londen et al: Wiley.

Putten, J. van (1982) 'Policy Styles in the Netherlands: Negotiation and Conflict', in Richardson, J. (ed.), *Policy Styles in Western Europe,* London: George Allen & Unwin, pp.168-196.

Rathje, W. and G. Murphy (1994) *Müll. Eine archäologische Reise durch die Welt des Abfalls,* München: Goldmann.

Rehbinder, E. and R. Stewart (1985) *Environmental Protection Policy,* Berlin and New York: De Gruyter.

Richardson, J. (ed.), (1982) *Policy Styles in Western Europe,* London: George Allen & Unwin.

Richardson, J. and N.S.J. Watts (1985) *National Policy Styles and the Environment. Britain and West Germany Compared,* Berlin: Wissenschaftszentrum Berlin, Discussion Papers IIUG dp 85-16.

Richardson, J., G. Gustafsson, and G. Jordan (1982) 'The concept of policy style', in Richardson, J. (ed.), *Policy Styles in Western Europe,* London: George Allen & Unwin, pp.1-16.

Ricken, C (1995) 'Nationaler Politikstil, Netzwerkstrukturen sowie ökonomischer Entwicklungsstand als Determinanten einer effektiven Umweltpolitik. Ein empirischer Industrieländervergleich', *Zeitschrift für Umweltpolitik,* no. 4, pp. 481-501.

Risse-Kappen, T. (1994) 'Ideas do not float freely: transnational coalitions, domestic structures, and the end of the cold war', *International Organizations,* vol. 48, pp. 185-214.

Rijksinstituut voor Volksgezondheid en Milieuhygiëne (1991) *Nationale Milieuverkenning 2, 1990-2010,* Alphen a.d. Rijn: Samsom H.D. Tjeenk Willink.

Rijksinstituut voor Volksgezondheid en Milieuhygiëne (1993) *Nationale Milieuverkenning 3, 1993-2015;.* Alphen a.d. Rijn: Samsom H.D. Tjeenk Willink.

Rose, R. (1993) *Lesson-drawing in public policy. A guide to learning across time and space,* Chatham, New Jersey: Chatham House.

Rosenthal, U. (1989) 'De Departementen', in Andeweg, R.B., A. Hoogerwerf and J.J.A Thomassen (eds) *Politiek in Nederland,* Alphen a.d. Rijn: Samsom H.D. Tjeenk WillinkH, pp. 239-263.

Royal Commission on Environmental Pollution, (1985) *Eleventh Report. Managing Waste: The Duty of Care,* London: Her Majesty's Stationery Office, Cmnd 9675.

Rüdig, W., M.N. Franklin, and L.G. Bennie (1993) 'Green Blues. The Rise and Decline of the British Green Party', Glasgow: Strathclyde Papers on Government and Politics no. 95.

Sabatier, P.A. and H.C. Jenkins-Smith (eds), (1993) *Policy change and learning: An advocacy coalition approach,* Boulder, CO: Westview Press.

Sachverständigenrat für Umweltfragen (1991) *Abfallwirtschaft. Sondergutachten,* Stuttgart: Metzler-Poeschel.

Sandholtz, W. (1993) 'Choosing Union: monetary politics and Maastricht', *International Organisation,* vol. 47, pp. 1-21.

Sbragia, A. (1996) 'Environmental Policy', in Wallace, H. and W. Wallace (eds), *Policy-Making in the European Union,* Oxford: Oxford UP, pp.235-255.

Scharpf, F.W. (1987) *Sozialdemokratische Krisenpolitik in Europa,* Frankfurt/M: Campus.

Scharpf. F.W. (1995) 'Negative and positive integration in the political economy of European welfare states', Florence: European University Institute, Jean Monnet Chair Papers, no.28.

Schattschneider, E.E. (1960) *The Semi-Sovereign People,* New York: Holt, Rinehart and Winston.

Schmidt, M.G. (1989a) 'Social policy in rich and poor countries: Socio-economic trends and political-institutional determinants', *European Journal of Political Research,* vol. 17, 641-659.

Schmidt, M.G. (1989b) 'Learning from Catastrophes. West Germany's Public Policy', in Castles, F.G. (ed.), *The Comparative History of Public Policy,* Cambridge: Polity Press, pp. 56-99.

Schmidt, M.G. (1993) 'Theorien in der international vergleichenden Staatstätigkeitsforschung', in Héritier, A. (ed.), *Policy-Analyse. Kritik und Neuorientierung*, Opladen: Westdeutscher Verlag, pp. 371-393.

Schuddeboom (1994) *Milieubeleid in de praktijk,* Aalphen aan den Rhijn: Samson Tjeenk Willink.

Sevenster, H.G. (1992) 'De geoorloofheid van milieubeleidsafspraken in Europees perspectief', in Aalders M.V.C en R.J.J. Van Acht (eds), *Afspraken in het milieurecht,* Zwolle: Tjeenk Willink, pp. 73-93.

Sewandono, I. (1993) 'Implementatie van EEG-richtlijnen met behulp van convenanten', in Ommeren F.J. van and H.F. Ru (eds), *Convenanten tussen overheid en maatschappelijke organisaties,* s'-Gravenhage: SDU, pp. 81-104.

Shonfield, A. (1965) *Modern capitalism*, London: Oxford UP.

Siedentopf, H. and J. Ziller (1988) *Making European policies work. The implementation of Community legislation in the Member States*, two volumes, London *et al.*: Sage.

Simonsson, E. (1995) 'The German Packaging Ordinance and EU environmental policy: subsidiarity, trade barriers, and harmonizations', Georgetown: University Center for German and European Studies, unpublished manuscript.

Skocpol, T. (ed.), (1984a) *Vision and method in historical sociology,* Cambridge *et al.*: Cambridge UP.

Skocpol, T. (1984b) 'Emerging agendas and recurrent strategies in historical sociology', in Skocpol, T. (ed.), *Vision and method in historical sociology,* Cambridge *et al.*: Cambridge UP.

Skocpol, T. (1985) 'Bringing the state back in: strategies of analysis in current research', in P.B. Evans, D. Rueschemeyer, and T. Skocpol (eds), *Bringing the state back in*, Cambridge: Cambridge UP, pp. 3-44.

Spies, R. (1994) 'Von der Kooperation in der Abfallpolitik zur staatlich flankierten Selbstorganisation: Das Duale System', *Staatswissenschaft & Staatspraxis,* vol. 4, pp. 267-302.

Steinmo, S., K. Thelen, and F. Longstreth, (1992) *Stucturing politics. Historical institutionalism in comparative analysis,* Cambridge *et al.*: Cambridge UP.

Sterkenburg, P.G.J van (1994) *Van Dale Handwoordenboek Hedendaags Nederlands,* second print, Utrecht and Antwerpen: Van Dale Lexicografie.

Steyger, E (1993) 'European Community law and the self-regulatory capacity of society', *Journal of Common Market Studies*, vol. 31, pp. 171-190.

Stichting Natuur en Milieu (1991) 'Brief aan de Tweede Kamer', 30.5.1991.

Stichting Verpakking en Milieu (1994a) 'SVM-Verslag Milieu- en Markt-Economische Analyses', in Commissie Verpakkingen, *Toetsing Milieu-analyses*, Utrecht: Commissie Verpakkingen, pp. 29-64.

Stichting Verpakking en Milieu (1994b) *Verpakkingsontwikkelingen. Packaging Developments. Implementation of the Packaging Covenant, illustrated,* Utrecht and Den Haag: SVM.

Stiching Verpakking en Milieu (1995) *The packaging chain organized,* Den Haag: SVM.

Stone Sweet, A. and W. Sandholtz (1997) 'European integration and supranational governance', *Journal of European Public Policy*, vol. 4, pp. 297-317.

Storm, P.-C. (1989) 'Einführung', in *Umweltrecht,* München: Beck-Texte im DTV, pp. 9-15.

Strange, S. (1995) 'The limits of politics', *Government and Opposition,* vol. 30, pp. 291-311.

Stupples, L (1993) 'Challenging the throwaway society in the UK', in Institut für Ökologisches Recycling (ed.), *Neue Wege ohne Abfall*, Fachkongreß Ökologische Abfallwirtschaft III, Berlin: Institut für Ökologisches Recycling, pp. 59-65.

Sturm, Roland (1994) 'Staatsordnung und politisches System', in Kastendiek, H., K. Rohe, and A. Volle (eds), *Länderbericht Großbritannien. Geschichte-Politik-Wirtschaft-Gesellschaft*, Bonn: Bundeszentrale für politische Bildung, pp 185-212.

Stuurgroep Milieu-Analyses Verpakkingen (1994) 'Eenmaal...andermal. Algemeen Rapport van Bevindingen van de Stuurgroep Milieu-Analyses Verpakkingen. Delft: Stafbureau Verpakkingen, TU Delft, Faculteit der Technische Bestuurskunde.

Tarrow, S. (1995) 'Bridging the Quantitative - Qualitative Divide in Political Science', *American Political Science Review*, vol. 89, no.2, pp. 471-475.

Tellegen, E. (1983) *Milieubeweging. Aula 734*, Utrecht and Antwerpen: Het Spectrum.

Thelen, K. and S. Steinmo (1992) 'Historical institutionalism in comparative politics', in S. Steinmo, K. Thelen, and F. Longstreth (eds), *Structuring Politics. Historical Institutionalism in Comparative Analysis*, Cambridge: Cambridge UP, pp. 1-32.

(The) Times (1996) *The Times 1000- 1996 - The definitive reference to business today,* London: Times Books.

Tsebelis, G. (1990) *Nested Games*, Berkely: University of California Press.

Tweede Kamer (1973-1974), Motie van het lid Vellekamp, 12304, no. 1.

Tweede Kamer (1979-1980), Motie van het lid Lansink., 15800, XVII, no. 21.

Tweede Kamer (1987-1988), Motie Preventie en Hergebruik (parlementaire discussie), 20.200 XI, nr. 69.

Tweede Kamer (1988-1989a) Nationaal Milieubeleidsplan. Kiezen of verliezen, 1988-1989, 21-137, nos 1-2.

Tweede Kamer (1990-1991) Brief van de Minister aan de Tweede Kamer, 21137, no. 49.

Umweltbundesamt (1992) *Life Cycle Analysis*, Berlin: UBA.

Unger, B. and F. van Waarden (1995) 'Introduction: An interdisciplinary approach to convergence', in Unger, B. en F. van Waarden, (eds), *Convergence or diversity,* Aldershot: Avebury, pp. 1-35.

Vogel, D. (1986) *National Styles of Regulation. Environmental Policy in Great Britain and the United States*, Ithaca and London: Cornell UP.

Vogel, D. (1993) 'Representing diffuse interests in environmental policymaking', in Weaver, R.K. and B.A. Rockman (eds), *Do institutions matter? Government capabilities in the United States and abroad,* Washington: D.C Brookings Institution, pp. 237-271.

Vogel, D. (1995) *Trading Up. Consumer and Environmental Regulation in a Global Economy*, Cambridge, MA: Harvard University Press.

Vogel, D., and V. Kun (1987) 'The comparative study of environmental policy: a review of the literature', in M. Dierkes, H.N. Weiler, and A.B. Antal (eds) *Comparative Policy Research. Learning from Experience,* Aldershot: Gower, pp. 99-170.

Vogel, S. (1996) *Freer markets, more rules. Regulatory reform in advanced industrial countries*, Ithaca and London: Cornell UP.

VROM (1988) 'Notitie Preventie en Hergebruik', Tweede Kamer, 1988-89, 20 877, no. 2.

VROM (1991) *The Packaging Covenant,* Den Haag: VROM.

VROM (1994) *Towards a sustainable Netherlands. Environmental policy development and implementation,* Den Haag: VROM.

VROM (1995) *Ontwerp-regeling verpakking en verpakkingsafval*, Staatscourant. No. 249, pp. 20-24.

VROM (1996) *Ontwerp-regeling verpakking and verpakkingsafval*, 20.6.1996. Den Haag: VROM.

VROM (1997a) *Huishoudelijke afvalcijvers van diverse Europese landen vergeleken*, Zoetermeer: VROM.

VROM (1997b) 'EU-Directive. The Dutch approach', presentation by VROM official at the EU Packaging & Waste Law conference at Brussels, 19.-20.3. 1997.

VROM (1997c) Brief aan de Vaste Commissie voor Vokshuisvesting, Ruimtelijke Ordening en Milieubeheer van de Tweede Kamer, 22.10.1997.

Waarden, F. van (1985) 'Varieties of collective self-regulation of business: the example of the Dutch dairy industry', in Streeck, W. and P.C. Schmitter (eds), *Private interest government. Beyond market and state*, London, Beverly Hills, and New Dehli: Sage, pp. 197-220.

Waarden, F. van (1992) 'Dimensions and types of policy networks', *European Journal of Political Research,* vol. 21, pp. 29-52.

Waarden, F. van (1995) 'Persistence of national policy styles: A study in their institutional foundations', in Unger, B. en F. van Waarden (eds), *Convergence or diversity,* Aldershot: Avebury, pp. 333-372.

Wallace, H. (1996) 'The institutions of the EU: experience and experiment', in Wallace, H. and W. Wallace (eds), *Policy-Making in the European Union*, Oxford: Oxford UP, pp.37-68.

Wallace, H. and W. Wallace (eds), (1996) *Policy-Making in the European Union*, Oxford: Oxford UP.

Weale, A. (1992) *The New Politics of Pollution*, Manchester and New York: Manchester UP.

Weale, A. (1992a) 'Vorsprung durch Technik? The Politics of German Environmental Regulation', in Dyson, K. (ed), *The Politics of German Regulation,* Bodmin, Cornwall: Hartnolls, pp. 159-183.

Weale, A., G. Pridham, A. Williams, and M.Porter (1996) 'Environmental administration in six European states. Secular convergence or national distinctiveness?' *Public Administration*, vol. 74, pp. 255-274.

Weber, H. (1994) 'Recht und Gerichtsbarkeit', in Kastendiek, H. K. Rohe and A. Volle (eds), *Länderbericht Großbritannien. Geschichte-Politik-Wirtschaft-Gesellschaft*, Bonn: Bundeszentrale für politische Bildung, pp 185-212.

Weidner, H. (1995) '25 Years of Modern Environmental Policy in Germany. Treading a Well-Worn Path to the Top of the International Field', Berlin: WZB papers FS II 95-301.

Weir, M. (1992) 'Ideas and the politics of bounded innovation', in Steinmo, S. K. Thelen, and F. Longstreth (eds), *Structuring Politics. Historical Institutionalism in Comparative Analysis*, Cambridge: Cambridge UP, pp. 188-213.

Weir. M. and T. Skocpol (1985)'State structures and the possibilities for "Keynesian" responses to the Great Depression in Sweden, Britain and the United States', in Evans, P.B., D. Rueschemeyer, and T. Skocpol (eds), *Bringing the state back in,* Cambridge: Cambridge UP, pp. 107-168.

Weßels, B. (1989) 'Politik, Industrie und Umweltschutz in der Bundesrepublik: Konsens und Konflikt in einem Politikfeld 1960-1986', in Herzog, D. and B. Weßels (eds), *Konfliktpotentiale und Konsensstrategien. Beiträge zur politischen Soziologie der Bundesrepublik,* Opladen: Westdeutscher Verlag, pp. 269-306.

Wicke, L. (1993) *Umweltökonomie*, fourth edition, München: Vahlen.

Wilensky, H.T. (1975) *The welfare state and equality*, Berkely: University of California Press.

Wilks, S. and M. Wright (1987) 'Conclusions: Comparing Government-Industry Relations: States, Sectors and Networks', in Wilks, S. and M. Wright (eds), *Comparative Government-Industry Relations. Western Europe, the United States, and Japan*, Oxford: Clarendon Press, pp. 274-313.

Windhoff-Héritier, A. (1991) 'Institutions, Interests and Political Choice', in R. Czada, and A. Windhoff-Héritier (ed.) *Political Choice. Institutions, Rules, and the Limits of Rationality*, Frankfurt/M and Boulder,Colorado: Campus and Westview, pp. 27-52.

World Commission on Environment and Development (1987) *Our Common Future*, Oxford: Clarendon Press.

Yin, R.K. (1994) *Case Study Research. Design and Method*, second edition, Thousand Oaks, London, and New Dehli: Sage.

Zahn, E (1989) *Regenten, rebellen en reformatoren, Een visie op Nederland en de Nederlanders*, Amsterdam: Contact.

Newspapers and specialized journals

Das Parlament	German fortnightly
De Financiële Telegraaf	section of the Dutch daily De Telegraaf
De Volkskrant	Dutch daily
Die Zeit	German weekly
ENDS Report	British specialized monthly newsletter
Environment Watch Western Europe (EWWE)	Brussels based fortnightly newsletter
Entsorga Magazin	newsletter from BDE
European Chemical News	fortnightly
European Packaging & Waste Law (former EC Packaging Report, EU Packaging Report)	Brussel based monthly newsletter
Financial Times (FT)	British daily
Financieel Dagblad	Dutch daily
Frankfurter Allgemeine Zeitung (FAZ)	German daily
Handelsblatt (HB)	German daily
INCPEN Factsheet	newsletter of INCPEN
Lebensmittelzeitung	German specialized fortnightly journal
Milieutechnologie	Dutch specialized monthly journal
Müllmagazin	German specialized monthly journal
NRC Handelsblad	Dutch daily
Ökologische Briefe	German specialized weekly newsletter
Pak Aan	newsletter of the SVM
Politische Ökologie	German monthly
Spiegel	German weekly
Süddeutsche Zeitung (SZ)	German daily
TAZ tageszeitung	German daily
The Economist	British weekly
Trouw	Dutch daily
Umwelt	German specialized monthly journal
Verpakken	Dutch specialized monthly journal
WARMER bulletin	British based specialized monthly
World Drink Report	Brussels based specialized monthly journal

Interviewees

Germany

12.11.1996	Official at the BMU
5.12.1996	Two respresentatives of the DSD
5.12.1996	Representative of BUND
6.12.1996	Official at the Deutsche Städtetag (German Association of Municipalities)
6.12.1996	Official at the BDI
11.12.1996	Official at the Environment Ministry of Northrine Westfalia (German Land)
10.12.1996	Two officials at the BMWi

Netherlands

3.11.1995	Official at the Packaging Commission
9.11.1995	Representative of SVM
20.11. 1995	Representative of Milieudefensie
4.12.1995	Official at VROM
7.12.1995	Professor Teuninga, Member of the LCA advisory board (expert)
8. 9.1997	Representative of VMK
11. 9.1997	Ten Heuvelhof, Chairman of the LCA steering committee (expert)

United Kingdom

15.5.1996	Official at the Department of Trade and Industry
20.5.1996	Professor Grant, University of Warwick (expert)
24.6.1996	British Industrialist, from multinational packer/filler, Vice Chairman of INCPEN
12.7.1996	Official at the Department of the Environment
23.7.1996	Trollope, specialized journalist (expert)
24.7.1996	Recycling Manager West Sussex County Council, Member of LARAC

Curriculum Vitae

Markus Haverland was born in Biedenkopf (Germany). He received his *Abitur* (A-Level) at the Städtisches Gymnasium Bad Laasphe in 1986. After having completed his alternative service, *Zivildienst,* in 1988, he studied public administration and political science at the Universities of Konstanz and Rotterdam. In 1993, he graduated with a thesis on the implementation of the European Fund for Regional Development in Germany and the United Kingdom. In the same year he became a doctoral student, *Assistent in Opleiding,* at the Center for Policy and Management Studies, University of Utrecht and the Netherlands School of Social and Economic Policy Research. He is currently holding a post-doc position at the School for Public Affairs, Faculty for Policy Science at the University of Nijmegen. This position will be interrupted by a visit at the Robert Schuman Centre of the European University Institute in Florence as a Jean Monnet Fellow.